Recent Progress in Bunyavirus Research

Special Issue Editors

Jane Tao
Pierre-Yves Lozach

Special Issue Editors
Jane Tao
Rice University
USA

Pierre-Yves Lozach
University Hospital Heidelberg
Germany

Editorial Office
MDPI AG
St. Alban-Anlage 66
Basel, Switzerland

This edition is a reprint of the Special Issue published online in the open access journal *Viruses* (ISSN 1999-4915) from 2015–2016 (available at: http://www.mdpi.com/journal/viruses/special_issues/bunyavirus-research).

For citation purposes, cite each article independently as indicated on the article page online and as indicated below:

Author 1; Author 2; Author 3 etc. Article title. *Journal Name*. **Year**. Article number/page range.

Photo courtesy of Dr. Psylvia Léger

ISBN 978-3-03842-392-8 (Pbk)
ISBN 978-3-03842-393-5 (PDF)

Articles in this volume are Open Access and distributed under the Creative Commons Attribution license (CC BY), which allows users to download, copy and build upon published articles even for commercial purposes, as long as the author and publisher are properly credited, which ensures maximum dissemination and a wider impact of our publications. The book taken as a whole is © 2017 MDPI, Basel, Switzerland, distributed under the terms and conditions of the Creative Commons license CC BY-NC-ND (http://creativecommons.org/licenses/by-nc-nd/4.0/).

Table of Contents

About the Guest Editors v

Preface to "Recent Progress in Bunyavirus Research" vii

Alain Kohl, Benjamin Brennan and Friedemann Weber
Homage to Richard M. Elliott
Reprinted from: *Viruses* **2016**, *8*(8), 224; doi: 10.3390/v8080224
http://www.mdpi.com/1999-4915/8/8/224 1

Jens H. Kuhn, Michael R. Wiley, Sergio E. Rodriguez, Yīmíng Bào, Karla Prieto, Amelia P. A. Travassos da Rosa, Hilda Guzman, Nazir Savji, Jason T. Ladner, Robert B. Tesh, Jiro Wada, Peter B. Jahrling, Dennis A. Bente and Gustavo Palacios
Genomic Characterization of the Genus *Nairovirus* (Family *Bunyaviridae*)
Reprinted from: *Viruses* **2016**, *8*(6), 164; doi: 10.3390/v8060164
http://www.mdpi.com/1999-4915/8/6/164 2

Alexey M. Shchetinin, Dmitry K. Lvov, Petr G. Deriabin, Andrey G. Botikov, Asya K. Gitelman, Jens H. Kuhn and Sergey V. Alkhovsky
Genetic and Phylogenetic Characterization of Tataguine and Witwatersrand Viruses and Other Orthobunyaviruses of the Anopheles A, Capim, Guamá, Koongol, Mapputta, Tete, and Turlock Serogroups
Reprinted from: *Viruses* **2015**, 7(11), 5987–6008; doi: 10.3390/v7112918
http://www.mdpi.com/1999-4915/7/11/2918 29

Inaia Phoenix, Nandadeva Lokugamage, Shoko Nishiyama and Tetsuro Ikegami
Mutational Analysis of the Rift Valley Fever Virus Glycoprotein Precursor Proteins for Gn Protein Expression
Reprinted from: *Viruses* **2016**, *8*(6), 151; doi: 10.3390/v8060151
http://www.mdpi.com/1999-4915/8/6/151 51

Inaia Phoenix, Shoko Nishiyama, Nandadeva Lokugamage, Terence E. Hill, Matthew B. Huante, Olga A.L. Slack, Victor H. Carpio, Alexander N. Freiberg and Tetsuro Ikegami
N-Glycans on the Rift Valley Fever Virus Envelope Glycoproteins Gn and Gc Redundantly Support Viral Infection via DC-SIGN
Reprinted from: *Viruses* **2016**, *8*(5), 149; doi: 10.3390/v8050149
http://www.mdpi.com/1999-4915/8/5/149 65

Martin Spiegel, Teresa Plegge and Stefan Pöhlmann
The Role of Phlebovirus Glycoproteins in Viral Entry, Assembly and Release
Reprinted from: *Viruses* **2016**, *8*(7), 202; doi: 10.3390/v8070202
http://www.mdpi.com/1999-4915/8/7/202 79

Sylvia Rothenberger, Giulia Torriani, Maria U. Johansson, Stefan Kunz and Olivier Engler
Conserved Endonuclease Function of Hantavirus L Polymerase
Reprinted from: *Viruses* **2016**, *8*(5), 108; doi: 10.3390/v8050108
http://www.mdpi.com/1999-4915/8/5/108 99

Amelina Albornoz, Anja B. Hoffmann, Pierre-Yves Lozach and Nicole D. Tischler
Early Bunyavirus-Host Cell Interactions
Reprinted from: *Viruses* **2016**, *8*(5), 143; doi: 10.3390/v8050143
http://www.mdpi.com/1999-4915/8/5/143 114

Myriam Ermonval, Florence Baychelier and Noël Tordo
What Do We Know about How Hantaviruses Interact with Their Different Hosts?
Reprinted from: *Viruses* **2016**, *8*(8), 223; doi: 10.3390/v8080223
http://www.mdpi.com/1999-4915/8/8/223 136

Marko Zivcec, Florine E. M. Scholte, Christina F. Spiropoulou, Jessica R. Spengler and Éric Bergeron
Molecular Insights into Crimean-Congo Hemorrhagic Fever Virus
Reprinted from: *Viruses* **2016**, *8*(4), 106; doi: 10.3390/v8040106
http://www.mdpi.com/1999-4915/8/4/106 153

William C. Wilson, A. Sally Davis, Natasha N. Gaudreault, Bonto Faburay, Jessie D. Trujillo, Vinay Shivanna, Sun Young Sunwoo, Aaron Balogh, Abaineh Endalew, Wenjun Ma, Barbara S. Drolet, Mark G. Ruder, Igor Morozov, D. Scott McVey and Juergen A. Richt
Experimental Infection of Calves by Two Genetically-Distinct Strains of Rift Valley Fever Virus
Reprinted from: *Viruses* **2016**, *8*(5), 145; doi: 10.3390/v8050145
http://www.mdpi.com/1999-4915/8/5/145 174

Jennifer Deborah Wuerthand Friedemann Weber
Phleboviruses and the Type I Interferon Response
Reprinted from: *Viruses* **2016**, *8*(6), 174; doi: 10.3390/v8060174
http://www.mdpi.com/1999-4915/8/6/174 191

Amber M. Riblett and Robert W. Doms
Making Bunyaviruses Talk: Interrogation Tactics to Identify Host Factors Required for Infection
Reprinted from: *Viruses* **2016**, *8*(5), 130; doi: 10.3390/v8050130
http://www.mdpi.com/1999-4915/8/5/130 208

About the Guest Editors

Pierre-Yves Lozach obtained his Ph.D degree in Virology at the Pasteur Institute (France) in 2004. He joined the lab of Ari Helenius (ETH Zurich, Switzerland) as a Marie-Curie postdoc fellow in 2007. He was appointed as tenure-track Assistant-Professor at the Armand Frappier Institute (Canada) in 2011, and then, granted the CellNetworks group leader position at the University of Heidelberg (Germany) in 2013.

Yizhi Jane Tao, Ph.D., is a biochemist, structural biologist, and Professor of Biochemistry and Cell Biology at Rice University in Houston, Texas. Born in China, Tao received a B.Sc. degree in Biology from Peking University in Beijing, China, in 1992. She later moved to West Lafayette, Indiana, where she received her Ph.D. in Biological Sciences while studying under Michael Rossmann. She completed a postdoctoral fellowship under Stephen Harrison at Harvard University in 2002. Upon completing her postdoctoral studies, Tao joined the faculty of Rice University, where she has made important contributions to the study of several RNA viruses, including influenza, hepatitis, astro, and birnaviruses.

Preface to "Recent Progress in Bunyavirus Research"

Over the last 25 years, scientific and public attention to bunyaviruses has grown considerably. There have been an increasing number of reports on new emerging bunyaviruses and infection episodes, including those causing Crimean–Congo hemorrhagic fever, Rift Valley fever, and severe fever with thrombocytopenia syndrome in humans, but also the Schmallenberg virus that infects cattle in Europe. With over 350 isolates distributed worldwide, the *Bunyaviridae* is the largest family of RNA viruses and is grouped into five genera, namely *Hantavirus*, *Orthobunyavirus*, *Nairovirus*, *Tospovirus*, and *Phlebovirus*. The genome of bunyaviruses contains three negative-sense, single-stranded RNA segments that encode a total of five to six proteins. Many bunyaviruses, which are carried and transmitted by either arthropods or rodents, are significant human or domestic animal pathogens. With international trade, travel, and climate change favoring the spread of vectors to new areas, bunyaviruses are emerging and re-emerging agents of disease that represent a global threat for agricultural productivity and public health.

Thus far, only a limited number of bunyaviruses have been investigated, with most of the available information coming from studies of a sprinkling of isolates. However, it is apparent that there is a wide variety of isolates, vectors, hosts, diseases, and geographical distributions. This diversity is also manifested at the genetic, cellular and molecular levels, as substantial differences are observed in the genomic organization, virion structure and architecture, transmission, tropism, host recognition, and cell entry mechanisms. However, the bunyavirus field has witnessed many exciting new findings and breakthroughs in recent years. These discoveries span a wide spectrum of research areas, which we intend to highlight in this book through several reviews and original research articles.

Briefly, genome-based analysis of *Nairo-* and *Orthobunyavirus* by **Kuhn et al.** and **the group of Sergei Alkhovsky** have led to the identification of new viruses and shed light on the phylogenetic lineages within these genera. The work by **Wilson and colleagues** on calf infection by Rift Valley fever virus (RVFV) opens larger perspectives for future investigations in vivo. The new molecular insights into the endonuclease activity of the hantavirus polymerase L from the study by **Rothenberger et al.** improve our understanding of hantavirus replication. Furthermore, different steps of the bunyavirus life cycle are documented here with, for instance, research on the role of *N*-glycans in the RVFV glycoprotein G_N expression, and also, in the RVFV infection via the receptor DC-SIGN (**Phoenix et al.**).

All of these new findings are further discussed through several thorough reviews, covering many topics such as the different hosts of hantaviruses (**Ermonval et al.**), the nairovirus Crimean Hemorrhagic Congo (**Zivcec et al.**), the molecular interplay between bunyaviruses and innate immunity (**Wuerth et al.**), the early bunyavirus–host cell interactions (**Albornoz et al.**), and the role of viral glycoproteins in viral entry, assembly, and release (**Spiegel et al.**). The review by **Riblett and Doms**, which discusses high-throughput screening approaches and the hundreds of cellular factors with a potential role in the bunyavirus life cycle, perfectly illustrates the recent research achievements made in the field.

Lastly, we would like to thank all contributing authors for their participation. Without their hard work, this book would have not been possible. We are also indebted to Dr. Delphine Guérin, who is the *Viruses* Managing Editor, for her patience and help along the way. Through collective efforts, we hope this book will provide the bunyavirus field an informed perspective of future research directions and also stimulate research in some of the understudied areas.

Pierre-Yves Lozach and Jane Tao

Guest Editors

Obituary

Homage to Richard M. Elliott

Alain Kohl [1], Benjamin Brennan [1] and Friedemann Weber [2,*]

1 MRC-University of Glasgow Centre for Virus Research, Glasgow G61 1QH, Scotland, UK; alain.kohl@glasgow.ac.uk (A.K.); Ben.Brennan@glasgow.ac.uk (B.B.)

2 Institute for Virology, FB10—Veterinary Medicine, Justus-Liebig University, Gießen 35392, Germany

* Correspondence: friedemann.weber@vetmed.uni-giessen.de; Tel.: +49-641-99-38351

Academic Editors: Jane Tao and Pierre-Yves Lozach

Received: 3 August 2016; Accepted: 8 August 2016; Published: 10 August 2016

In the last 25 years, the scientific and public attention paid to bunyaviruses has increased considerably. This has many reasons (one of them being that new family members are constantly emerging) and many drivers, but there was one man whose name will be forever connected with the *Bunyaviridae* family. Richard M. Elliott passed away in 2015 at the age of 61. With his unstoppable enthusiasm, strong vision, perseverance, and his keen interest in almost every aspect of bunyavirus replication, he greatly contributed to the progress in the field [1]. While his most prominent achievement may be the first rescue of a segmented negative-strand RNA virus (the type species Bunyamwera) from plasmid cDNA, he also studied glycoprotein processing, particle assembly, anti-interferon mechanisms, phylogeny, vaccine development and host cell factors. Yes, these are the main topics of this special issue of *Viruses*, and Richard could have chosen any of them to write a competent review himself. Lamentably, this is not possible any more, but we will always remember the man who helped to foster so much of the past research and train the next generation of scientists studying the "Cinderellas of virology" (in his own words) which in reality constitute one of the biggest virus families ever known.

References

1. Brennan, B.; Weber, F.; Kormelink, R.; Schnettler, E.; Bouloy, M.; Failloux, A.B.; Weaver, S.C.; Fazakerley, J.K.; Fragkoudis, R.; Harris, M.; et al. In memoriam—Richard M. Elliott (1954–2015). *J. Gen. Virol.* **2015**, *96*, 1975–1978. [CrossRef] [PubMed]

© 2016 by the authors. Licensee MDPI, Basel, Switzerland. This article is an open access article distributed under the terms and conditions of the Creative Commons Attribution (CC BY) license (http://creativecommons.org/licenses/by/4.0/).

Article

Genomic Characterization of the Genus *Nairovirus* (Family *Bunyaviridae*)

Jens H. Kuhn [1], Michael R. Wiley [2], Sergio E. Rodriguez [3], Yīmíng Bào [4], Karla Prieto [2], Amelia P. A. Travassos da Rosa [3], Hilda Guzman [3], Nazir Savji [5], Jason T. Ladner [2], Robert B. Tesh [3], Jiro Wada [1], Peter B. Jahrling [1], Dennis A. Bente [3] and Gustavo Palacios [2,*]

1 Integrated Research Facility at Fort Detrick, Division of Clinical Research, National Institute of Allergy and Infectious Diseases, National Institutes of Health, Frederick, MD 21702, USA; kuhnjens@mail.nih.gov (J.H.K.); wadaj@mail.nih.gov (J.W.); jahrlingp@niaid.nih.gov (P.B.J.)

2 Center for Genome Sciences, United States Army Medical Research Institute of Infectious Diseases, Fort Detrick, Frederick, MD 21702, USA; michael.r.wiley19.ctr@mail.mil (M.R.W.); karla.prieto.ctr@mail.mil (K.P.); jason.t.ladner.ctr@mail.mil (J.T.L.)

3 Galveston National Laboratory, Institute for Human Infection and Immunity, Department of Microbiology & Immunology, University of Texas Medical Branch, Galveston, TX 77555, USA; seerodri@utmb.edu (S.E.R.); aptravas@utmb.edu (A.P.A.T.d.R.); hguzman@utmb.edu (H.G.); rtesh@utmb.edu (R.B.T.); dabente@utmb.edu (D.A.B.)

4 Information Engineering Branch, National Center for Biotechnology Information, National Library of Medicine, National Institutes of Health, Bethesda, MD 20892, USA; bao@ncbi.nlm.nih.gov (Y.B.)

5 School of Medicine, New York University, New York, NY 10016, USA; nazir.savji@gmail.com (N.S.)

* Correspondence: gustavo.f.palacios.ctr@mail.mil; Tel.: +1-301-619-8732

Academic Editors: Jane Tao and Pierre-Yves Lozach
Received: 2 April 2016; Accepted: 26 May 2016; Published: 10 June 2016

Abstract: *Nairovirus*, one of five bunyaviral genera, includes seven species. Genomic sequence information is limited for members of the *Dera Ghazi Khan*, *Hughes*, *Qalyub*, *Sakhalin*, and *Thiafora nairovirus* species. We used next-generation sequencing and historical virus-culture samples to determine 14 complete and nine coding-complete nairoviral genome sequences to further characterize these species. Previously unsequenced viruses include Abu Mina, Clo Mor, Great Saltee, Hughes, Raza, Sakhalin, Soldado, and Tillamook viruses. In addition, we present genomic sequence information on additional isolates of previously sequenced Avalon, Dugbe, Sapphire II, and Zirqa viruses. Finally, we identify Tunis virus, previously thought to be a phlebovirus, as an isolate of Abu Hammad virus. Phylogenetic analyses indicate the need for reassignment of Sapphire II virus to *Dera Ghazi Khan nairovirus* and reassignment of Hazara, Tofla, and Nairobi sheep disease viruses to novel species. We also propose new species for the Kasokero group (Kasokero, Leopards Hill, Yogue viruses), the Ketarah group (Gossas, Issyk-kul, Keterah/soft tick viruses) and the Burana group (Wēnzhōu tick virus, Huángpí tick virus 1, Tǎchéng tick virus 1). Our analyses emphasize the sister relationship of nairoviruses and arenaviruses, and indicate that several nairo-like viruses (Shāyáng spider virus 1, Xīnzhōu spider virus, Sānxiá water strider virus 1, South Bay virus, Wǔhàn millipede virus 2) require establishment of novel genera in a larger nairovirus-arenavirus supergroup.

Keywords: *Bunyaviridae*; bunyavirus; nairovirus; Dera Ghazi Khan virus; Erve virus; Ganjam virus; Hughes virus; Qalyub virus; Sakhalin virus; Tunis virus; virus classification; virus taxonomy

---

1. Introduction

With over 530 members, *Bunyaviridae* is one of the largest virus families [1]. Bunyaviruses are characterized by single-stranded RNA genomes that typically consist of separate small (S), medium (M), and large (L) segments, all of which have complementary 3′ and 5′ termini. Most bunyavirus genomes are of negative polarity, but some viruses use ambisense strategies to express their proteins [1,2]. The S, M, and L segments encode the structural nucleoprotein (NP), glycoprotein precursor (GPC), and RNA-dependent RNA polymerase (L) proteins, respectively [1]. Nonstructural proteins are encoded by several, but not all bunyaviruses, by either the S or M or by both S and M segments. Bunyavirions enter host cells by engaging cell-surface receptors with their glycoproteins followed by endocytosis and release of genomes. The viruses typically replicate in the cytosol of infected cells and produce progeny virions that bud from cellular membranes derived from the Golgi apparatus via exocytosis [3].

The family *Bunyaviridae* currently includes five recognized genera: *Hantavirus*, *Nairovirus*, *Orthobunyavirus*, *Phlebovirus*, and *Tospovirus* [1]. Family members have been assigned to these genera, and within genera to species, based primarily on serological cross-reactions, characteristic genus-specific genome segment termini sequences, host association (invertebrates, vertebrates or plants), transmission pathways (arthropod-borne *versus* vertebrate excreta-driven) and, until recently, very limited genomic sequence information [1].

The genus *Nairovirus* includes seven species that are accepted by the International Committee on Taxonomy of Viruses (ICTV) [1]. Most of these species have several distinct members, all of which are either maintained in arthropods or transmitted by ticks among bats, birds, eulipotyphla, or rodents. The most important nairovirus with public-health impact is the tick-borne Crimean-Congo hemorrhagic fever virus (CCHFV), which causes a frequently lethal viral hemorrhagic fever in Western Asia, the Balkans, Southern Europe, and most of Africa [3]. The most important nairoviruses of veterinary importance are the tick-borne Nairobi sheep disease and Ganjam viruses (NSDV and GANV, respectively), which are known to cause lethal hemorrhagic gastroenteritis in small ruminants in Africa and India [4].

The typical nairovirus genome is approximately 18.8 kb in length (S: $\approx$1.7 kb; M: $\approx$4.9 kb; L: $\approx$12.2 kb) and characterized by the genus-specific 3′ segment terminus AGAGUUUCU and 5′ segment terminus AGAAACUCU. Classical nairovirions are enveloped spheres (80–120 nm in diameter) spiked with heterodimeric glycoprotein projections consisting of the cleavage products of the glycoprotein precursor (Gn and Gc) [3].

Next-generation sequencing followed by coding-complete or complete genomic sequence assembly (see [5] for sequencing nomenclature) is increasingly used to classify previously uncharacterized phleboviruses [6–14] and orthobunyaviruses [15–21] and to characterize novel bunyavirus clades, such as "goukoviruses," "herbeviruses," "phasmaviruses," and the Ferak and Jonchet virus groups [22–24]. Several unclassified bunyaviruses and viruses assigned to bunyaviral genera other than *Nairovirus* have been identified as bona fide nairoviruses [25–34]. At least one classified nairovirus was identified as an actual phlebovirus [14]. Novel nairoviruses have been discovered in bats [25,27,29,35,36], and in arachnids, millipedes, and water striders [37–40]. Even more interestingly, at least two nairo-like viruses with only bisegmented genomes have been reported [37,41]. Shortly before this manuscript was submitted, Walker *et al.* reported the coding-complete sequences of 11 nairoviruses (Abu Hammad virus (AHV), Avalon virus (AVAV), Bandia virus (BDAV), Dera Ghazi Khan virus (DGKV), Erve virus (ERVEV), Farallon virus (FARV), GANV, Punta Salinas virus (PSV), Qalyub virus (QYBV), Taggert virus (TAGV), and Zirqa virus (ZIRV)) [42]. An overview of all viruses currently thought to be nairoviruses or nairo like-viruses, and their relationships based on data prior to this study are provided in Table S1.

As is evident from the table, genomic sequence information for nairoviruses is still limited. Here we report either the coding-complete or complete genomic sequences of 23 nairoviruses (Table 1). Ten of these sequences have also been determined by Walker *et al.* [42]. Four of the 23 sequences are for novel strains of previously sequenced nairoviruses. Nine of the 23 sequences are new from previously unsequenced viruses. We extended 14 sequences to include all of the 3′ and 5′ genome segment termini. Our subsequent phylogenetic analyses indicate a number of changes in the organization of nairoviruses. At least five new nairovirus species ought to be established. GANV should be considered an isolate of NSDV, and soft tick bunyavirus should be considered an isolate of Keterah virus (KRTV). Tunis virus (TUNV), which was serologically identified as a phlebovirus, is an isolate of AHV in the *Dera Ghazi Khan nairovirus* species. At least seven nairo-like viruses should be classified into novel genera, and these genera and all nairoviruses are more closely related to arenaviruses than to other bunyaviruses.

Table 1. Viruses sequenced for this study. NCR, noncoding regions; RSFSR, Russian Soviet Federated Socialist Republic; USSR, United Soviet Socialist Republic.

Virus Name (Abbreviation)	Strain Designation	Source	Date; Place of Isolation	Ref.	BioSampleID GenBank Accession Numbers	L 5′ NCR	L 3′ NCR	M 5′ NCR	M 3′ NCR	S 5′ NCR	S 3′ NCR
Abu Hammad virus (AHV)	Eg ArT 1194	Ticks (*Argas hermanni*) collected from pigeon	7 June 1971; Abu Hammad, al-Sharqia Governorate, Egypt	[43]	Re-sequenced [42]: SAMN04530531 S: KU925436 M: KU925435 L: KU925434	Yes	Yes	Yes	Yes	Yes	Yes
Abu Mina virus (AMV)	Eg An 4996-63	European turtle dove (*Streptopelia turtur*) and associated ticks (*Argas streptopelia*)	1 May 1963; Abu Mina, Matrouh Governorate, Egypt	[43]	Newly sequenced: SAMN04530533 S: KU925439 M: KU925438 L: KU925437	Yes	Yes	Yes	Yes	Yes	Yes
Avalon virus (AVAV)	Brest/Ar T261	Ticks (*Ixodes uriae*)	1979; Brittany, France	[44]	Newly sequenced: SAMN04530548 S: KU925445 M: KU925444 L: KU925443	Yes	Yes	Yes	Yes	Yes	Yes
Avalon virus (AVAV)	CanAr 173	Ticks (*Ixodes uriae*) from European herring gull (*Larus argentatus*)	31 July 1972; Great Island, Newfoundland and Labrador, Canada	[45]	Re-sequenced [42]: SAMN04530547 S: KU925442 M: KU925441 L: KU925440	Yes	Yes	Yes	Yes	Yes	Yes
Bandia virus (BDAV)	IPD/A 611	Rodent (*Mastomys* sp.) and ticks (*Ornithodoros sonrai*) collected from rodent burrow	26 February 1965; Bandia Forest, Thiès Region, Senegal	[46]	Re-sequenced [42]: SAMN04530545 S: KU925448 M: KU925447 L: KU925446	No	No	No	No	No	No
Clo Mor virus (C[L]MV)	ScotAr 7	Ticks (*Ixodes uriae*) in nesting sites of common murees (*Uria aalge*)	15 June 1973; Clo Mor, Cape Wrath, Scotland, UK	[45]	Newly sequenced: SAMN04530553 S: KU925451 M: KU925450 L: KU925449	No	No	No	No	No	No
Dera Ghazi Khan virus (DGKV)	JD 254	Ticks (*Hyalomma dromedarii*) collected from a camelid	4 April 1966; Dera Ghazi Khan District, Punjab Province, Pakistan	[47]	Re-sequenced [42]: SAMN04530534 S: KU925454 M: KU925453 L: KU925452	Yes	Yes	Yes	Yes	Yes	Yes
Dugbe virus (DUGV)	IbAr 1792	Ticks (*Amblyomma variegatum*) collected from cattle	14 October 1964; Ibadan, Oyo State, Nigeria	[48]	Newly sequenced: SAMN04530543 S: KU925457 M: KU925456 L: KU925455	Yes	Yes	Yes	Yes	Yes	Yes
Erve virus (ERVEV)	Brest/An 221 (TVP21049)	Greater white-toothed shrew (*Crocidura russula*)	5 May 1982; Saulges, Mayenne Département, France	[49]	Re-sequenced [42]: SAMN04530552 S: KU925460 M: KU925459 L: KU925458	No	No	No	No	Yes	Yes
Farallon virus (FARV)	Cal Ar846	Ticks (*Carios denmarki*)	20 July 1965; Farallon Islands, California, USA	[50]	Re-sequenced [42]: SAMN04530536 S: KU925463 M: KU925462 L: KU925461	Yes	Yes	Yes	Yes	Yes	Yes
Ganjam virus (GANV)	G 619 (TVP20486)	Ticks (*Haemaphysalis intermedia*) collected from a domestic goat	6 November 1954; Bhanjanagar, Ganjam District, Orissa, India	[51]	Re-sequenced (Yadav *et al.*, unpublished) SAMN04530544 S: KU925466 M: KU925465 L: KU925464	No	Yes	No	No	No	No

Table 1. *Cont.*

Virus Name (Abbreviation)	Strain Designation	Source	Date; Place of Isolation	Ref.	BioSampleID GenBank Accession Numbers	L 5′ NCR	L 3′ NCR	M 5′ NCR	M 3′ NCR	S 5′ NCR	S 3′ NCR
Great Saltee virus (GRSV)	RML 59972	Ticks (*Carios maritimus*) collected from a seabird nest	1972; Great Saltee Island, County Wexford, Ireland	[52]	Newly sequenced: SAMN04530537 S: KU925469 M: KU925468 L: KU925467	Yes	Yes	Yes	Yes	Yes	Yes
Hughes virus (HUGV)	G2126	Ticks (*Carios denmarki*)	January, 1962; Bush Key, Dry Tortugas, Florida, USA	[53,54]	Newly sequenced: SAMN04530538 S: KU925472 M: KU925471 L: KU925470	Yes	Yes	Yes	Yes	Yes	Yes
Punta Salinas virus (PSV)	Cal Ar888	Ticks (*Carios amblus*)	14 October 1967; Punta Salinas, Huaura Province, Lima Region, Peru	[55]	Re-sequenced [42]: SAMN04530539 S: KU925475 M: KU925474 L: KU925473	No	No	No	No	No	No
Qalyub virus (QYBV)	Eg Ar 370	Ticks (*Carios erraticus*) collected from a rat nest	28 August 1952; Qalyub, al-Qalyubiyah Governorate, Egypt (British Protectorate)	[56]	Re-sequenced [42]: SAMN04530546 S: KU925478 M: KU925477 L: KU925476	Yes	Yes	Yes	Yes	Yes	Yes
Raza virus (RAZAV)	829	Ticks (*Carios denmarki*)	20 May 1962; Raza Island, Baja California, Mexico	[57]	Newly sequenced: SAMN04530540 S: KU925481 M: KU925480 L: KU925479	Yes	Yes	Yes	No	Yes	Yes
Sakhalin virus (SAKV)	LEIV-71C	Ticks (*Ixodes uriae*) collected from nesting sites of common murees (*Uria aalge*)	21 November 1969; Tyuleniy Island, Sea of Okhotsk, Sakhalin Oblast, RSFSR, USSR	[58]	Newly sequenced: SAMN04530549 S: KU925484 M: KU925483 L: KU925482	No	No	No	Yes	No	No
Sapphire II virus (SAPV)	RML 52323-14	Ticks (*Argas cooleyi*) collected from a cliff swallow nest	August 1969; Garza County, Texas, USA	[59]	Newly sequenced: SAMN04530535 S: KU925487 M: KU925486 L: KU925485	Yes	Yes	Yes	Yes	Yes	Yes
Soldado virus (SOLV)	TRVL 52214	Ticks (*Carios capensis*)	16 June 1963; Soldado Rock, Trinidad and Tobago	[60]	Newly sequenced: SAMN04530541 S: KU925490 M: KU925489 L: KU925488	Yes	Yes	Yes	Yes	Yes	Yes
Taggert virus (TAGV)	MI14850	Ticks (*Ixodes uriae*) from seabird rookery	1 January 1972; Macquarie Island, Tasmania, Australia	[61]	Re-sequenced [42]: SAMN04530550 S: KU925493 M: KU925492 L: KU925491	No	Yes	No	Yes	Yes	Yes
Tillamook virus (TILLV)	RML 86	Ticks (*Ixodes uriae*)	1970; Oregon, USA	[62]	Newly sequenced: SAMN04530551 S: KU925496 M: KU925495 L: KU925494	Yes	Yes	Yes	Yes	Yes	Yes
Tunis virus (TUNV)	Brest/Ar/T2756	Ticks (*Argas hermanni*)	October 1989; El Kef, Kef Governorate, Tunisia	[63]	Newly sequenced: SAMN04530532 S: KU925499 M: KU925498 L: KU925497	Yes	Yes	Yes	Yes	Yes	Yes
Zirqa virus (ZIRV)	POR7866	Ticks (*Carios muesebecki*)	2 November 1969; Zirku (Zirqa/Zarrakuh) Island, Abu Dhabi, United Arab Emirates	[64]	Newly sequenced: SAMN04530542 S: KU925502 M: KU925501 L: KU925500	Yes	Yes	No	Yes	No	Yes

2. Materials and Methods

2.1. Viruses

The viruses used in this study were obtained from the World Reference Center for Emerging Viruses and Arboviruses at the University of Texas Medical Branch, Galveston, TX, USA. All of these viruses have been described before. Table 1 provides specifics about the viruses and GenBank accession numbers for all newly sequenced and re-sequenced viruses.

2.2. Genome Sequencing

Viral stocks were obtained in TRIzol LS (Invitrogen, Carlsbad, CA, USA), and RNAs were extracted using the Direct-zol™ RNA MiniPrep kit (Zymo, Irvine, CA, USA). RNAs were converted to cDNAs and amplified using sequence-independent single primer amplification as described previously [65] with some modifications to resolve the 5′ and 3′ ends. An oligonucleotide containing three ribonucleotides (rGTP) at the 3′ end (GCCGGAGCTCTGCAGATATCGGCCATTAT GGCCrGrGrG) and the FR40RV-T primer [65] were added during first-strand cDNA synthesis. The reverse transcriptase was changed to Maxima H Minus reverse transcriptase (Thermo Fisher Scientific, Waltham, MA, USA), which has terminal transferase activity that adds the rGTP-containing oligonucleotide to the 5′ end during cDNA synthesis. cDNA was sheared to ≈400 bp in length and used as starting material for creation of Illumina TRUseq DNA libraries. Sequencing was performed on either an Illumina MiSeq or NextSeq desktop sequencer using 300-cycle kits (2 × 150). Open-source Cutadapt [66] and Prinseq-lite [67] were used to trim primers and remove poor quality reads, respectively. Reads were assembled into contigs using open-source Ray Meta [68]. Annotation was determined using basic local alignment search tool (BLAST) in combination with custom scripts. Contigs related to nairovirus sequences were used as references.

2.3. Phylogenetic Analysis

A set of nairovirus sequences (252 for the N gene of the S segment, 111 for the M segment, and 93 for the L segment) comprising the majority of the nucleotide (nt) sequences from GenBank available on 1 March 2016, were aligned using the CLUSTAL algorithm. Because the nairovirus sequences of all analyzed nairoviruses were so different that the alignment reached substitution saturation (no detection of signal), alignments were instead implemented at the amino acid (aa) level (using MEGA Version 5 [69]). Non-coding regions of S segments therefore had to be excluded. Additional manual editing was performed to ensure the highest possible quality of alignments. Neighbor-joining (NJ) analysis at the aa level was performed due to the observed high variability of the underlying nt sequences. The statistical significance of tree topology was evaluated by bootstrap re-sampling of the sequences 1000 times. Phylogenetic analyses were performed using MEGA Version 5.

2.4. Detection of Reassortant Events

Systematic screening for the presence of recombination patterns was pursued by using the nt alignments and the Recombination Detection Program (RDP [70]), Bootscan [71], maximum chi-square (MaxChi) [72], Chimaera [73], Likelihood Analysis of Recombination in DNA (LARD) [74], and Phylip Plot [75].

2.5. Sequence Analysis

Geneious 4.8.3 (Biomatters Inc., Newark, NJ, USA) was used for sequence assembly and analysis. Topology, sizes, and targeting predictions were generated by employing SignalP 4.1, NetOGlyc 4.0, NetNGlyc 1.0, Prop 1.0, tied mixture hidden Markov model (TMHMM) 2.0 [76], SnapGene Viewer 2.82 [77], the web-based version of TopPred2 [78], and integrated predictions in Geneious [79–83].

3. Results

3.1. Genomic Characterization and Phylogenetic Analysis

Consistent with the genomic organization characteristic for already sequenced nairoviral genomes, each of the 23 viral genomes sequenced during this study is comprised of three RNA segments including (a) a small (S) segment encoding the NP and, in an ambisense orientation, a non-structural protein (NSs); (b) a medium (M) segment encoding a GPC; and (c) a large (L) segment encoding an RNA-dependent RNA polymerase. Fourteen nairovirus genomes were completely characterized. The 3′ terminal sequences were obtained for 57 segments, and the 5′ terminal sequences were obtained for 51 segments (Table 1). For most of the viral genomes sequenced in this study, the nine most terminal nucleotides of each segment were identical to those previously reported for nairoviruses (3′ segment terminus AGAGUUUCU and 5′ segment terminus AGAAACUCU) [1,3]. However, the Abu Hammad virus (AHV), Abu Mina virus (AMV), Dera Ghazi Khan virus (DGKV), Sapphire II virus (SAPV), and Tunis virus (TUNV) genome segments have termini that differ by one nt (AGAGUUUCA and TGAAACUCU). Likewise, the Qalyub virus (QYBV) genomic segments termini differ from the consensus sequences by one nt (AGAGAUUCU and AGAATCUCU).The results of phylogenetic analyses of the newly obtained L, M, and S segment sequences are shown in Figures 1–3.

The phylogenetic placement of the newly sequenced viruses is largely consistent with their previous serological classification, including recent amendments [42] (Table S1). However, Hazara virus (HAZV) and Tofla virus (TOFV) clustered with each other but not with CCHFV and, therefore, should not be classified in the species *Crimean-Congo hemorrhagic fever nairovirus*. Likewise, both Kupe virus (KUPEV) and Nairobi sheep disease virus (NSDV) did not cluster with Dugbe virus (DUGV), and, therefore, should be removed from the species *Dugbe nairovirus* and re-assigned to new species (here proposed as "*Hazara nairovirus*" (HAZV, TOFV)) and "*Nairobi sheep disease virus*" (NSDV), respectively). Ganjam virus (GANV) is clearly identified as an isolate of NSDV. Our analysis confirm that Leopards Hill virus (LPHV), Kasokero virus (KAS(O)V), and Yogue virus (YOGV) form a novel nairovirus genogroup (proposed species "*Kasokero nairovirus*"), as do Keterah virus (KRTV) and Issyk-kul virus (ISKV) (proposed species "*Keterah virus*") [29,42]. The recently described soft tick bunyavirus [38] is identified as an isolate of KRTV. Genetic characterization of TUNV clearly demonstrates that this virus is a nairovirus and not a phlebovirus as previously described by serological analysis [63]. The TUNV genome represents an isolate of AHV, indicating necessary classification into the species *Dera Ghazi Khan nairovirus*. Another new species, proposed to be named "*Burana nairovirus*" should be established for Wēnzhōu tick virus, Huángpí tick virus 1, and Tǎchéng tick virus 1. Finally, the phylogenetic trees demonstrate that several nairo-like viruses with three (Shāyáng spider virus 1, Xīnzhōu spider virus (XSV), Sānxiá water strider virus 1 (SWSV-1), South Bay virus (SBV)) or two genomic segments (SBV, Wǔhàn millipede virus 2) should not be classified in the genus *Nairovirus*.

Although genomic segment reassortment has been found very frequently among CCHFV strains and lineages [84–88], we were unable to detect any instance of reassortment among the other nairoviruses using RDP, Bootscan, MaxChi, LARD and Phylip Plot. Phylogenetic incongruence was only detected in the case of HAZV: whereas the HAZV M and N open reading frames (ORFs) cluster together with those of NSDV/KUPEV, the HAZV L ORF does not. However, given the genetic distance between these sequences, whether this distance is the result of reassortment or saturation of the phylogenetic signal is not clear.

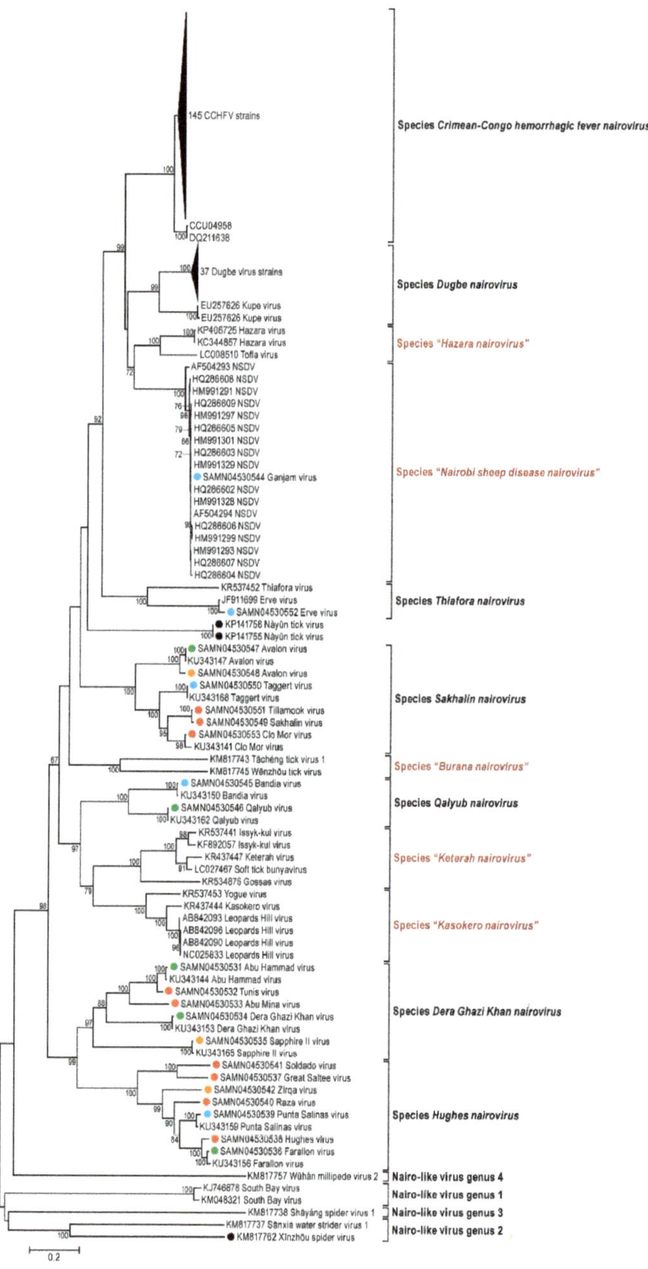

Figure 1. Phylogenetic analysis of nairovirus and nairo-like virus S segment N gene sequences, including newly determined virus sequences (red dots), newly determined virus isolate sequences (orange dots), re-sequenced genomes (blue dots), and re-sequenced genomes with genomic termini determined for the first time (green dots). Sequences marked with black dots correspond to partial sequences. Nairovirus sequences comprise all partial or complete sequences from GenBank available on 1 March 2016. Proposed new taxa are highlighted in red and placed in quotation marks.

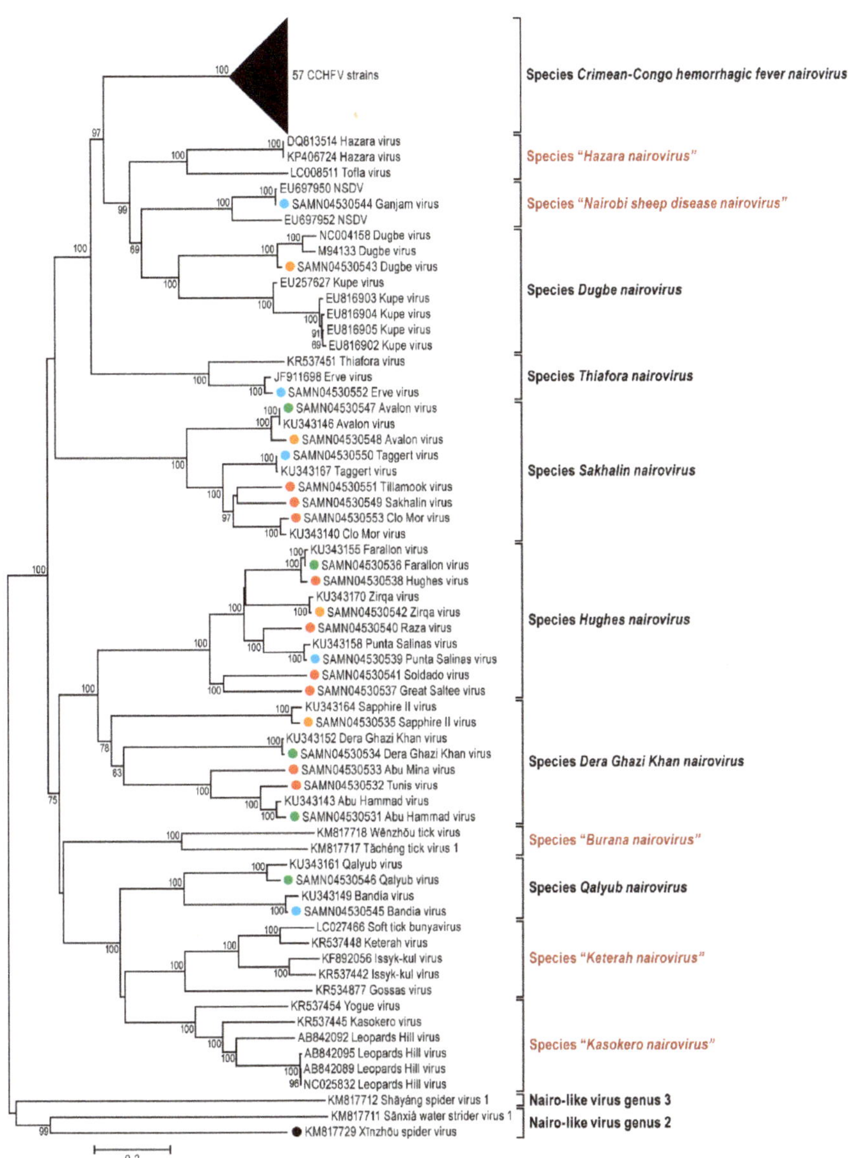

Figure 2. Phylogenetic analysis of nairovirus and nairo-like virus M segment sequences. Analysis was performed as outlined for Figure 1.

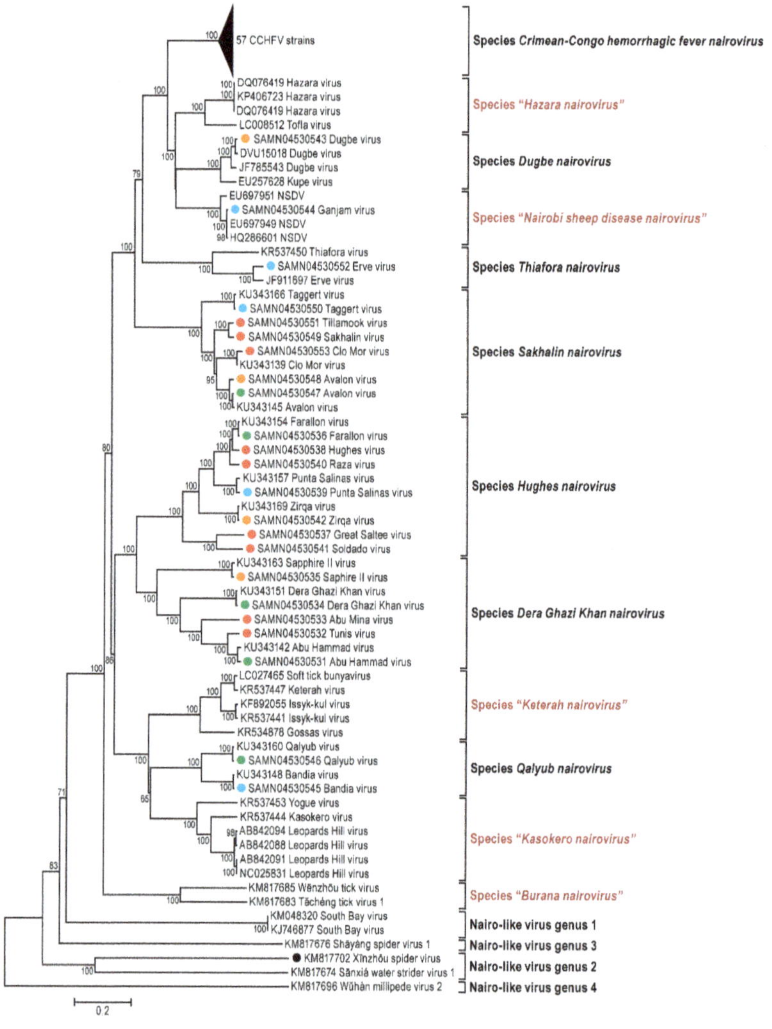

Figure 3. Phylogenetic analysis of nairovirus and nairo-like virus L segment sequences. Analysis was performed as outlined for Figure 1.

3.2. *Open Reading Frames*

3.2.1. Small (S) Segment—Nucleocapsid Protein

Nairoviral NPs may recognize specific nairoviral RNA sequences, bind non-specifically to single-stranded RNA, and form the ribonucleoprotein (RNP) complex [89–94]. The structure of NP has been determined for CCHFV, ERVEV, HAZV, and KUPEV [89,92–94]. All nairovirus NPs assume a racket-shaped structure with distinct "head" and "stalk" domains that are typical for bunyaviruses and unique among other negative-sense single-stranded RNA viruses.

In the case of CCHFV NP, two positively charged regions are responsible for RNA binding [92]. One region forms a large positively charged crevice (residues K339, K343, K346, R384, K411, H453, and Q457), of which two residues contribute to a conserved nairovirus motif (EH_{453}(L/M), (L/F)HQ_{457}). The other region is delineated by residues R134, R140, and Q468. The CCHFV NP stalk region also contains a positively charged region consisting of residues R195, H197, K222, R225, R282, and R286. Only three of the positively charged residues (R134, K222 and K343) are absolutely conserved among all nairovirus NPs, although most of the substitutions observed maintain the overall hydrophobicity profile.

CCHFV NP interacts with the antiviral defense factor MxA [95] and the apoptosis mediator caspase-3 [96]. Thus, we expected some degree of conservation of the NP areas mediating those functions. Using a CCHFV minireplicon system [89], three separate NP residues (K132, Q300, and K411) were identified to be essential for replicon activity, and mutation of another two residues (K90 and H456) resulted in significantly reduced NP functionality. However, only H456 and Q300 are completely conserved among nairovirus NPs.

Because protein structure is evolutionarily conserved to a higher degree compared to the primary aa sequence, we used homology modeling principles and techniques to identify conserved structures among nairovirus proteins. We used the Phyre2 (Protein homology/analogy recognition engine) server to model the structure of the proteins and to align remotely related sequences based on hidden Markov models (HMMs) (Figure 4).

Figure 4. Nairoviral nucleoproteins (NPs) similarity plot comparing typical features of nairoviral NPs. The NP sequence of Crimean-Congo hemorrhagic fever virus (CCHFV) is taken as the reference sequence. Ovals in orange and blue highlight two regions responsible for RNA binding.

In addition, upon identification of a conserved nairovirus protein structure, we performed mutational sensitivity analysis using the Disease-Susceptibility-based SAV Phenotype Prediction (SuSPect) tool [97], to predict whether missense mutations in a nairovirus protein are likely to functionally affect structure.

Finally, the mutational sensitivity score was plotted for each position against the average percent identity for that position in the nairovirus protein alignment. We assumed that if a mutation would have an influence on the structure of a conserved domain, the frequency of that mutation would be diminished. In summary, we expected that the positions with a higher score in the mutational sensitivity analysis should directly correlate with positions of higher conservation. The validity of this approach was tested initially with nairovirus NP sequences (Figure 5A,B). As expected, all nairovirus NPs were found to be structurally homologous to CCHFV NP, and the positional analysis revealed a direct correlation between positions of high mutational sensitivity with those of high identity (Figure 5A).

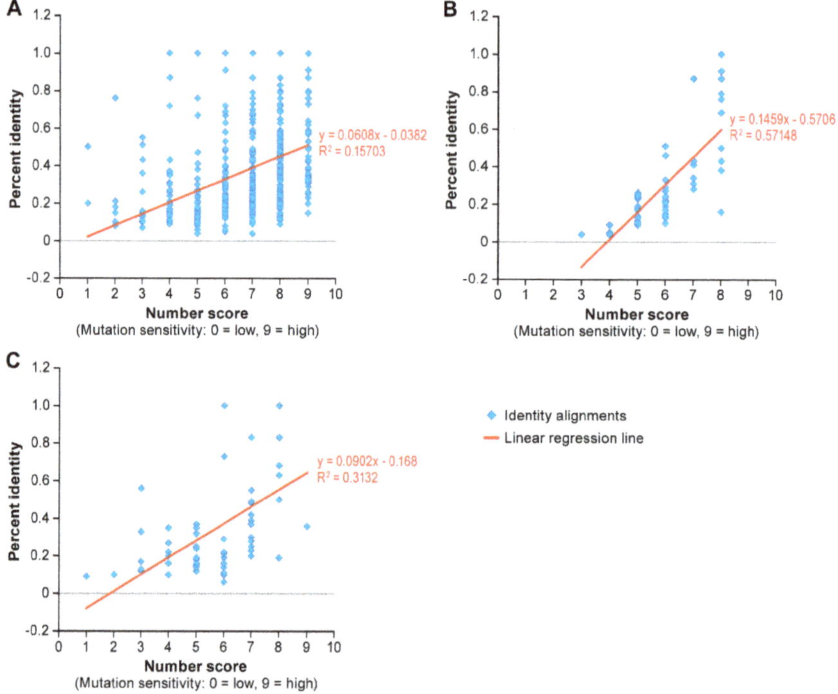

Figure 5. Mutational Sensitivity Analysis. The mutational sensitivity number score for each nairovirus NP amino-acid position (SuSPect) was plotted against the percent identity for that position in the nairovirus NP protein sequence alignment. (**A**) Full-length nairovirus NP; (**B**) myosin-4 motor protein-like NP domain; (**C**) ovarian tumor (OTU) domain of nairovirus L.

Interestingly, we identified two other nairovirus NP domains that were structurally similar to other known, non-viral structures. The first domain, at approximate position 150–200 of the nairovirus alignment, is similar to the globular tail of myosin-4 motor protein (type V myosin; confidence 60%; Protein Data Bank (PDB) ID: 3mmi). Myosin-4 motor protein is a monomeric myosin with motility uniquely adapted to transport mRNA [98]. Interestingly, SuSPect analysis suggested that the domain was sensitive to aa changes (Figure 5B). The second domain, located at the C-terminal NP domain (approximately at positions 460–481), is similar to the structures of the cholesterol-binding toxins intermedilysin (confidence 35%; PDB: 1s3r), perfringolysin (PDB: 1pfo) and pneumolysin (PDB: 4qqq), but SuSPect analysis did not show mutational sensitivity at these positions (data not shown).

3.2.2. Medium (M) Segment—Glycoprotein Precursor

ORF analysis of 36 sequenced nairovirus M segments generally yielded single unit polyprotein-encoding ORFs in each case (two units in the case of "*Burana nairovirus*" M segments). The predicted masses of the encoded unmodified polyproteins, the GPCs, ranged from 143 kDa (Thiafora virus (TFAV)) to 187 kDa (CCHFV). Each nairovirus polyprotein was approximated, via software modeling, to contain a signal peptide at the N-terminus. Experimental evidence garnered from CCHFV GPC processing [99–101] was used for functional element assignment in predicted nairovirus GPCs. Cleavage motifs for signal peptidases, furin (RSKR), subtilase SKI-I/S1P-like (RRLL and RKLL), and an unknown convertase that cleaves at the aa sequence RKPL, are highly conserved across CCHFV isolates and are critical post-translational motifs in the viral lifecycle [99–102]. Interestingly, the RSKR motif appears to be unique to CCHFV. The RKLL motif for SKI-I/S1P protease is conserved among nairovirus GPCs and is found in most members of all established and putative nairoviruses except in AVAV, DUGV, ERVEV, KAS(O)V, KUPEV, NSDV/GANV, and YOGV. The RKLL motif is not confined to any specific region/domain and occurs throughout the nairovirus GPCs—in some instances more than once (e.g., DUGV and members of *Qalyub nairovirus* and "*Keterah nairovirus*"). The RRLL motif is the second most prevalent cleavage site with the exception of viruses belonging to "*Keterah nairovirus*" and *Qalyub nairovirus*, which do not contain the motif. The RKPL and RRLL motifs are also conserved across nairoviruses of many species.

Using the four cleavage motifs (RSKR, RKPL, RRLL, RKLL), the nairovirus GPs stemming from GPC processing were predicted and annotated numerically, starting at the N-termini and depicted as colored arrows (Figure 6). We predicted the synthesis of two to five GPs depending on the examined nairovirus (Table S2 and Figure 6). To further explore GPC processing phenotypes, proprotein convertase prediction software was used to identify additional cleavage patterns. The most prevalent motifs predicted with a high degree of probability to mediate post-translation cleavage along the GPC were R-X-X-L, G-X-X-R, Q-X-X-C, and R-X-X-K, (data not shown and Figure 6). Other predicted motifs (with varying degrees of probability) are shown in sky blue and pink boxes within Figure 6.

Other post-translational modifications, such as, *N*- and *O*-linked glycosylations, were predicted for all analyzed nairoviruses using software modeling (NetOGlyc 4.0, NetNGlyc 1.0). Averages of *N*-linked glycosylation sites between viruses of different species ranged from seven to twenty sites along polyproteins. The most *N*-linked glycosylations were found for members of the "*Keterah nairovirus*" and *Sakhalin nairovirus* species, whereas the members of the *Dugbe* and *Hazara nairovirus* species had the fewest. The extent of nairovirus glycoprotein *O*-linked glycosylation ranged from four to over 130 sites (Table S2). The fewest *O*-glycans were predicted for members of the *Hughes nairovirus* and *Dera Ghazi Khan nairovirus* species, ranging from four to 17 sites. Intermediate *O*-linked glycosylation was predicted for *Hazara*, *Dugbe*, *Qalyub*, *Sakhalin*, and *Thiafora nairovirus* species members, with averages that ranged from 29 to 48 sites. The highest number of *O*-glycans (100 and 135 sites, respectively), were predicted for viruses of the *Crimean-Congo hemorrhagic fever nairovirus* and "*Keterah nairovirus*" species. Relatively few sites were predicted towards the C-termini of nairovirus GPC, with the exception of the "*Keterah nairovirus*" species members, which were predicted to contain a small cluster of *O*-linked sites between residues 994 and 1033.

Heavily *O*-glycosylated GPCs were typically characterized by site clustering towards the N-termini. O-glycosylated members of the *Dugbe*, "*Keterah*," and *Qalyub nairovirus* species contain RKPL and RKLL proteolytic motifs in these areas. If used by proteases, these motifs could mediate the production of separate, stand-alone peptides that are *O*-glycosylated. Such peptides have been identified in the cases of CCHFV and DUGV as mucin-like domains (MLDs) [103]. Additionally, viruses of the "*Leopards Hill nairovirus*" species encode GPC with shorter regions of *O*-linked glycosylation clustered towards the N-termini in the vicinity of predicted proteolytic cleavage sites. Of nairovirus genus members containing regions of O-glycosylation/MLDs, unmodified averaged masses ranged from eight to 37 kDa. Notably, of the GPC of seven viruses belonging to the *Hughes nairovirus* species, none were predicted to be *O*-glycosylated.

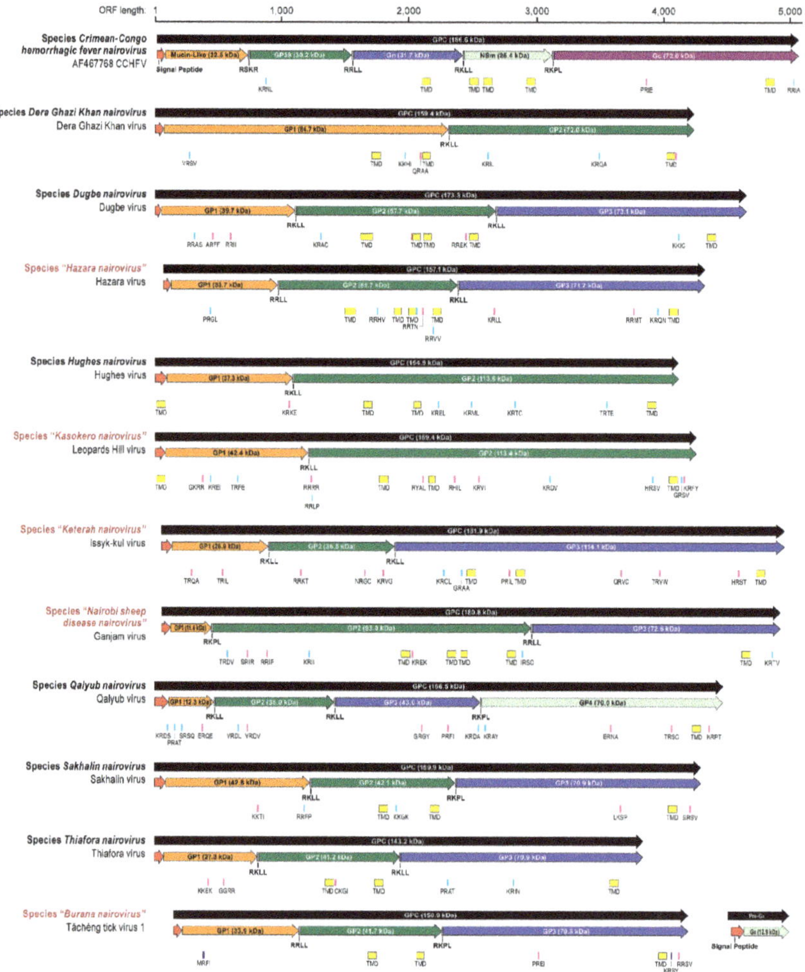

Figure 6. Nairovirus M segment-encoded polyprotein features and annotations. Type virus glycoprotein precursors (GPC) are represented for each species as black arrows. Putative glycoproteins (GP) are designated based on experimentally proven cleavages of CCHFV GPC (indicated in grey boxes separating glycoproteins). GPs are designated as colored arrows and numerically annotated beginning from the N-termini. The predicted molecular weights (kDa) of GPCs and putative GPs are annotated within colored arrows. Molecular weights were predicted without glycosylation. No designations of structural *versus* non-structural (Gn/Gc *vs.* NSm) proteins are listed due to lack of available experimental evidence across the genus. Signal peptides were predicted using posterior probability thresholds of 1.0 and 0.1 (SignalP 4.1) and are annotated with red arrows. Proprotein cleavage predictions (Prop 1.0) were analyzed for general convertase and furin predictions. Proprotein cleavage prediction thresholds of 0.3–0.49 are annotated with pink boxes, and thresholds of 0.5 or higher are annotated in sky blue. For both predictions, four-letter amino-acid sequences are provided below each colored box. Transmembrane domains (TMD) are annotated with yellow boxes and predicted using TMHMM 2.0. Two members of the species "*Burana nairovirus*" were predicted to have two open reading frames on the M-segment encoding a separate stand-alone glycoprotein (Pre-Gx) next to the polyprotein (GPC).

We also analyzed the nairovirus GPC for the occurrence, location, and topology of transmembrane regions using software modeling (TopPred2). The number of TMDs varied between one and five domains. Interestingly, viruses of the *Qalyub nairovirus* species had the fewest transmembrane regions in their glycoprotein precursors (e.g., QYBV has only one such region). By comparison, CCHFV and DUGV GPCs have five TMDs. All nairovirus GPCs were predicted to have at least a single conserved C-terminal transmembrane region approximately 40–60 residues prior to the C-termini of the GPC (Figure 6 and Table S2).

3.2.3. Large (L) Segment—RNA-Dependent RNA Polymerase

Nairovirus RNA-dependent RNA polymerases (Ls) are substantially larger than other bunyavirus L homologs. All nairovirus L sequences maintain the characteristic RNA-dependent RNA polymerase core motifs described by Poch [104] and Muller *et al.* [105] comprising residues 2361–2669 of the L gene alignment (domain A), and therefore include the polymerase module pre-A motif through motif E (Figure 7).

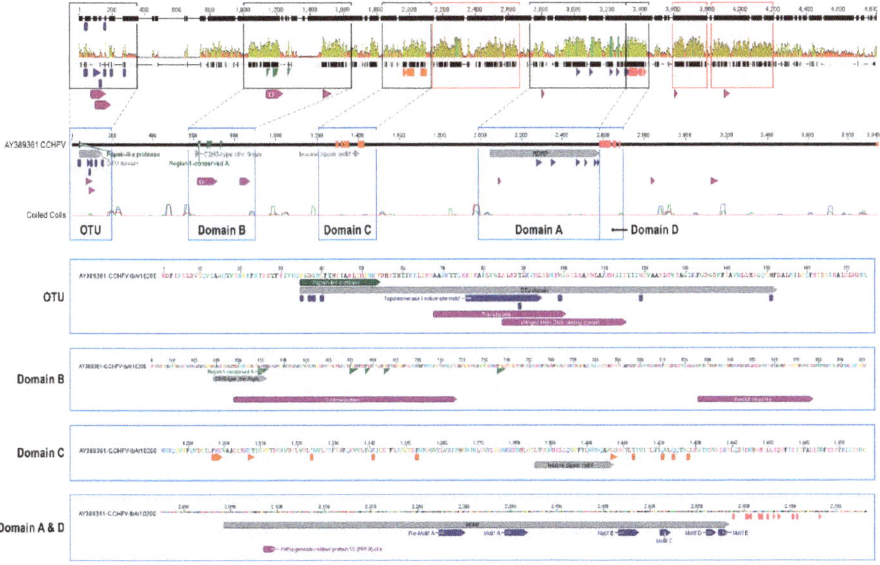

Figure 7. Cartoon showing conserved regions in nairovirus RNA-dependent RNA polymerases (Ls) using CCHFV as a reference. OTU, ovarian tumor family-like domain.

Moreover, inter-motif regions are moderately conserved, suggesting structural constrains on their three-dimensional arrangements. The invariant sequences DXX KW and SDD of motifs A and C, respectively, may have metal-binding activities necessary for catalytic functions [106,107] (Figure S1). A phylogenetic analysis of the nairovirus and nairo-like virus core polymerase modules with the corresponding regions of other bunyaviruses (hantaviruses, orthobunyaviruses, phleboviruses), mammarenaviruses, and orthomyxoviruses is shown in Figure 8.

Our analysis demonstrates that the nairovirus and nairo-like virus RNA-dependent-RNA polymerase core domain is more closely related to the arenavirus domain than to any other bunyaviral domain, supporting the existence of an arenavirus-nairovirus supergroup.

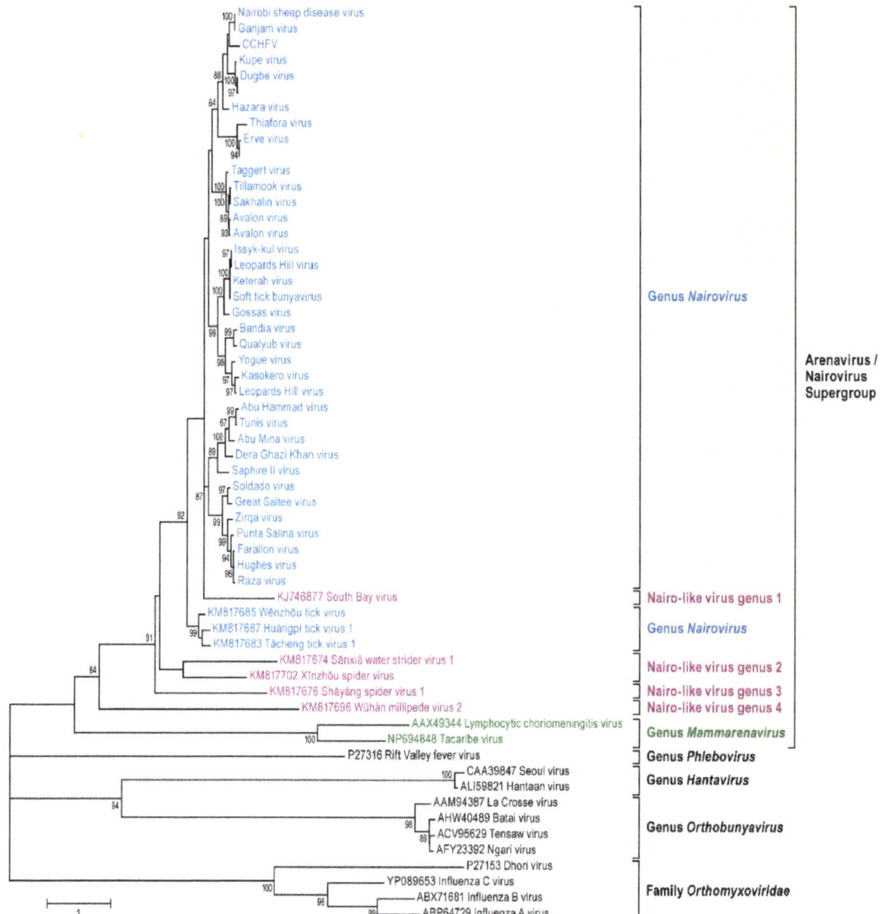

Figure 8. Phylogenetic analysis of the RNA-dependent-RNA polymerase core domain. A set of 58 core domains, comprising motifs A through E for representative viruses were analyzed by maximum likelihood method at the amino acid level using PHYML.

CCHFV encodes a deubiquitinase (DUB) of the OTU family, which unlike eukaryotic OTU DUBs, also targets interferon-stimulated gene 15 (ISG15) modifications [108,109]. The catalytic motifs characteristics of OTU-like cysteine proteases are clearly detected in all nairoviruses, but not in any of the nairo-like viruses. The role of the CCHFV OTU-like cysteine protease in the cleavage of host cell proteins, and specifically on the ubiquitin- and ISG15-dependent innate immune response, has been widely investigated [108,109]. Several OTU-characteristic residues (P35, D37, G38, C40, Y89, W99, W119, and H151) are highly conserved among all nairoviruses (Figure S7-2). Interestingly, these residues are not conserved in any of the nairo-like viruses [110]. P35, D37, G38, C40, and H151 are part of OTU's catalytic site (Figure S2). Y89 is the key aa residue of a conserved site resembling a topoisomerase motif (SXXXY), but its serine residue is not conserved among nairoviruses. Interestingly, the conserved site Y89 is located very close to P77, E78, and R80, which are key residues for the interaction with ubiquitin and ISG15 [111]. Moreover, this area of the nairovirus OTU domain is structurally similar to the catalytic domain of a transferase-like protein (PDB ID: 1k98; methionine synthase protein, confidence 60.8%) and to the DNA/RNA-binding 3-helical bundle fold of "winged helix" DNA-binding domain proteins. Interestingly, the region in the methionine synthase protein identified as structurally similar to the nairovirus OTU domain is the binding domain for vitamin B12. As expected, positional analysis revealed a correlation between positions with high mutational sensitivity and high identity (Figure 5C). The presence of the conserved OTU motif in all nairovirus Ls and the structural similarities between topoisomerases and strand-specific recombinases could indicate a role of the OTU domain in RNA strand manipulation.

The majority of nairoviruses pathogenic to humans (CCHFV, DUGV virus, ERVEV, ISKV, KAS(O)V, and NSDV) or other mammals (NSDV/GANV) contain an identifiable topoisomerase I active site motif, whereas most nairoviruses without that domain are only known to infect arthropods or to establish subclinical infections in vertebrates. The deubiquitinylation and deISGylation activities of CCHFV L have been proposed as a mechanism of virus evasion from the innate immune response via efficient interference with antiviral signaling pathways mediated by nuclear factor (NF)-kB, interferon regulatory factor 3 (IRF3), and type 1 interferon (IFN-α/β) [108,109]. These pathways rely on protein ubiquitinylation for their activation—one outcome is the modification of the pathway factor with ISG15. Thus, researchers posit that the CCHFV OTU domain might be an important virulence determinant, as some differences were observed in the functionality of the domain in very virulent CCHFV and the less virulent DUGV [112]. The observation of a potential structural and phylogenetic difference between the nairovirus OTUs of pathogenic and non-pathogenic nairoviruses is therefore intriguing.

Position-specific iterated (PSI)-BLAST searches had previously demonstrated that the region between the nairovirus OTU-like cysteine protease domain and the core region of the RNA-dependent polymerase (domain A) is conserved [42,113] (Figure 7). A region, including the C_2H_2-type zinc finger domain and several aa positions conserved among all RNA-dependent RNA polymerase modules of segmented negative-sense RNA viruses, was named domain B [106,107] (Figure 7 and Figure S3). Interestingly, this area of L is structurally similar to an oxidoreductase (PDB: 1mv8, a GDP-mannose 6-hydrogenase; positions 328–433; confidence 86.5%: identity 64%). Although several aa residues essential for the proper structure and function of this domain are conserved among nairovirus and other bunyavirus Ls, no correlation between mutational sensitivity and conservation was detected (data not shown). Downstream, Phyre2 analysis identified an area with similarity to the mammalian suppressor of yeast Sec 4 (Mss4)-like superfamily of proteins, of the family Rab guanine nucleotide exchange factor (GEF) Mss4 (PDB: 2fu5, confidence 67.4%, identity 44%). Interestingly, this area also contains several aa residues that are highly conserved among all nairoviruses (Figure 7 and Figure S3). However, no correlation with mutational sensitivity was detected (data not shown).

A region immediately upstream of the core polymerase region is highly conserved among arenaviruses and bunyaviruses (domain C) (Figure 7 and Figure S4; residues conserved among families are highlighted in orange in the figure) [114]. In the case of CCHFV, domain C includes a leucine zipper motif, which is found in all nairovirus genomes. Leucine zipper domains are a common three-dimensional structural motif of transcription factors, characterized by a periodic repetition of leucine residues at every seventh position over a distance covering eight helical turns [115]. The polypeptide segments containing these periodic arrays of leucine residues were proposed to exist in an alpha-helical conformation, and the leucine side chains from one alpha helix interdigitate with those from the alpha helix of a second polypeptide, facilitating dimerization. Basic-region leucine zippers (bZIPs) are a class of eukaryotic transcription factors including leucine zipper domains of 60 to 80 residues in length with highly conserved DNA binding basic regions [116]. The nairovirus leucine zipper domain surrounds a highly conserved NRRQ domain in the center of the 7 moderately conserved leucine positions. Of note, this motif is also structurally similar to human immunodeficiency virus 1 Nef (PDB: 2xi1, positions 97–110, confidence 11%, identify 43%), which includes the functional conserved motif XR [117,118]. Interestingly, structural analysis of nairovirus domain C revealed a similarity to RNA-binding an endoribonuclease VapD (PDB: 3ui3, positions 69–132, confidence 45.6%, identity 40%).

Finally, domain D, which has been previously described only for orthobunyaviruses (Figure 7 and Figure S5; residues conserved among family members are highlighted in orange in the figure), could also be identified as a conserved feature of nairovirus Ls, but structural similarities could not be detected.

Three additional, new conserved domains (Figure 7; orange boxes) were found in the nairovirus L alignment (1424–1605; 1700–1898; and 2763–3368). Only the third (and largest) domain (aa 3122–3155) was found to be structurally similar to another domain, namely the homodimerization domain of the female germline-specific tumor suppressor protein gld-1 (PDB: 3kbl; confidence 59%; identity 27%).

4. Discussion

The family *Bunyaviridae* currently has 530 putative members [1]. These members are either classified in the five recognized bunyaviral genera *Hantavirus*, *Nairovirus*, *Orthobunyavirus*, *Phlebovirus*, and *Tospovirus*, or remain to be classified into these existing or novel genera. Until recently, bunyaviral classification predominantly relied on antigenic relationships as genomic sequence information on individual bunyaviruses was scarce [1]. Classical and next-generation sequencing is now increasingly applied to historical isolates of presumed or newly discovered putative bunyaviruses [6–42]. These studies demonstrated that bunyaviral diversity is far broader than previously appreciated, probably necessitating the establishment of additional bunyaviral genera [22–24]. Other studies revealed that arenaviruses, emaraviruses, tenuiviruses, and Mourilyan virus may have to be included among bunyaviruses despite having genomes with more or less than the bunyavirus-typical three segments [119–126]. Finally, although sequencing of putative bunyaviruses by and large confirmed historical antigenic classifications, several bunyaviruses were assigned to wrong genera [14,25–34].

Broadening the bunyaviral sequence space to encompass as many bunyaviruses and bunya-like viruses as possible is necessary to elucidate their true phylogenetic relationships and to establish a modern comprehensive bunyaviral taxonomy that adequately reflects evolution. Here we reported 23 bunyaviral genome sequences (Table 1) with the goal of understanding the relationships of nairoviruses and nairo-like viruses—a group of bunyaviruses for which almost no sequence information was available until very recently [29,42]. While we confirmed the overall monophyly of the current genus *Nairovirus*, our results indicated that (a) novel species will have to be established; (b) that some nairoviruses will have to be moved between species; (c) that a long-thought phlebovirus is actually a nairovirus; and (d) that several newly discovered bunyaviruses do not directly fall into the nairovirus clade but are nevertheless more closely related to nairoviruses than to other bunyaviruses (Figures 1–3). We propose here that in the absence of at least coding-complete genomic information, "bunyaviruses" should not be classified into bunyaviral taxa but rather be seen as putative members of the family *Bunyaviridae*. Applying this stringent criterium to the list of "nairoviruses" (Table S1), we propose a new taxonomic organization for the genus *Nairovirus* in Table 2. This organization is overall in line with a similar, very recent, proposal by Walker *et al.* [42].

Coding-complete genome sequences of at least five additional nairoviruses (Burana virus, Caspiyi virus, Chim virus, Geran virus, and Tamdy virus) have been determined but are not yet available for analysis [30–34] (Table S1). Therefore, the taxonomy proposed in Table 2 should be considered preliminary until these sequences become available and incorporated. In particular, the Burana virus and Tamdy virus sequences may help to further refine the Burana and Qalyub genogroups.

Genomic reassortment (*i.e.*, the relatively free swapping of S, M, and L segments, between taxonomically diverse bunyaviruses simultaneously infecting the same host and consequently resulting in novel viruses) has been described for orthobunyaviruses, phleboviruses, and tospoviruses [127]. Interestingly, our analysis did not reveal any signs of inter-nairovirus reassortment. This result suggests that distinct nairoviruses may rarely have the opportunity in nature to infect the same host at the same time or that molecular-biological constraints prevent inter-nairovirus reassortment. *In vitro* experiments should be performed to evaluate these hypotheses.

Table 2. Proposed new taxonomy of the genus *Nairovirus* based on genomic data. Viruses mentioned in Table S1 but not here ought to be considered putative nairoviruses that based on current data cannot/should not be classified. Proposed new taxa are highlighted in red and placed in quotation marks.

Species	Virus Members
"Burana Nairovirus"	Huángpí tick virus 1 (HTV-1) Tǎchéng tick virus 1 (TTV-1) Wēnzhōu tick virus (WTV)
Crimean-Congo hemorrhagic fever nairovirus	Crimean-Congo hemorrhagic fever virus (CCHFV)
Dera Ghazi Khan nairovirus	Abu Hammad virus (AHV) including Tunis isolate Abu Mina virus (AMV) Dera Ghazi Khan virus (DGKV) Sapphire II virus (SAPV)
Dugbe nairovirus	Dugbe virus (DUGV) Kupe virus (KUPEV)
"Hazara nairovirus"	Hazara virus (HAZV) Tofla virus (TFLV)
Hughes nairovirus	Caspiy virus (CASV) Farallon virus (FARV) Great Saltee virus (GRSV) Hughes virus (HUGV) Punta Salinas virus (PSV) Raza virus (RAZAV) Soldado virus (SOLV) Zirqa virus (ZIRV)
"Keterah nairovirus"	Gossas virus (GOSV) Issyk-kul virus (ISKV) Keterah virus (KTRV) including soft tick isolate Uzun-Agach virus (UZAV)
"Kasokero nairovirus"	Kasokero virus (KAS(O)V) Leopards Hill virus (LPHV) Yogue virus (YOGV)
"Nairobi sheep disease virus nairovirus"	Ganjam virus (GANV) Nairobi sheep disease virus (NSDV) including Ganjam isolate
Qalyub nairovirus	Bandia virus (BDAV) Qalyub virus (QYBV)
Sakhalin nairovirus	Avalon virus (AVAV) Clo Mor virus (C(L)MV) Sakhalin virus (SAKV) Taggert virus (TAGB) Tillamook virus (TILLV)
Thiafora nairovirus	Erve virus (ERVEV) Thiafora virus (TFAV)

Two recent studies on arthropod samples revealed the existence of at least five bunyaviruses (SWSV-1, Shāyáng spider virus 1, SBV, Wǔhàn millipede virus 2, and XSV) that appeared to be closely related to nairoviruses [37,41]. Interestingly, SBV and Wǔhàn millipede virus 2 genomes were found to consist of only two segments, rather than the bunyavirus/nairovirus-typical three segments. Our phylogenetic analyses (Figures 1–3 and Figure 8) indicate that all five viruses should not be classified in the genus *Nairovirus*, but confirm that these viruses are more closely related to nairoviruses than to all other bunyaviruses. To further resolve phylogenetic placement of these five viruses, we performed PAirwise Sequence Comparison (PASC) of bunyavirus sequences using the National Center for Biotechnology Information's (NCBIs) PASC tool. Histograms from this analysis demonstrates the distribution of the number of virus genome pairs at each identity percentage. Histogram peaks and valleys can be used to differentiate taxon ranks and to establish taxon demarcation criteria using objective criteria [128,129]. A preliminary PASC analysis with representative bunyaviral genomes indicate that cut-offs of ≈26% and 31%–34% identity for the M and L segments, respectively, ought to be used to uphold the current bunyaviral division into *Hantavirus*, *Nairovirus*, *Orthobunyavirus*, *Phlebovirus*, and *Tospovirus* genera (data not shown). Using these cut-offs, PASC confirms the monophyly of the genus *Nairovirus* as outlined in Table 2 and indicates the need to establish four nairovirus-like genera for (1) SBV; (2) SWSV-1 and XSV; (3) Shāyáng spider virus 1; and (4) Wǔhàn millipede virus 2 (Figures 1–3; Figure 8: nairoviruses in blue and nairo-like viruses in purple). Interestingly, phylogenetic analyses also indicate that the bisegmented arenaviruses, currently not part of the *Bunyaviridae*, are more related to these viruses than they are to other bunyaviruses, suggesting the need to establish an arenavirus/nairovirus supergroup within the family (Figure 8).

Supplementary Materials: The following are available online at www.mdpi.com/link, Table S1: All known and putative nairoviruses and nairo-like viruses and their deduced relationships prior to this study; Table S2: M-segment polyprotein descriptions. Figures S1 through S5: conserved features of nairovirus RNA-dependent RNA polymerases (Ls).

Acknowledgments: The content of this publication does not necessarily reflect the views or policies of the US Department of Health and Human Services, or the institutions and companies affiliated with the authors. This work was supported by the Defense Threat Reduction Agency #1881290, and the United States Department of Defense, Defense Biological Products Assurance Office. A.P.A.T.d.R., H.G., and R.B.T. were supported by US National Institutes of Health (NIH) contract HHSN272201000040I/HHSN27200004/D04. Y.B. was supported by the Intramural Research Program of the NIH, National Library of Medicine. This work was supported in part through Battelle Memorial Institute's prime contract with the US National Institute of Allergy and Infectious Diseases (NIAID) under Contract No. HHSN272200700016I. A Battelle employee involved in this work is: J.W. A subcontractor to Battelle Memorial Institute who performed this work is: J.H.K., an employee of Tunnell Government Services, Inc. We thank Laura Bollinger (Battelle Memorial Institute) for editing this manuscript.

Author Contributions: M.R.W., S.E.R., and G.P. conceived and designed the experiments; M.R.W., S.E.R., K.P., A.P.A.T.d.R., H.G., N.S., and G.P. performed the experiments; J.H.K., M.R.W., S.E.R., J.T.L., Y.B., P.B.J., and G.P. analyzed the data; R.B.T. and D.A.B. contributed reagents/materials/analysis tools; J.H.K., M.R.W., S.E.R., J.W., and G.P. wrote the paper and/or helped prepare figures.

Conflicts of Interest: The authors declare no conflict of interest. The funding sponsors had no role in the design of the study; in the collection, analyses, or interpretation of data; in the writing of the manuscript; and in the decision to publish the results.

References

1. Plyusnin, A.; Beaty, B.J.; Elliott, R.M.; Goldbach, R.; Kormelink, R.; Lundkvist, A.; Schmaljohn, C.S.; Tesh, R.B. Family *Bunyaviridae*. In *Virus Taxonomy—Ninth Report of the International Committee on Taxonomy of Viruses*; King, A.M.Q., Adams, M.J., Carstens, E.B., Lefkowitz, E.J., Eds.; Elsevier/Academic Press: London, UK, 2011; pp. 725–741.
2. De Haan, P.; Wagemakers, L.; Peters, D.; Goldbach, R. The S RNA segment of tomato spotted wilt virus has an ambisense character. *J. Gen. Virol.* **1990**, *71*, 1001–1007. [CrossRef] [PubMed]
3. Elliott, R.M.; Schmaljohn, C.S. Bunyaviridae. In *Fields Virology*, 7th ed.; Knipe, D.M., Howley, P.M., Eds.; Wolters Kluwer/Lippincott Williams & Wilkins: Philadelphia, PA, USA, 2013; pp. 1244–1282.
4. Davies, F.G.; Terpstra, C. Nairobi sheep disease. In *Infectious Diseases of Livestock*; Coetzer, J.A.W., Tustin, R.C., Eds.; Oxford University Press: Oxford, UK, 2004; Volume 2, pp. 1071–1076.
5. Ladner, J.T.; Beitzel, B.; Chain, P.S.; Davenport, M.G.; Donaldson, E.F.; Frieman, M.; Kugelman, J.R.; Kuhn, J.H.; O'Rear, J.; Sabeti, P.C.; *et al.* Standards for sequencing viral genomes in the era of high-throughput sequencing. *MBio* **2014**, *5*, e01360-14. [CrossRef] [PubMed]
6. Palacios, G.; Wiley, M.R.; Travassos da Rosa, A.P.A.; Guzman, H.; Quiroz, E.; Savji, N.; Carrera, J.-P.; Bussetti, A.V.; Ladner, J.T.; Lipkin, W.I.; *et al.* Characterization of the Punta Toro species complex (genus *Phlebovirus*, family *Bunyaviridae*). *J. Gen. Virol.* **2015**, *96*, 2079–2085. [CrossRef] [PubMed]

7. Palacios, G.; Tesh, R.B.; Savji, N.; Travassos da Rosa, A.P.A.; Guzman, H.; Bussetti, A.V.; Desai, A.; Ladner, J.; Sanchez-Seco, M.; Lipkin, W.I. Characterization of the sandfly fever Naples species complex and description of a new Karimabad species complex (genus *Phlebovirus*, family *Bunyaviridae*). *J. Gen. Virol.* **2014**, *95*, 292–300. [CrossRef] [PubMed]
8. Palacios, G.; Tesh, R.; Travassos da Rosa, A.; Savji, N.; Sze, W.; Jain, K.; Serge, R.; Guzman, H.; Guevara, C.; Nunes, M.R.T.; *et al.* Characterization of the Candiru antigenic complex (*Bunyaviridae*: *Phlebovirus*), a highly diverse and reassorting group of viruses affecting humans in tropical America. *J. Virol.* **2011**, *85*, 3811–3820. [CrossRef] [PubMed]
9. Palacios, G.; Savji, N.; Travassos da Rosa, A.; Guzman, H.; Yu, X.; Desai, A.; Rosen, G.E.; Hutchison, S.; Lipkin, W.I.; Tesh, R. Characterization of the Uukuniemi virus group (*Phlebovirus*: *Bunyaviridae*): Evidence for seven distinct species. *J. Virol.* **2013**, *87*, 3187–3195. [CrossRef] [PubMed]
10. Palacios, G.; Savji, N.; Travassos da Rosa, A.; Desai, A.; Sanchez-Seco, M.P.; Guzman, H.; Lipkin, W.I.; Tesh, R. Characterization of the Salehabad virus species complex of the genus *Phlebovirus* (*Bunyaviridae*). *J. Gen. Virol.* **2013**, *94*, 837–842. [CrossRef] [PubMed]
11. Palacios, G.; Travassos da Rosa, A.; Savji, N.; Sze, W.; Wick, I.; Guzman, H.; Hutchison, S.; Tesh, R.; Lipkin, W.I. *Aguacate virus*, a new antigenic complex of the genus *Phlebovirus* (family *Bunyaviridae*). *J. Gen. Virol.* **2011**, *92*, 1445–1453. [CrossRef] [PubMed]
12. Matsuno, K.; Weisend, C.; Travassos da Rosa, A.P.A.; Anzick, S.L.; Dahlstrom, E.; Porcella, S.F.; Dorward, D.W.; Yu, X.-J.; Tesh, R.B.; Ebihara, H. Characterization of the Bhanja serogroup viruses (*Bunyaviridae*): A novel species of the genus *Phlebovirus* and its relationship with other emerging tick-borne phleboviruses. *J. Virol.* **2013**, *87*, 3719–3728. [CrossRef] [PubMed]
13. Matsuno, K.; Weisend, C.; Kajihara, M.; Matysiak, C.; Williamson, B.N.; Simuunza, M.; Mweene, A.S.; Takada, A.; Tesh, R.B.; Ebihara, H. Comprehensive molecular detection of tick-borne phleboviruses leads to the retrospective identification of taxonomically unassigned bunyaviruses and the discovery of a novel member of the genus *Phlebovirus*. *J. Virol.* **2015**, *89*, 594–604. [CrossRef] [PubMed]
14. Альховский, С.В.; Львов, Д.К.; Щелканов, М.Ю.; Щетинин, А.М.; Дерябин, П.Г.; Самохвалов, Е.И.; Гительман, А.К.; Ботиков, А.Г. Таксономия вируса Хасан (Khasan, KHAV)—нового вируса рода *Phlebovirus* (сем. *Bunyaviridae*), изолированного из клещей *Haemaphysalis longicornis* (Neumann, 1901) в Приморском крае (Россия). Al'hovskij, S.V.; L'vov, D.K.; Ŝelkanov, M.Û.; Ŝetinin, A.M.; Derŝbin, P.G.; Samohvalov, E.I.; Gitel'man, A.K.; Botikov, A.G. The taxonomy of the Khasan virus (KHAV), a new representative of the *Phlebovirus* genus (*Bunyaviridae*), isolated from *Haemaphysalis longicornis* (Neumann, 1901) ticks in the Maritime Territory. *Vopr. Virusol.* **2013**, *58*, 15–18. (In Russian)
15. Stockwell, T.B.; Heberlein-Larson, L.A.; Tan, Y.; Halpin, R.A.; Fedorova, N.; Katzel, D.A.; Smole, S.; Unnasch, T.R.; Kramer, L.D.; Das, S.R. First complete genome sequences of two Keystone viruses from Florida. *Genome Announc.* **2015**, *3*, e01255-15. [CrossRef] [PubMed]
16. Savji, N.; Palacios, G.; Travassos da Rosa, A.; Hutchison, S.; Celone, C.; Hui, J.; Briese, T.; Calisher, C.H.; Tesh, R.B.; Lipkin, W.I. Genomic and phylogenetic characterization of Leanyer virus, a novel orthobunyavirus isolated in northern Australia. *J. Gen. Virol.* **2011**, *92*, 1676–1687. [CrossRef] [PubMed]
17. Ladner, J.T.; Savji, N.; Lofts, L.; Travassos da Rosa, A.; Wiley, M.R.; Gestole, M.C.; Rosen, G.E.; Guzman, H.; Vasconcelos, P.F.C.; Nunes, M.R.T.; *et al.* Genomic and phylogenetic characterization of viruses included in the Manzanilla and Oropouche species complexes of the genus *Orthobunyavirus*, family *Bunyaviridae*. *J. Gen. Virol.* **2014**, *95*, 1055–1066. [CrossRef] [PubMed]
18. Hontz, R.D.; Guevara, C.; Halsey, E.S.; Silvas, J.; Santiago, F.W.; Widen, S.G.; Wood, T.G.; Casanova, W.; Vasilakis, N.; Watts, D.M.; *et al.* Itaya virus, a novel *Orthobunyavirus* associated with human febrile illness, Peru. *Emerg. Infect. Dis.* **2015**, *21*, 781–788. [CrossRef] [PubMed]
19. Groseth, A.; Vine, V.; Weisend, C.; Ebihara, H. Complete genome sequence of Trivittatus virus. *Arch. Virol.* **2015**, *160*, 2637–2639. [CrossRef] [PubMed]
20. Groseth, A.; Mampilli, V.; Weisend, C.; Dahlstrom, E.; Porcella, S.F.; Russell, B.J.; Tesh, R.B.; Ebihara, H. Molecular characterization of human pathogenic bunyaviruses of the Nyando and Bwamba/Pongola virus groups leads to the genetic identification of Mojuí dos Campos and Kaeng Khoi virus. *PLoS Negl. Trop. Dis.* **2014**, *8*, e3147. [CrossRef] [PubMed]

21. Shchetinin, A.M.; Lvov, D.K.; Deriabin, P.G.; Botikov, A.G.; Gitelman, A.K.; Kuhn, J.H.; Alkhovsky, S.V. Genetic and phylogenetic characterization of Tataguine and Witwatersrand viruses and other orthobunyaviruses of the Anopheles A, Capim, Guama, Koongol, Mapputta, Tete, and Turlock serogroups. *Viruses* **2015**, *7*, 5987–6008. [CrossRef] [PubMed]
22. Marklewitz, M.; Zirkel, F.; Kurth, A.; Drosten, C.; Junglen, S. Evolutionary and phenotypic analysis of live virus isolates suggests arthropod origin of a pathogenic RNA virus family. *Proc. Natl. Acad. Sci. USA* **2015**, *112*, 7536–7541. [CrossRef] [PubMed]
23. Marklewitz, M.; Zirkel, F.; Rwego, I.B.; Heidemann, H.; Trippner, P.; Kurth, A.; Kallies, R.; Briese, T.; Lipkin, W.I.; Drosten, C.; *et al.* Discovery of a unique novel clade of mosquito-associated bunyaviruses. *J. Virol.* **2013**, *87*, 12850–12865. [CrossRef] [PubMed]
24. Ballinger, M.J.; Bruenn, J.A.; Hay, J.; Czechowski, D.; Taylor, D.J. Discovery and evolution of bunyavirids in arctic phantom midges and ancient bunyavirid-like sequences in insect genomes. *J. Virol.* **2014**, *88*, 8783–8794. [CrossRef] [PubMed]
25. Альховский, С.В.; Львов, Д.К.; Щелканов, М.Ю.; Дерябин, П.Г.; Щетинин, А.М.; Самохвалов, Е.И.; Аристова, В.А.; Гительман, А.К.; Ботиков, А.Г. Генетическая характеристика вируса Узун-Агач (UZAV—Uzun-Agach virus) (*Bunyaviridae, Nairovirus*), изолированного в Казахстане от остроухой ночницы *Myotis blythii oxygnathus* Monticelli, 1885 (*Chiroptera; Vespertilionidae*). Al'hovskij, S.V.; L'vov, D.K.; Ŝelkanov, M.Û.; Derâbin, P.G.; Ŝetinin, A.M.; Samohvalov, E.I.; Aristova, V.A.; Gitel'man, A.K.; Botikov, A.G. Genetic characterization of Uzun-Agach virus (UZAV, *Bunyaviridae, Nairovirus*), isolated from *Myotis blythii oxygnathus* Monticelli, 1885 bats (*Chiroptera; Vespertilionidae*) in Kazakhstan. *Vopr. Virusol.* **2014**, *59*, 23–26. (In Russian)
26. Альховский, С.В.; Львов, Д.К.; Щелканов, М.Ю.; Щетинин, А.М.; Дерябин, П.Г.; Гительман, А.К.; Ботиков, А.Г.; Самохвалов, Е.И.; Закарян, В.А. Таксономия вируса Арташат (ARTSV—Artashat virus) (*Bunyaviridae, Nairovirus*), изолированного из клещей *Ornithodoros alactagalis* Issaakjan, 1936 и *O. verrucosus* Olenev, Sassuchin et Fenuk, 1934 (*Argasidae* Koch, 1844), собранных в Закавказье. Al'hovskij, S.V.; L'vov, D.K.; Ŝelkanov, M.Û.; Ŝetinin, A.M.; Derâbin, P.G.; Gitel'man, A.K.; Botikov, A.G.; Samohvalov, E.I.; Zakarân, V.A. Taxonomic status of Artashat virus (ARTSV) (*Bunyaviridae, Nairovirus*) isolated from *Ornithodoros alactagalis* Issaakjan, 1936 and *O. verrucosus* Olenev, Sassuchin et Fenuk, 1934 ticks (*Argasidae* Koch, 1844) collected in Transcaucasia. *Vopr. Virusol.* **2014**, *59*, 24–28. (In Russian)
27. Альховский, С.В.; Львов, Д.К.; Щелканов, М.Ю.; Щетинин, А.М.; Дерябин, П.Г.; Самохвалов, Е.И.; Гительман, А.К.; Ботиков, А.Г. Таксономия вируса Иссык-Куль (Issyk-kul virus, ISKV; *Bunyaviridae, Nairovirus*), возбудителя Иссык-Кульской лихорадки, изолированного от летучих мышей (*Vespertilionidae*) и клещей *Argas* (*Carios*) *vespertilionis* (Latreille, 1796). Al'hovskij, S.V.; L'vov, D.K.; Ŝelkanov, M.Û.; Ŝetinin, A.M.; Derâbin, P.G.; Samohvalov, E.I.; Gitel'man, A.K.; Botikov, A.G. Taxonomy of Issyk-kul virus (ISKV, *Bunyaviridae, Nairovirus*), the etiologic agent of Issyk-kul fever isolated from bats (*Vespertilionidae*) and *Argas* (*Carios*) *vespertilionis* (Latreille, 1796) ticks. *Vopr. Virusol.* **2013**, *58*, 11–15. (In Russian)
28. Atkinson, B.; Marston, D.A.; Ellis, R.J.; Fooks, A.R.; Hewson, R. Complete genomic sequence of Issyk-kul virus. *Genome Announc.* **2015**, *3*, e00662-15. [CrossRef] [PubMed]
29. Walker, P.J.; Widen, S.G.; Firth, C.; Blasdell, K.R.; Wood, T.G.; Travassos da Rosa, A.P.A.; Guzman, H.; Tesh, R.B.; Vasilakis, N. Genomic characterization of Yogue, Kasokero, Issyk-Kul, Keterah, Gossas, and Thiafora viruses: Nairoviruses naturally infecting bats, shrews, and ticks. *Am. J. Trop. Med. Hyg.* **2015**, *93*, 1041–1051. [CrossRef] [PubMed]
30. Львов, Д.К.; Альховский, С.В.; Щелканов, М.Ю.; Дерябин, П.Г.; Щетинин, А.М.; Самохвалов, Е.И.; Аристова, В.А.; Гительман, А.К.; Ботиков, А.Г. Генетическая характеристика вируса Герань (GERV - Geran virus) (*Bunyaviridae, Nairovirus*, группа Кальюб), изолированного в Азербайджане от клещей *Ornithodoros verrucosus* Olenev, Zasukhin and Fenyuk, 1934 (*Argasidae*), собранных в норе краснохвостой песчанки (*Meriones erythrourus* Grey, 1842). L'vov, D.K.; Al'hovskij, S.V.; Ŝelkanov, M.Û.; Derâbin, P.G.; Ŝetinin, A.M.; Samohvalov, E.I.; Aristova, V.A.; Gitel'man, A.K.; Botikov, A.G. Genetic characterization of Geran virus (GERV, *Bunyaviridae, Nairovirus*, Qalyub group) isolated from *Ornithodoros verrucosus* Olenev, Zasukhin and Fenyuk, 1934 ticks (*Argasidae*) collected in the burrow of *Meriones erythrourus* Grey, 1842 in Azerbaijan. *Vopr. Virusol.* **2014**, *59*, 13–18. (In Russian)

31. Львов, Д.К.; Альховский, С.В.; Щелканов, М.Ю.; Щетинин, А.М.; Аристова, В.А.; Гительман, А.К.; Дерябин, П.Г.; Ботиков, А.Г. Таксономия ранее негруппированного вируса Тамды (TAMV-Tamdy virus) (*Bunyaviridae, Nairovirus*), изолированного от иксодовых клещей *Hyalomma asiaticum asiaticum* Schülce et Schlottke, 1929 (*Ixodidae, Hyalomminae*) в Средней Азии и Закавказье. Lvov, D.K.; Alkhovsky, S.V.; Shchelkanov, M.Yu.; Shchetinin, A.M.; Aristova, V.A.; Gitelman, A.K.; Deryabin, P.G.; Botikov, A.G. Taxonomy of previously unclassified Tamdy virus (TAMV) (*Bunyaviridae, Nairovirus*) isolated from *Hyalomma asiaticum asiaticum* Schülce et Schlottke, 1929 (*Ixodidae, Hyalomminae*) ticks in the Middle East and Transcaucasia. *Vopr. Virusol.* **2014**, *59*, 15–22. (In Russian)
32. Львов, Д.К.; Альховский, С.В.; Щелканов, М.Ю.; Щетинин, А.М.; Аристова, В.А.; Морозова, Т.Н.; Гительман, А.К.; Дерябин, П.Г.; Ботиков, А.Г. Таксономия ранее не классифицированного вируса ЧИМ (CHIMV - Chim virus) (*Bunyaviridae, Nairovirus*, группа Кальюб), изолированного в Узбекистане и Казахстане из иксодовых (*Acari*: *Ixodidae*) и аргасовых (*Acari*: *Argasidae*) клещей, собранных в норах больших песчанок *Rhombomys opimus* Lichtenstein, 1823 (*Muridae, Gerbillinae*). L'vov, D.K.; Al'hovskij, S.V.; Shhelkanov, M.J.; Shhetinin, A.M.; Aristova, V.A.; Morozova, T.N.; Gitel'man, A.K.; Derjabin, P.G.; Botikov, A.G. Taxonomic status of Chim virus (CHIMV) (*Bunyaviridae, Nairovirus*, Qalyub group) isolated from *Ixodidae* and *Argasidae* ticks collected from great gerbil (*Rhombomys opimus* Lichtenstein, 1823) (*Muridae, Gerbillinae*) burrows in Uzbekistan and Kazakhstan. *Vopr. Virusol.* **2014**, *59*, 18–23. (In Russian)
33. Львов, Д.К.; Альховский, С.В.; Щелканов, М.Ю.; Щетинин, А.М.; Дерябин, П.Г.; Гительман, А.К.; Аристова, В.А.; Ботиков, А.Г. Таксономический статус вируса Бурана (BURV—Burana virus) (*Bunyaviridae, Nairovirus*, группа Тамды), изолированного из клещей *Haemaphysalis punctata* Canestrini et Fanzago, 1877 и *Haem. concinna* Koch, 1844 (*Ixodidae, Haemaphysalinae*) в Кыргызстане. L'vov, D.K.; Al'hovskij, S.V.; Shhelkanov, M.J.; Shhetinin, A.M.; Derjabin, P.G.; Gitel'man, A.K.; Aristova, V.A.; Botikov, A.G. Taxonomic status of Burana virus (BURV) (*Bunyaviridae, Nairovirus*, Tamdy group) isolated from *Haemaphysalis punctata* Canestrini et Fanzago, 1877 and *Haem. concinna* Koch, 1844 ticks (*Ixodidae, Haemaphysalinae*) in Kyrgyzstan. *Vopr. Virusol.* **2014**, *59*, 10–15. (In Russian)
34. Львов, Д.К.; Альховский, С.В.; Щелканов, М.Ю.; Щетинин, А.М.; Дерябин, П.Г.; Самохвалов, Е.И.; Гительман, А.К.; Ботиков, А.Г. Генетическая характеристика вируса Каспий (CASV - *Caspiy virus*) (*Bunyaviridae, Nairovirus*), изолированного от чайковых (Laridae Vigors, 1825) и крачковых (*Sternidae* Bonaparte, 1838) птиц и аргасовых клещей *Ornithodoros capensis* Neumann, 1901 (*Argasidae* Koch, 1844) на западном и восточном побережьях Каспийского моря. L'vov, D.K.; Al'hovskij, S.V.; Shhelkanov, M.Ju.; Shhetinin, A.M.; Derjabin, P.G.; Samohvalov, E.I.; Gitel'man, A.K.; Botikov, A.G. Genetic characterization of *Caspiy virus* (CASV) (*Bunyaviridae, Nairovirus*) isolated from Laridae (Vigors, 1825) and *Sternidae* (Bonaparte, 1838) birds and *Argasidae* (Koch, 1844) *Ornithodoros capensis* Neumann, 1901, ticks form western and eastern coasts of the Caspian Sea. *Vopr. Virusol.* **2014**, *59*, 24–29. (In Russian)
35. Dacheux, L.; Cervantes-Gonzalez, M.; Guigon, G.; Thiberge, J.-M.; Vandenbogaert, M.; Maufrais, C.; Caro, V.; Bourhy, H. A preliminary study of viral metagenomics of French bat species in contact with humans: Identification of new mammalian viruses. *PLoS ONE* **2014**, *9*, e87194. [CrossRef] [PubMed]
36. Ishii, A.; Ueno, K.; Orba, Y.; Sasaki, M.; Moonga, L.; Hang'ombe, B.M.; Mweene, A.S.; Umemura, T.; Ito, K.; Hall, W.W.; *et al.* A nairovirus isolated from African bats causes haemorrhagic gastroenteritis and severe hepatic disease in mice. *Nat. Commun.* **2014**, *5*, 5651. [CrossRef] [PubMed]
37. Li, C.X.; Shi, M.; Tian, J.H.; Lin, X.D.; Kang, Y.J.; Chen, L.J.; Qin, X.C.; Xu, J.; Holmes, E.C.; Zhang, Y.Z. Unprecedented genomic diversity of RNA viruses in arthropods reveals the ancestry of negative-sense RNA viruses. *Elife* **2015**, *4*, e05378. [CrossRef] [PubMed]
38. Oba, M.; Omatsu, T.; Takano, A.; Fujita, H.; Sato, K.; Nakamoto, A.; Takahashi, M.; Takada, N.; Kawabata, H.; Ando, S.; *et al.* A novel Bunyavirus from the soft tick, *Argas vespertilionis*, in Japan. *J. Vet. Med. Sci.* **2016**, *78*, 443–445. [CrossRef] [PubMed]
39. Xia, H.; Hu, C.; Zhang, D.; Tang, S.; Zhang, Z.; Kou, Z.; Fan, Z.; Bente, D.; Zeng, C.; Li, T. Metagenomic profile of the viral communities in *Rhipicephalus* spp. ticks from Yunnan, China. *PLoS ONE* **2015**, *10*, e0121609. [CrossRef] [PubMed]
40. Shimada, S.; Aoki, K.; Nabeshima, T.; Fuxun, Y.; Kurosaki, Y.; Shiogama, K.; Onouchi, T.; Sakaguchi, M.; Fuchigami, T.; *et al.* Tofla virus: A newly identified *Nairovirus* of the Crimean-Congo hemorrhagic fever group isolated from ticks in Japan. *Sci. Rep.* **2016**, *6*, 20213. [CrossRef] [PubMed]

41. Tokarz, R.; Williams, S.H.; Sameroff, S.; Sanchez Leon, M.; Jain, K.; Lipkin, W.I. Virome analysis of *Amblyomma americanum*, *Dermacentor variabilis*, and *Ixodes scapularis* ticks reveals novel highly divergent vertebrate and invertebrate viruses. *J. Virol.* **2014**, *88*, 11480–11492. [CrossRef] [PubMed]
42. Walker, P.J.; Widen, S.G.; Wood, T.G.; Guzman, H.; Tesh, R.B.; Vasilakis, N. A global genomic characterization of nairoviruses identifies nine discrete genogroups with distinctive structural characteristics and host-vector associations. *Am. J. Trop. Med. Hyg.* **2016**, *94*, 1107–1122. [CrossRef] [PubMed]
43. Darwish, M.A.; Imam, I.Z.E.; Omar, F.M. Complement-fixing antibodies against Abu Hammad and Abu Mina viruses in mammalian sera from Egypt. *J. Egypt. Public Health Assoc.* **1976**, *51*, 51–54. [PubMed]
44. Quillien, M.C.; Monnat, J.Y.; Le Lay, G.; Le Goff, F.; Hardy, E.; Chastel, C. Avalon virus, Sakhalin group (*Nairovirus, Bunyaviridae*) from the seabird tick *Ixodes* (*Ceratixodes*) *uriae* White 1852 in France. *Acta Virol.* **1986**, *30*, 418–427. [PubMed]
45. Main, A.J.; Downs, W.G.; Shope, R.E.; Wallis, R.C. Avalon and Clo Mor: Two new Sakhalin group viruses from the North Atlantic. *J. Med. Entomol.* **1976**, *13*, 309–315. [CrossRef] [PubMed]
46. Brès, P.; Cornet, M.; Robin, Y. Le virus de la Forêt de Bandia (IPD/A 611), nouveau prototype d'arbovirus isolé au Sénégal [The Bandia Forest virus (IPD/A 611), a new arbovirus prototype isolated in Senegal]. *Ann. Inst. Pasteur (Paris)* **1967**, *113*, 739–747. (In French) [PubMed]
47. Begum, F.; Wisseman, C.L., Jr.; Casals, J. Tick-borne viruses of West Pakistan. III. *Dera Ghazi Khan* virus, a new agent isolated from *Hyalomma dromedarii* ticks in the D.G.Khan District of West Pakistan. *Am. J. Epidemiol.* **1970**, *92*, 195–196. [PubMed]
48. The Subcommitee on Information Exchange of the American Commitee on Arthropod-borne Viruses, No. 226. Dugbe (DUG). Strain: AR 1792. *Am. J. Trop. Med. Hyg.* **1970**, *19*, 1123–1124.
49. Chastel, C.; Main, A.J.; Richard, P.; le Lay, G.; Legrand-Quillien, M.C.; Beaucournu, J.C. Erve virus, a probable member of *Bunyaviridae* family isolated from shrews (*Crocidura russula*) in France. *Acta Viro* **1989**, *33*, 270–280.
50. Radovsky, F.J.; Stiller, D.; Johnson, H.N.; Clifford, C.M. Descriptive notes on *Ornithodoros* ticks from gull nests on the Farallon Islands and isolation of a variant of Hughes virus. *J. Parasitol.* **1967**, *53*, 890–892. [CrossRef]
51. Dandawate, C.N.; Shah, K.V. Ganjam virus: A new arbovirus isolated from Ticks *Haemaphysalis Intermedia* Warburton and Nuttall, 1909 in Orissa, India. *Indian J. Med. Res.* **1969**, *57*, 799–804. [PubMed]
52. Keirans, J.E.; Yunker, C.E.; Clifford, C.M.; Thomas, L.A.; Walton, G.A.; Kelly, T.C. Isolation of a Soldado-like virus (Hughes group) from *Ornithodorus maritimus* ticks in Ireland. *Experientia* **1976**, *32*, 453–454. [CrossRef] [PubMed]
53. Hughes, L.E.; Clifford, C.M.; Thomas, L.A.; Denmark, H.A.; Philip, C.B. Isolation and characterization of a virus from soft ticks (*Ornithodoros capensis* group) collected on Bush Key, Dry Tortugas, Florida. *Am. J. Trop. Med. Hyg.* **1964**, *13*, 118–122. [PubMed]
54. Philip, C.B. Hughes virus, a new arboviral agent from marine bird ticks. *J. Parasitol.* **1965**, *51*, 252. [CrossRef] [PubMed]
55. Converse, J.D.; Moussa, M.I.; Easton, E.R.; Casals, J. Punta Salinas virus (Hughes group) from *Argas arboreus* (Ixodoidea: Argasidae) in Tanzania. *Trans. R. Soc. Trop. Med. Hyg.* **1981**, *75*, 755–756. [CrossRef]
56. The Subcommitee on Information Exchange of the American Commitee on Arthropod-borne Viruses, No. 222. Qalyub (QYB). Strain: Ar 370. *Am. J. Trop. Med. Hyg.* **1970**, *19*, 1115–1116.
57. Clifford, C.M.; Thomas, L.A.; Hughes, L.E.; Kohls, G.M.; Philip, C.B. Identification and comparison of two viruses isolated from ticks of the genus *Ornithodoros*. *Am. J. Trop. Med. Hyg.* **1968**, *17*, 881–885. [PubMed]
58. Lvov, D.K.; Timofeeve, A.A.; Gromashevski, V.L.; Chervonsky, V.I.; Gromov, A.I.; Tsynkin, Y.M.; Pogrebenko, A.G.; Kostyrko, I.N. "Sakhalin" virus—A new arbovirus isolated from *Ixodes* (*Ceratixodes*) *putus* Pick.-Camb. 1878 collected on Tuleniy Island, Sea of Okhotsk. *Arch. Gesamte Virusforsch.* **1972**, *38*, 133–138. [CrossRef] [PubMed]
59. Yunker, C.E.; Clifford, C.M.; Thomas, L.A.; Cory, J.; George, J.E. Isolation of viruses from swallow ticks, *Argas cooleyi*, in the southwestern United States. *Acta Virol.* **1972**, *16*, 415–421. [PubMed]
60. Jonkers, A.H.; Casals, J.; Aitken, T.H.G.; Spence, L. Soldado virus, a new agent from Trinidadian *Ornithodoros* ticks. *J. Med. Entomol.* **1973**, *10*, 517–519. [CrossRef] [PubMed]
61. Doherty, R.L.; Carley, J.G.; Murray, M.D.; Main, A.J., Jr.; Kay, B.H.; Domrow, R. Isolation of arboviruses (Kemerovo group, Sakhalin group) from *Ixodes uriae* collected at Macquarie Island, Southern Ocean. *Am. J. Trop. Med. Hyg.* **1975**, *24*, 521–526. [PubMed]

62. Thomas, L.A.; Clifford, C.M.; Yunker, C.E.; Keirans, J.E.; Patzer, E.R.; Monk, G.E.; Easton, E.R. Tickborne viruses in western North America. I. Viruses isolated from *Ixodes uriae* in coastal Oregon in 1970. *J. Med. Entomol.* **1973**, *10*, 165–168. [CrossRef] [PubMed]
63. Chastel, C.; Bach-Hamba, D.; Karabatsos, N.; Bouattour, A.; le Lay, G.; le Goff, F.; Vermeil, C. Tunis virus: A new *Phlebovirus* from *Argas reflexus hermanni* ticks in Tunisia. *Acta Virol.* **1994**, *38*, 285–289. [PubMed]
64. Varma, M.G.; Bowen, E.T.; Simpson, D.I.; Casals, J. Zirga virus, a new arbovirus isolated from bird-infesting ticks. *Nature* **1973**, *244*, 452. [CrossRef] [PubMed]
65. Djikeng, A.; Halpin, R.; Kuzmickas, R.; Depasse, J.; Feldblyum, J.; Sengamalay, N.; Afonso, C.; Zhang, X.; Anderson, N.G.; Ghedin, E.; *et al.* Viral genome sequencing by random priming methods. *BMC Genom.* **2008**, *9*, 5. [CrossRef] [PubMed]
66. Martin, M. Cutadapt removes adapter sequences from high-throughput sequencing reads. *EMBnet.J.* **2011**, *17*, 10–12. [CrossRef]
67. Schmieder, R.; Edwards, R. Quality control and preprocessing of metagenomic datasets. *Bioinformatics* **2011**, *27*, 863–864. [CrossRef] [PubMed]
68. Boisvert, S.; Raymond, F.; Godzaridis, E.; Laviolette, F.; Corbeil, J. Ray Meta: Scalable de novo metagenome assembly and profiling. *Genome Biol.* **2012**, *13*, R122. [CrossRef] [PubMed]
69. Kumar, S.; Tamura, K.; Nei, M. MEGA3: Integrated software for molecular evolutionary genetics analysis and sequence alignment. *Brief. Bioinform.* **2004**, *5*, 150–163. [CrossRef] [PubMed]
70. Martin, D.; Rybicki, E. RDP: Detection of recombination amongst aligned sequences. *Bioinformatics* **2000**, *16*, 562–563. [CrossRef] [PubMed]
71. Salminen, M.O.; Carr, J.K.; Burke, D.S.; McCutchan, F.E. Identification of breakpoints in intergenotypic recombinants of HIV type 1 by bootscanning. *AIDS Res. Hum. Retrovir.* **1995**, *11*, 1423–1425. [CrossRef] [PubMed]
72. Smith, J.M. Analyzing the mosaic structure of genes. *J. Mol. Evol.* **1992**, *34*, 126–129. [CrossRef] [PubMed]
73. Posada, D.; Crandall, K.A. Evaluation of methods for detecting recombination from DNA sequences: Computer simulations. *Proc. Natl. Acad. Sci. USA* **2001**, *98*, 13757–13762. [CrossRef] [PubMed]
74. Holmes, E.C. Molecular epidemiology of dengue virus—The time for big science. *Trop. Med. Int. Health* **1998**, *3*, 855–856. [CrossRef] [PubMed]
75. Felsenstein, J. PHYLIP—Phylogeny inference package (version 3.2). *Cladistics* **1989**, *5*, 164–166.
76. Bendtsen, J.D.; Nielsen, H.; von Heijne, G.; Brunak, S. Improved prediction of signal peptides: SignalP 3.0. *J. Mol. Biol.* **2004**, *340*, 783–795. [CrossRef] [PubMed]
77. GSL Biotech LLC. SnapGene: Software for Everyday Molecular Biology. Available online: http://www.snapgene.com/ (accessed on 31 May 2016).
78. Von Heijne, G. Membrane protein structure prediction. Hydrophobicity analysis and the positive-inside rule. *J. Mol. Biol.* **1992**, *225*, 487–494. [CrossRef]
79. Bendtsen, J.D.; Nielsen, H.; von Heijne, G.; Brunak, S. Improved prediction of signal peptides: SignalP 3.0. *J. Mol. Biol.* **2004**, *340*, 783–795. [CrossRef] [PubMed]
80. Claros, M.G.; von Heijne, G. TopPred II: An improved software for membrane protein structure predictions. *Comput. Appl. Biosci.* **1994**, *10*, 685–686. [CrossRef] [PubMed]
81. Kahsay, R.Y.; Gao, G.; Liao, L. An improved hidden Markov model for transmembrane protein detection and topology prediction and its applications to complete genomes. *Bioinformatics* **2005**, *21*, 1853–1858. [CrossRef] [PubMed]
82. Kall, L.; Krogh, A.; Sonnhammer, E.L. A combined transmembrane topology and signal peptide prediction method. *J. Mol. Biol.* **2004**, *338*, 1027–1036. [CrossRef] [PubMed]
83. Krogh, A.; Larsson, B.; von Heijne, G.; Sonnhammer, E.L. Predicting transmembrane protein topology with a hidden Markov model: Application to complete genomes. *J. Mol. Biol.* **2001**, *305*, 567–580. [CrossRef] [PubMed]
84. Burt, F.J.; Paweska, J.T.; Ashkettle, B.; Swanepoel, R. Genetic relationship in southern African Crimean-Congo haemorrhagic fever virus isolates: Evidence for occurrence of reassortment. *Epidemiol. Infect.* **2009**, *137*, 1302–1308. [CrossRef] [PubMed]
85. Hewson, R.; Gmyl, A.; Gmyl, L.; Smirnova, S.E.; Karganova, G.; Jamil, B.; Hasan, R.; Chamberlain, J.; Clegg, C. Evidence of segment reassortment in Crimean-Congo haemorrhagic fever virus. *J. Gen. Virol.* **2004**, *85*, 3059–3070. [CrossRef] [PubMed]

86. Lukashev, A.N. Evidence for recombination in Crimean-Congo hemorrhagic fever virus. *J. Gen. Virol.* **2005**, *86*, 2333–2338. [CrossRef] [PubMed]
87. Goedhals, D.; Bester, P.A.; Paweska, J.T.; Swanepoel, R.; Burt, F.J. Next-generation sequencing of southern African Crimean-Congo haemorrhagic fever virus isolates reveals a high frequency of M segment reassortment. *Epidemiol. Infect.* **2014**, *142*, 1952–1962. [CrossRef] [PubMed]
88. Zhou, Z.; Deng, F.; Han, N.; Wang, H.; Sun, S.; Zhang, Y.; Hu, Z.; Rayner, S. Reassortment and migration analysis of Crimean-Congo haemorrhagic fever virus. *J. Gen. Virol.* **2013**, *94*, 2536–2548. [CrossRef] [PubMed]
89. Carter, S.D.; Surtees, R.; Walter, C.T.; Ariza, A.; Bergeron, É.; Nichol, S.T.; Hiscox, J.A.; Edwards, T.A.; Barr, J.N. Structure, function, and evolution of the Crimean-Congo hemorrhagic fever virus nucleocapsid protein. *J. Virol.* **2012**, *86*, 10914–10923. [CrossRef] [PubMed]
90. Dayer, M.R.; Dayer, M.S.; Rezatofighi, S.E. Mechanism of preferential packaging of negative sense genomic RNA by viral nucleoproteins in Crimean-Congo hemorrhagic Fever virus. *Protein J.* **2015**, *34*, 91–102. [CrossRef] [PubMed]
91. Fajs, L.; Resman, K.; Avšič-Županc, T. Crimean-Congo hemorrhagic fever virus nucleoprotein suppresses IFN-beta-promoter-mediated gene expression. *Arch. Virol.* **2014**, *159*, 345–348. [CrossRef] [PubMed]
92. Guo, Y.; Wang, W.; Ji, W.; Deng, M.; Sun, Y.; Zhou, H.; Yang, C.; Deng, F.; Wang, H.; Hu, Z.; *et al.* Crimean-Congo hemorrhagic fever virus nucleoprotein reveals endonuclease activity in bunyaviruses. *Proc. Natl. Acad. Sci. USA* **2012**, *109*, 5046–5051. [CrossRef] [PubMed]
93. Wang, W.; Liu, X.; Wang, X.; Dong, H.; Ma, C.; Wang, J.; Liu, B.; Mao, Y.; Wang, Y.; Li, T.; *et al.* Structural and functional diversity of nairovirus-encoded nucleoproteins. *J. Virol.* **2015**, *89*, 11740–11749. [CrossRef] [PubMed]
94. Wang, Y.; Dutta, S.; Karlberg, H.; Devignot, S.; Weber, F.; Hao, Q.; Tan, Y.J.; Mirazimi, A.; Kotaka, M. Structure of Crimean-Congo hemorrhagic fever virus nucleoprotein: Superhelical homo-oligomers and the role of caspase-3 cleavage. *J. Virol.* **2012**, *86*, 12294–12303. [CrossRef] [PubMed]
95. Andersson, I.; Bladh, L.; Mousavi-Jazi, M.; Magnusson, K.-E.; Lundkvist, Å.; Haller, O.; Mirazimi, A. Human MxA protein inhibits the replication of Crimean-Congo hemorrhagic fever virus. *J. Virol.* **2004**, *78*, 4323–4329. [CrossRef] [PubMed]
96. Karlberg, H.; Tan, Y.J.; Mirazimi, A. Induction of caspase activation and cleavage of the viral nucleocapsid protein in different cell types during Crimean-Congo hemorrhagic fever virus infection. *J. Biol. Chem.* **2011**, *286*, 3227–3234. [CrossRef] [PubMed]
97. Yates, C.M.; Filippis, I.; Kelley, L.A.; Sternberg, M.J. SuSPect: Enhanced prediction of single amino acid variant (SAV) phenotype using network features. *J. Mol. Biol.* **2014**, *426*, 2692–2701. [CrossRef] [PubMed]
98. Dunn, B.D.; Sakamoto, T.; Hong, M.S.; Sellers, J.R.; Takizawa, P.A. Myo4p is a monomeric myosin with motility uniquely adapted to transport mRNA. *J. Cell Biol.* **2007**, *178*, 1193–1206. [CrossRef] [PubMed]
99. Bergeron, É.; Zivcec, M.; Chakrabarti, A.K.; Nichol, S.T.; Albariño, C.G.; Spiropoulou, C.F. Recovery of Recombinant Crimean Congo Hemorrhagic Fever Virus Reveals a Function for Non-structural Glycoproteins Cleavage by Furin. *PLoS Pathog.* **2015**, *11*, e1004879. [CrossRef] [PubMed]
100. Sanchez, A.J.; Vincent, M.J.; Erickson, B.R.; Nichol, S.T. Crimean-Congo hemorrhagic fever virus glycoprotein precursor is cleaved by furin-like and SKI-1 proteases to generate a novel 38-kilodalton glycoprotein. *J. Virol.* **2006**, *80*, 514–525. [CrossRef] [PubMed]
101. Vincent, M.J.; Sanchez, A.J.; Erickson, B.R.; Basak, A.; Chretien, M.; Seidah, N.G.; Nichol, S.T. Crimean-Congo hemorrhagic fever virus glycoprotein proteolytic processing by subtilase SKI-1. *J. Virol.* **2003**, *77*, 8640–8649. [CrossRef] [PubMed]
102. Deyde, V.M.; Khristova, M.L.; Rollin, P.E.; Ksiazek, T.G.; Nichol, S.T. Crimean-Congo hemorrhagic fever virus genomics and global diversity. *J. Virol.* **2006**, *80*, 8834–8842. [CrossRef] [PubMed]
103. Sanchez, A.J.; Vincent, M.J.; Nichol, S.T. Characterization of the glycoproteins of Crimean-Congo hemorrhagic fever virus. *J. Virol.* **2002**, *76*, 7263–7275. [CrossRef] [PubMed]
104. Poch, O.; Sauvaget, I.; Delarue, M.; Tordo, N. Identification of four conserved motifs among the RNA-dependent polymerase encoding elements. *EMBO J.* **1989**, *8*, 3867–3874. [PubMed]
105. Müller, R.; Poch, O.; Delarue, M.; Bishop, D.H.L.; Bouloy, M. Rift Valley fever virus L segment: Correction of the sequence and possible functional role of newly identified regions conserved in RNA-dependent polymerases. *J. Gen. Virol.* **1994**, *75*, 1345–1352. [CrossRef] [PubMed]

106. Das, K.; Arnold, E. Negative-strand RNA virus L proteins: one machine, many activities. *Cell* **2015**, *162*, 239–241. [CrossRef] [PubMed]
107. Reguera, J.; Gerlach, P.; Cusack, S. Towards a structural understanding of RNA synthesis by negative strand RNA viral polymerases. *Curr. Opin. Struct. Biol.* **2016**, *36*, 75–84. [CrossRef] [PubMed]
108. Frias-Staheli, N.; Giannakopoulos, N.V.; Kikkert, M.; Taylor, S.L.; Bridgen, A.; Paragas, J.; Richt, J.A.; Rowland, R.R.; Schmaljohn, C.S.; Lenschow, D.J.; *et al.* Ovarian tumor domain-containing viral proteases evade ubiquitin- and ISG15-dependent innate immune responses. *Cell Host Microbe* **2007**, *2*, 404–416. [CrossRef] [PubMed]
109. James, T.W.; Frias-Staheli, N.; Bacik, J.P.; Levingston Macleod, J.M.; Khajehpour, M.; García-Sastre, A.; Mark, B.L. Structural basis for the removal of ubiquitin and interferon-stimulated gene 15 by a viral ovarian tumor domain-containing protease. *Proc. Natl. Acad. Sci. USA* **2011**, *108*, 2222–2227. [CrossRef] [PubMed]
110. Capodagli, G.C.; Deaton, M.K.; Baker, E.A.; Lumpkin, R.J.; Pegan, S.D. Diversity of ubiquitin and ISG15 specificity among nairoviruses' viral ovarian tumor domain proteases. *J. Virol.* **2013**, *87*, 3815–3827. [CrossRef] [PubMed]
111. Capodagli, G.C.; McKercher, M.A.; Baker, E.A.; Masters, E.M.; Brunzelle, J.S.; Pegan, S.D. Structural analysis of a viral ovarian tumor domain protease from the Crimean-Congo hemorrhagic fever virus in complex with covalently bonded ubiquitin. *J. Virol.* **2011**, *85*, 3621–3630. [CrossRef] [PubMed]
112. Bakshi, S.; Holzer, B.; Bridgen, A.; McMullan, G.; Quinn, D.G.; Baron, M.D. Dugbe virus ovarian tumour domain interferes with ubiquitin/ISG15-regulated innate immune cell signalling. *J. Gen. Virol.* **2013**, *94*, 298–307. [CrossRef] [PubMed]
113. Van Kasteren, P.B.; Beugeling, C.; Ninaber, D.K.; Frias-Staheli, N.; van Boheemen, S.; García-Sastre, A.; Snijder, E.J.; Kikkert, M. Arterivirus and nairovirus ovarian tumor domain-containing Deubiquitinases target activated RIG-I to control innate immune signaling. *J. Virol.* **2012**, *86*, 773–785. [CrossRef] [PubMed]
114. Kinsella, E.; Martin, S.G.; Grolla, A.; Czub, M.; Feldmann, H.; Flick, R. Sequence determination of the Crimean-Congo hemorrhagic fever virus L segment. *Virology* **2004**, *321*, 23–28. [CrossRef] [PubMed]
115. Landschulz, W.H.; Johnson, P.F.; McKnight, S.L. The leucine zipper: A hypothetical structure common to a new class of DNA binding proteins. *Science* **1988**, *240*, 1759–1764. [CrossRef] [PubMed]
116. Hurst, H.C. Transcription factors. 1: bZIP proteins. *Protein Profile* **1994**, *1*, 123–168. [PubMed]
117. Shugars, D.C.; Smith, M.S.; Glueck, D.H.; Nantermet, P.V.; Seillier-Moiseiwitsch, F.; Swanstrom, R. Analysis of human immunodeficiency virus type 1 *nef* gene sequences present *in vivo*. *J. Virol.* **1993**, *67*, 4639–4650. [PubMed]
118. Singh, P.; Yadav, G.P.; Gupta, S.; Tripathi, A.K.; Ramachandran, R.; Tripathi, R.K. A novel dimer-tetramer transition captured by the crystal structure of the HIV-1 Nef. *PLoS ONE* **2011**, *6*, e26629. [CrossRef] [PubMed]
119. Cowley, J.A.; McCulloch, R.J.; Spann, K.M.; Cadogan, L.C.; Walker, P.J. Preliminary molecular and biological characterization of Mourilyan virus (MoV): A new bunya-related virus of penaeid prawns. In *Diseases in Asian Aquaculture V. Proceedings of the 5th Symposium on Diseases in Asian Aquaculture*; Walker, P.J., Lester, R.G., Bondad-Reantaso, M.G., Eds.; Fish Health Section, Asian Fisheries Society: Manila, The Philippines, 2005; pp. 113–124.
120. Mielke, N.; Muehlbach, H.P. A novel, multipartite, negative-strand RNA virus is associated with the ringspot disease of European mountain ash (Sorbus aucuparia L.). *J. Gen. Virol.* **2007**, *88*, 1337–1346. [CrossRef] [PubMed]
121. Mielke-Ehret, N.; Mühlbach, H.-P. *Emaravirus*: A novel genus of multipartite, negative strand RNA plant viruses. *Viruses* **2012**, *4*, 1515–1536. [CrossRef] [PubMed]
122. Kakutani, T.; Hayano, Y.; Hayashi, T.; Minobe, Y. Ambisense segment 4 of rice stripe virus: Possible evolutionary relationship with phleboviruses and uukuviruses (*Bunyaviridae*). *J. Gen. Virol.* **1990**, *71*, 1427–1432. [CrossRef] [PubMed]
123. Bucher, E.; Sijen, T.; De Haan, P.; Goldbach, R.; Prins, M. Negative-strand tospoviruses and tenuiviruses carry a gene for a suppressor of gene silencing at analogous genomic positions. *J. Virol.* **2003**, *77*, 1329–1336. [CrossRef] [PubMed]
124. Falk, B.W.; Tsai, J.H. Biology and molecular biology of viruses in the genus *Tenuivirus*. *Annu. Rev. Phytopathol.* **1998**, *36*, 139–163. [CrossRef] [PubMed]

125. Garry, C.E.; Garry, R.F. Proteomics computational analyses suggest that the carboxyl terminal glycoproteins of Bunyaviruses are class II viral fusion protein (beta-penetrenes). *Theor. Biol. Med. Model.* **2004**, *1*, 10. [CrossRef] [PubMed]
126. Reguera, J.; Weber, F.; Cusack, S. *Bunyaviridae* RNA polymerases (L-protein) have an N-terminal, influenza-like endonuclease domain, essential for viral cap-dependent transcription. *PLoS Pathog.* **2010**, *6*, e1001101. [CrossRef] [PubMed]
127. Briese, T.; Calisher, C.H.; Higgs, S. Viruses of the family *Bunyaviridae*: Are all available isolates reassortants? *Virology* **2013**, *446*, 207–216. [CrossRef] [PubMed]
128. Bao, Y.; Chetvernin, V.; Tatusova, T. Improvements to pairwise sequence comparison (PASC): A genome-based web tool for virus classification. *Arch. Virol.* **2014**, *159*, 3293–3304. [CrossRef] [PubMed]
129. Bao, Y.; Kapustin, Y.; Tatusova, T. Virus classification by Pairwise Sequence Comparison (PASC). In *Encyclopedia of Virology*, 3rd ed.; Mahy, B.W.J., Regenmortel, M.H.V., Eds.; Academic Press: Oxford, UK, 2008; pp. 342–348.

© 2016 by the authors. Licensee MDPI, Basel, Switzerland. This article is an open access article distributed under the terms and conditions of the Creative Commons Attribution (CC BY) license (http://creativecommons.org/licenses/by/4.0/).

Article

Genetic and Phylogenetic Characterization of Tataguine and Witwatersrand Viruses and Other Orthobunyaviruses of the Anopheles A, Capim, Guamá, Koongol, Mapputta, Tete, and Turlock Serogroups

Alexey M. Shchetinin [1], Dmitry K. Lvov [1], Petr G. Deriabin [1], Andrey G. Botikov [1], Asya K. Gitelman [1], Jens H. Kuhn [2] and Sergey V. Alkhovsky [1,*]

[1] D.I. Ivanovsky Institute of Virology, Gamaleya Federal Research Center for Epidemiology and Microbiology, Ministry of Health of the Russian Federation, 123098, Moscow, Russia; shchetinin.alexey@yandex.ru (A.M.S.); dk_lvov@mail.ru (D.K.L.); pg_deryabin@mail.ru (P.G.D.); tessey@mail.ru (A.G.B.); gitelman_ak@mail.ru (A.K.G.)

[2] Integrated Research Facility at Fort Detrick, National Institute of Allergy and Infectious Diseases, National Institutes of Health, Fort Detrick, Frederick, MD 21702, USA; kuhnjens@mail.nih.gov

* Correspondence: salkh@yandex.ru; Tel.: +7-499-190-3043; Fax: +7-499-190-2867

Academic Editors: Jane Tao and Pierre-Yves Lozach
Received: 2 September 2015; Accepted: 7 November 2015; Published: 23 November 2015

Abstract: The family *Bunyaviridae* has more than 530 members that are distributed among five genera or remain to be classified. The genus *Orthobunyavirus* is the most diverse bunyaviral genus with more than 220 viruses that have been assigned to more than 18 serogroups based on serological cross-reactions and limited molecular-biological characterization. Sequence information for all three orthobunyaviral genome segments is only available for viruses belonging to the Bunyamwera, Bwamba/Pongola, California encephalitis, Gamboa, Group C, Mapputta, Nyando, and Simbu serogroups. Here we present coding-complete sequences for all three genome segments of 15 orthobunyaviruses belonging to the Anopheles A, Capim, Guamá, Kongool, Tete, and Turlock serogroups, and of two unclassified bunyaviruses previously not known to be orthobunyaviruses (Tataguine and Witwatersrand viruses). Using those sequence data, we established the most comprehensive phylogeny of the *Orthobunyavirus* genus to date, now covering 15 serogroups. Our results emphasize the high genetic diversity of orthobunyaviruses and reveal that the presence of the small nonstructural protein (NSs)-encoding open reading frame is not as common in orthobunyavirus genomes as previously thought.

Keywords: Anopheles A virus; bunyavirus; Capim virus; Guamá virus; Koongol virus; orthobunyavirus; Tataguine virus; Tete virus; Turlock virus; Witwatersrand virus

---

1. Introduction

The family *Bunyaviridae* ranks among the largest families of RNA viruses. The family has more than 530 named members that are either assigned to the five included genera *Hantavirus*, *Nairovirus*, *Orthobunyavirus*, *Phlebovirus*, or *Tospovirus*, or remain to be classified [1–3]. Many bunyaviruses can cause disease in humans. These diseases commonly manifest as arthritides/rashes, fevers/myalgias, pulmonary diseases, encephalitides, or viral hemorrhagic fevers [4]. In addition, bunyaviruses can cause severe disease in wild and domesticated animals and wild or cultivated crop or ornamental plants (tospoviruses only). The majority of bunyaviruses is transmitted among hosts by arthropods (predominantly mosquitoes, ticks, sandflies, biting midges, or thrips), whereas hantaviruses are transmitted by contaminated excreta or secreta or bites of infected rodents, eulipotyphla, or bats [1,3]. In addition, several genus-level clades of insect-specific bunyaviruses (Ferak and Jonchet virus clades, "goukoviruses", "herbeviruses", "phasmaviruses") have been described recently. Among those, "herbeviruses," *i.e.*, Herbert virus, Kibale virus, and Taï virus, form a deeply-rooted sister taxon to the genus *Orthobunyavirus* [5,6].

Bunyaviruses have tripartite, negative-sense, single-stranded RNA genomes [7,8]. The three linear genomic segments, commonly referred to as small (S), medium (M), and large (L) based on their overall lengths, are characterized by complementary 3′- and 5′-untranslated regions (UTRs) that form panhandle-like structures. The S segment encodes the nucleocapsid protein (N) and often a small nonstructural protein (NSs). The M segment encodes a glycoprotein precursor polyprotein, which after cleavage yields viral glycoproteins (Gn and Gc) and, in the case of some bunyaviruses, a small, nonstructural protein (NSm). The L segment encodes the RNA-dependent RNA polymerase (RdRp, L). The glycoproteins mediate virion entry into susceptible cells through endocytosis. L transcribes and replicates the individual segments in the cytosol. Newly synthesized N encapsidates progeny RNA genome segments to form ribonucleoprotein complexes that associate with a few molecules of RdRp. These complexes acquire glycoprotein-containing envelopes upon budding into Golgi-derived vesicles and egress from the host cell by exocytosis. The functions of the various nonstructural proteins are less clear, but NSs has been identified as an interferon antagonist [7,8].

The genus *Orthobunyavirus* is the most complex genus of the family *Bunyaviridae* and currently contains approximately 220 named viruses. The majority of these viruses were assigned to more than 18 different serogroups based on presence, lack, or extent of serological cross reactions using various assays, such as complement fixation (CF), hemagglutination inhibition (HI), and neutralization test (NT) [9–14]. Mostly within those serogroups, these orthobunyaviruses were assigned to 48 official species [2]. Until recently, genomic sequence information was limited to some orthobunyaviruses of the Bunyamwera, California encephalitis, Simbu, and Group C serogroups [15–23]. Recent advances in next-generation sequencing have led to an acquisition of coding-complete and sometimes complete genome sequences of some orthobunyaviruses from the Bwamba/Pongola, Gamboa, Mapputta, and Nyando serogroups, and of several ungrouped viruses [24–28]. However, viruses of more than 10 serogroups remain to be characterized on the genomic level.

Here we report the first coding-complete genomic sequences of 15 distinct orthobunyaviruses from the Anopheles A, Capim, Guamá, Koongol, Mapputta, Tete, and Turlock serogroups (Table 1).

Viruses of the Anopheles A serogroup have been isolated from *Aedes* and *Anopheles* mosquitoes mostly collected in South American and Caribbean countries. Eleven different viruses of this group have been assigned to the two species *Anopheles A orthobunyavirus* (Anopheles A, Arumateua, Caraipé, Las Maloyas, Lukuni, Trombetas, Tucuruí viruses) and *Tacaiuma orthobunyavirus* (CoAr 1071, CoAr 3627, Tacaiuma, Virgin River viruses) [1,2,29–32]. We present coding-complete sequence information for Lukuni virus, which was originally isolated from *Aedes* mosquitoes collected in 1955 in Trinidad and Brazil [33], but has not been associated with human disease.

Capim serogroup viruses are classified into five species: *Acara orthobunyavirus* (Acara, Moriche viruses), *Benevides orthobunyavirus* (Benevides virus), *Bushbush orthobunyavirus* (Benfica, Bushbush, Juan Diaz viruses), *Capim orthobunyavirus* (Capim virus), and *Guajará orthobunyavirus* (Guajará virus). These viruses are endemic to South and Northern America, where they are transmitted by mosquitoes among small vertebrates [1,2,32]. We determined the coding-complete sequence of Capim virus and Guajará virus. Capim virus was originally isolated from a trapped woolly opossum (*Caluromys philander*) in 1958 in Pará State, Brazil, and was also recovered from sentinel spiny rats (*Proechimys* spp.) and *Culex* mosquitoes. Antibodies to the virus were detected in spiny rat sera, but not in human sera in Pará State. Guajará virus was first isolated from a sentinel Swiss laboratory mouse in Pará State, Brazil. Additional isolates were obtained in South America from wild rodents and mosquitoes of different species, and antibodies against the virus were detected repeatedly in spiny rats [11,34,35].

Table 1. Newly sequenced orthobunyaviruses (this study).

Orthobunyavirus Serogroup Virus Name (Virus Abbreviation) Isolate Designation	Year of Virus Isolation	Country of Virus Isolation (Current Country Name)	Virus Source (Species)	Signs of Human Infection	Reference(s)	New GenBank Accession Numbers
Anopheles A serogroup						
Lukuni virus (LUKV)						
TRVL 10076	1955	Trinidad and Tobago Crown Colony of the British Empire (Trinidad and Tobago)	Mosquitoes (*Aedes* (*Ochlerotatus*) *scapularis*)	NR	[33]	KP792670-72
Capim serogroup						
Capim virus (CAPV)						
BeAn 8582	1958	Brazil	Woolly opossum (*Caluromys philander*)	NR	[34]	KT160026-28
Guajará virus (GJAV)						
BeAn 10615	1959	Brazil	Swiss laboratory mouse, sentinel	NR	[34,35]	KP792661-63
Guamá serogroup						
Bimiti virus (BIMV)						
TRVL 8362	1955	Trinidad and Tobago Crown Colony of the British Empire (Trinidad and Tobago)	Mosquitoes (*Culex* (*Melanoconion*) *spissipes*)	NR	[36,37]	KP792655-57
Catú virus (CATUV)						
BeH 151	1955	Brazil	Mosquitoes (*Culex* (*Melanoconion*) *spissipes*)	Fever, myalgia	[38]	KP792658-60
Guamá virus (GMAV)						
BeAn 277	1955	Brazil	Tufted capuchin (*Cebus apella*), sentinel	Fever, myalgia	[38]	KP792664-66
Mahogany Hammock virus (MHV)						
Fe4-2a	1964	USA	Mosquitoes (*Culex* (*Melanoconion*) sp.)	NR	[39]	KP835518-20
Moju virus (MOJUV)						
BeAr 12590	1959	Brazil	Mosquitoes (*Culex* sp.)	NR	[34]	KP792673-75
Koongol serogroup						
Koongol virus (KOOV)						
MRM31	1960	Australia	Mosquitoes (*Culex* (*Culex*) *annulirostris*)	NR	[40]	KP792667-69

Table 1. *Cont.*

Orthobunyavirus Serogroup Virus Name (Virus Abbreviation) Isolate Designation	Year of Virus Isolation	Country of Virus Isolation (Current Country Name)	Virus Source (Species)	Signs of Human Infection	Reference(s)	New GenBank Accession Numbers
Mapputta serogroup						
Mapputta virus (MAPV)						
MRM186	1960	Australia	Mosquitoes (*Anopheles* (*Cellia*) *meraukensis*)	NR	[40]	KP792694-96
Trubanaman virus (TRUV)						
MRM3630	1966	Australia	Mosquitoes (*Anopheles* (*Cellia*) *annulipes*)	Arthritis, rash	[41,42]	KP792682-84
Tete serogroup						
Bahig virus (BAHV)						
EgB 90	1966	Egypt	Eurasian golden oriole (*Oriolus oriolus*)	NR	[43]	KP792652-54
Matruh virus (MTRV)						
An-1047	1961	United Arab Republic (Egypt)	Lesser whitethroat (*Sylvia curruca*)	NR	[43]	KP792691-93
Tete virus (TETEV)						
SAAn 3518	1959	Union of South Africa (South Africa)	Village weaver (*Ploceus cucullatus*)	NR	[44]	KP792679-81
Turlock serogroup						
Umbre virus (UMBV)						
IG1424	1955	India	Mosquitoes (*Culex* (*Oculeomyia*) *bitaeniorhynchus*)	NR	[44]	KP792685-87
Unassigned						
Tataguine virus (TATV)						
Ib-H 9963	1968	Nigeria	Human (*Homo sapiens*)	Fever, myalgia	[45]	KP792676-78
Witwatersrand virus (WITV)						
SAAr 1062	1958	Union of South Africa (South Africa)	Mosquitoes (*Culex* (*Eumelanomyia*) *rubinotus*)	NR	[46]	KP792688-90

NR, none reported.

The Guamá serogroup viruses are distributed among five established species: *Bertioga orthobunyavirus* (Bertioga, Cananeia, Guaratuba, Itimirim, Mirim viruses), *Bimiti orthobunyavirus* (Bimiti virus), *Catú orthobunyavirus* (Catú virus), *Guamá orthobunyavirus* (Ananindeua, Guamá, Mahogany Hammock, Moju viruses), and *Timboteua orthobunyavirus* (Timboteua virus) [1,2,32]. Guamá serogroup viruses are mostly endemic to South America. The exceptions are Guamá and Mahogany Hammock viruses, which were isolated in North America. Catú virus and Guamá virus were isolated from humans with fever/myalgia. Viruses of the Guamá serogroup are usually transmitted by culicine mosquitoes among vertebrate hosts including bats, birds, marsupials, and rodents [34,36–39,47]. Here, we present coding-complete sequences for Bimiti, Catú, Guamá, Mahogany Hammock, and Moju viruses.

The Koongol serogroup currently consists of Koongol virus and Wongal virus, both of which are assigned to the species *Koongol orthobunyavirus* [1,2,32]. Both viruses were originally isolated in 1960 in Queensland, Australia, from *Culex* mosquitoes [40]. Koongol virus, newly sequenced here, was also obtained from *Ficalbia* mosquitoes in New Guinea [35]. HI tests suggested that both Koongol and Wongal viruses may be able to infect a range of mammals, marsupials, and possibly birds, and may be widespread throughout Australia. However, these results were not confirmed using NT [48–50].

The Mapputta serogroup currently has seven members (Buffalo Creek, Gan Gan, Mapputta, Maprik, Murrumbidgee, Salt Ash, Trubanaman viruses) that have not yet been assigned to species [1,24,51]. Buffalo Creek virus and Murrumbidgee viruses [24,52,53] were isolated from *Anopheles* mosquitoes collected in Northern Territory and Griffith, Australia, respectively. Mapputta and Trubanaman viruses, which we have sequenced during this study, were initially isolated from *Anopheles* mosquitoes, respectively, collected in Queensland, Australia in the 1960s [40,41]. Antibodies against Trubanaman virus have been detected in humans and domestic and wild animals in different parts of Queensland, and the virus is suspected to cause arthritis/rash in humans [42].

Viruses of the species *Batama orthobunyavirus* (Batama virus) and *Tete orthobunyavirus* (Bahig, Matruh, Tete, Tsuruse, Weldona viruses) comprise the Tete serogroup [1,2,32]. Together with Bakau virus (Bakau serogroup) and Estero Real virus (Patois serogroup), Tete serogroup viruses are the only currently known classified orthobunyaviruses transmitted by ticks. We sequenced Bahig, Matruh, and Tete viruses. Bahig and Matruh viruses, first discovered in 1966 and 1961, respectively [43], have been repeatedly isolated from birds and ticks collected in Egypt and Italy; successful isolations have also been made from birds trapped in Cyprus [54–56]. Tete virus, the prototype virus of the serogroup, was originally isolated from a spotted-backed weaver bird (*Ploceus cucullatus*) collected in South Africa in 1959 [44].

Finally, we sequenced Umbre virus, a virus of the Turlock serogroup that is represented by viruses belonging to the two species *M'Poko orthobunyavirus* (M'Poko virus, Yaba-1 virus) and *Turlock orthobunyavirus* (Lednice virus, Turlock virus, Umbre virus) [1,2]. Turlock serogroup viruses are distributed in Africa, Asia, Europe, and Northern and South America. Umbre virus was initially isolated from *Culex bitaeniorhynchus* mosquitoes collected in Poona (today Pune), India in 1955 [44]. Additional virus isolates were obtained from *Culex* mosquitoes and birds in India and Malaysia. Anti-Umbre virus antibodies could be detected in sera collected from wild birds and sentinel chickens from Malaysia [35].

In addition, we determined coding-complete genomic sequences of two unclassified members of the family *Bunyaviridae*, Tataguine virus and Witwatersrand virus (Table 1). Tataguine virus was initially recovered from pooled *Culex* and *Anopheles* mosquitoes collected in Senegal in 1962 [57]. This virus is widespread in Africa as evidenced by isolation from areas of today's Cameroon, Central African Republic, Ethiopia, Nigeria, and Senegal [45,58–64]. Tataguine virus is a known human pathogen; infected patients present with fever, headache, rash, and joint pain [63–65]. Witwatersrand virus was originally isolated from *Culex* mosquitoes collected in Germiston, South Africa, in 1958 [46]. Additional isolates were obtained from *Culex* mosquitoes, sentinel hamsters, and various rodents sampled in

Mozambique, South Africa, and Uganda. Antibodies to Witwatersrand virus also have been detected in human sera, although the virus has not been associated with human disease [66].

Our data unequivocally identify both viruses as members of the genus *Orthobunyavirus* and largely confirm the previously deduced relationships of the remaining 15 viruses using serological assays.

2. Materials and Methods

2.1. Viral Genomic RNA Isolation and Library Preparation

Orthobunyairuses were obtained from the Russian State Collection of Viruses in the form of lyophilized infected suckling mouse brains (Table 1). Total RNA was isolated from vials with 1 mL of TRI Reagent (Molecular Research Center, Cincinnati, OH, USA). Total RNA was additionally purified using the RNeasy MinElute Cleanup Kit (Qiagen, Hilden, Germany) followed by ribosomal RNA depletion using the GeneRead rRNA Depletion Kit (Qiagen) according to the manufacturers' instructions. Purified RNA was reverse-transcribed with RevertAid Reverse Transcriptase (Thermo Fisher Scientific, Grand Island, NY, USA) using hexameric random primers (Promega, Madison, WI, USA). First strand cDNA was converted to double-stranded cDNA using the NEBNext Second Strand Synthesis Module (New England BioLabs, Ipswich, MA, USA) according to the manufacturer's instructions. Resulting dsDNA was used to prepare next-generation sequencing libraries using the TruSeq DNA LT Library Prep Kit (Illumina, San Diego, CA, USA). A paired-end 250-bp protocol was used for sequencing indexed libraries on an Illumina MiSeq instrument.

2.2. Bioinformatic and Phylogenetic Analyses

Primary analysis of sequencing data and *de novo* genome assembly were performed with CLC Genomics Workbench 7.0 (CLC bio, Waltham, MA, USA). Open reading frame (ORF) analysis and general work with assembled contigs were performed using the Lasergene 11.0.0 (DNAStar, Madison, WI, USA) software package. After *de novo* assembly of trimmed reads, BLASTx (BLAST, basic local alignment search tool, open-source software) analysis was performed against orthobunyaviral sequences, and matching contigs were extracted. All resulting contigs contained parts of non-coding terminal regions and full-length ORFs corresponding to their matches in the BLASTx search. Identified ORFs were translated, and the resulting amino acid (aa) sequences were used in further analyses. Deduced aa sequences of the proteins encoded by sequenced orthobunyaviral genome segments and corresponding sequences of selected representatives of already characterized orthobunyaviruses were aligned using all available multiple sequence alignment methods implemented on M-Coffee server [67] and only columns with a score of 5 or higher were retained. N, glycoprotein precursor polyprotein, and RdRp final alignment lengths were 240, 1412, and 2331 aa residues, respectively. Most suitable models of protein evolution were predicted with open-source ProtTest 3.2 for three alignments [68].

Phylogenetic trees were inferred using MrBayes 3.2.4 [69] under LG + G + I model for N and LG + G + I + F model for glycoprotein precursor polyprotein and RdRp alignments, with 1,000,000 generations and a 25% burn-in value. Maximum likelihood (ML) phylogenies were inferred using MEGA6 with the same protein evolution models and 1000 bootstrap replicates. Consensus trees were visualized with TreeGraph 2.4 [70].

Signal peptide cleavage sites of orthobunyaviral glycoprotein precursors were determined from deduced aa sequences using SignalP 4.1 Server [71]. Transmembrane domains of glycoprotein precursors were predicted using the same sequences and open-source TMHMM Server v 2.0 [72]. Putative *N*-glycosylation sites were determined with the open-source NetNGlyc 1.0 Server [73].

3. Results

Segment-specific ORF-based phylogenies of the genus *Orthobunyavirus*, including previously and newly characterized viruses, are presented in Figure 1.

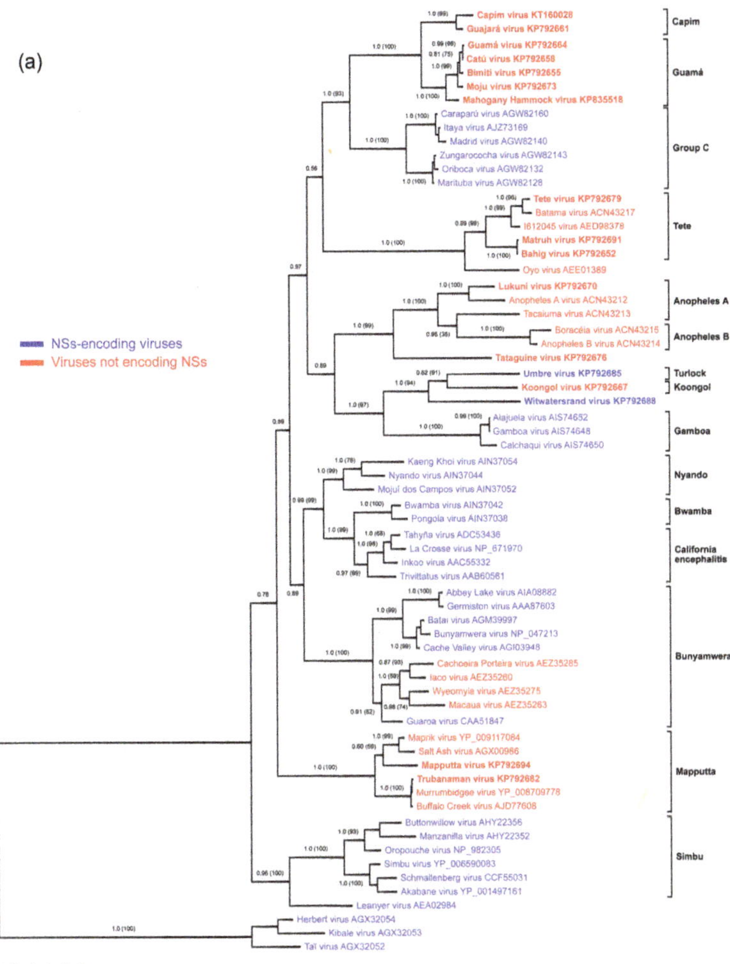

Figure 1. *Cont.*

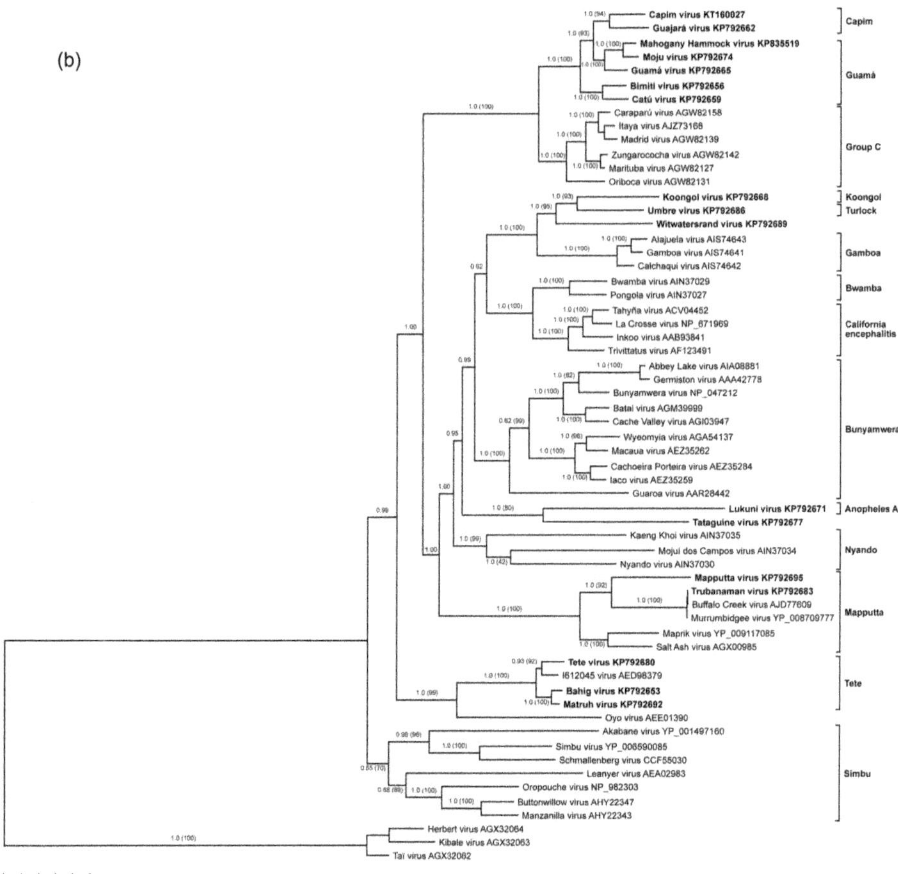

Figure 1. *Cont.*

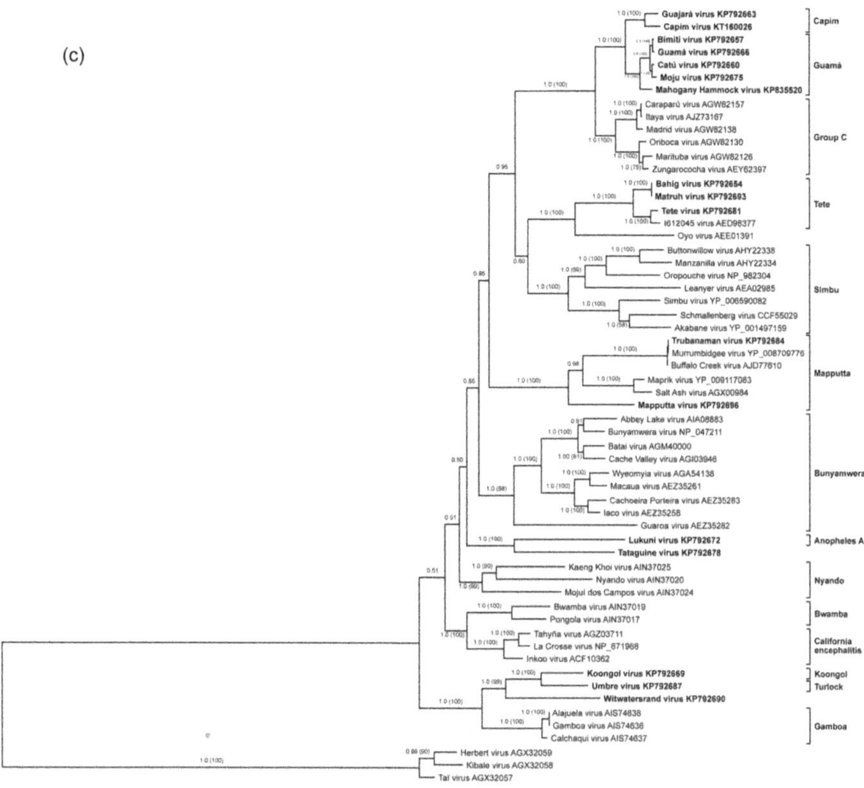

Figure 1. Protein sequence-based phylogenies of orthobunyaviruses. Phylogenies were inferred for (**a**) N; (**b**) glycoprotein precursor polyprotein; and (**c**) L proteins using Bayesian and Maximum Likelihood analyses. The LG + I + G model of amino acid (aa) substitution was used for inferring the N protein phylogeny, whereas the LG + I + G + F model was used to investigate glycoprotein precursor polyprotein and L protein phylogenetic relationships. Numbers represent Bayesian posterior probabilities (Maximum Likelihood bootstrap values). "Herbeviruses" (Herbert virus, Kibale virus, and Taï virus) [5,6] were used to root the phylograms. Trees are drawn to scale measured in substitutions per site. In (a), viruses that encode NSs proteins are marked in blue, whereas viruses that do not are marked in red. Viruses studied in the present work are depicted in bold. GenBank accession numbers and serogroups are to the right of virus names.

3.1. *Anopheles A Serogroup*

Genomic data of the Anopheles A serogroup orthobunyaviruses were thus far limited to S segment sequences of Anopheles A and Tacaiuma viruses [74]. Here, we expanded this sequence space by adding the complete coding sequences of Lukuni virus. Based on the obtained genomic data, Lukuni virus is closely related to Anopheles A virus (71.5% aa nucleocapsid sequence identity) and is more distantly related to Tacaiuma virus (53.9% nucleocapsid sequence identity).

3.2. *Guamá and Capim Serogroups*

All Guamá serogroup members studied in the present work (Bimiti, Catú, Guamá, Mahogany Hammock, Moju viruses) clustered together in all three phylogenetic trees (Figure 1), and formed a monophyletic group along with Capim [34,35] and Group C serogroup viruses. The Group C serogroup includes four species: *Caraparu orthobunyavirus* (Apeú, Bruconha, Caraparú, Ossa, Vinces viruses), *Madrid orthobunyavirus* (Madrid virus), *Marituba orthobunyavirus* (Gumbo Limbo, Marituba, Murutucú, Nepuyo, Restan, Zungarococha viruses), and *Oriboca orthobunyavirus* (Itaquí, Oriboca viruses) [1,2]. Group C serogroup viruses were originally thought not to be related to Guamá group viruses, but a low level of HI cross-reactivity between them links these groups of viruses [47].

The measured divergence of Guamá serogroup viruses is somewhat inconsistent among phylogenies inferred for different viral proteins (Figure 1). For instance, based on the phylogenies obtained for N and L proteins, Guamá, Catú, Moju, and Bimiti viruses are more closely related to each other than to Mahogany Hammock virus, although exhibiting different branching orders inside the group. Additionally, analysis of the glycoprotein precursor polyproteins pairs Moju virus with Guamá virus, Mahogany Hammock virus, and Capim serogroup viruses. Taken together, these discrepancies suggest that all examined Guamá group viruses may be reassortants.

The most obvious example of this observation is the relationship between Catú BeH 151 and Guamá BeAn 277 viruses, which were found to be almost identical when compared by their N protein sequences, differing only by two aa (99.2% sequence identity). However, their glycoprotein precursor polyprotein and L protein sequences are more divergent (64.5% polyprotein identity, 95.8% L protein identity). These data are in agreement with original studies that distinguished these two viruses in NT, but not in CF tests [37]. The observed relationships between Catú and Guamá viruses support the idea that one or both of the viruses may be reassortants.

Capim and Guajará viruses fall basal to viruses of the Guamá group phylogenies inferred for N and L proteins, but are placed inside the Guamá serogroup in the glycoprotein precursor polyprotein phylogeny (Figure 1). Capim and Guajará viruses are one-way reactive with Guamá, and Catú and Guamá viruses, respectively, in NT but not in CF assays. Interestingly, no reactivity between Capim and Guajará viruses was found in NT [35]. Bushbush virus, another member of the Capim serogroup, has similar cross-reactivity in NT but not in the CF assay with Bimiti and Catú viruses of the Guamá serogroup [35]. The observed phylogenetic and antigenic relationships between Guamá and Capim serogroup viruses might indicate that their ancestors were inter-group reassortant viruses that shared an M segment. If validated by further experiments and analyses, this finding may be of significance for orthobunyavirus taxonomy, as thus far reassortment has only been observed among viruses of the same orthobunyaviral species and this restriction has been used as one of the official orthobunyavirus species demarcation criteria [2].

3.3. *Mapputta Serogroup*

The obtained genomic sequences of Mapputta virus MRM186 is more than 99% identical to the sequences of the same isolate reported previously [24], and all identified nucleotide (nt) substitutions are synonymous. These substitutions likely arose due to different passaging and maintenance procedures. Our phylogenetic analysis placed Trubanaman virus firmly inside the Mapputta serogroup and revealed its close relationship to two previously characterized viruses: Buffalo Creek virus and Murrumbidgee virus [24,52,53]. The N protein divergence of Buffalo Creek, Murrumbidgee, and Trubanaman viruses does not exceed 2.1%, suggesting that Buffalo Creek and Murrumbidgee viruses are different isolates of Trubanaman virus rather than distinct viruses. An increasing need for further characterization of this group of viruses is indicated by evidence that Buffalo Creek virus [52] and Trubanaman virus are suspected as human pathogens [42].

3.4. *Tete Serogroup*

Sequence information for this serogroup was limited to the S segment of Tete and Batama viruses [74] and a partial Weldona virus M segment sequence. We provide complete coding sequences for three Tete serogroup viruses: Tete, Bahig, and Matruh viruses. Interestingly, the S segment sequence of Tete virus SAAn 3518 obtained here differs slightly from that previously reported [74]. The conflicting region is located closer to the C-terminus of the N protein and, in our case, is represented by a characteristic amino acid motif (G_{158}–S_{164}) found in all other N protein sequences belonging to Tete group viruses, but not in the Tete virus SAAn 3518 sequence reported earlier. Our analyses reveal Bahig and Matruh viruses to be closely related, with 98.8%, 99.1%, and 87.2% aa identities among their N, L, and glycoprotein precursor polyprotein sequences. This observed relationship is in agreement with the results of CF tests, which showed that these viruses were practically indistinguishable, and HI tests, which proved them to be easily distinguishable [35]. The obtained phylogenetic tree topology and branching order of Tete serogroup viruses (Figure 1) is consistent among the three segments.

Two unclassified bunyaviruses, I612045 virus (GenBank HM627179-81) and Oyo virus (GenBank HM639778-80) form a monophyletic group with Tete serogroup viruses, with Oyo virus falling basal to the other viruses in this group. While the measured nt and aa identities of I612045 virus with other orthobunyaviruses clearly indicate its taxonomic status as a member of Tete serogroup, Oyo virus is indeed distinct and may represent a distinct species in the genus *Orthobunyavirus*.

3.5. Koongol and Turlock Serogroups

Here we report coding-complete sequences of all three genomic segments of Koongol and Umbre viruses. The sequence of the Umbre virus IG1424 M segment is more than 99% identical to a previously published M segment of the same isolate [75]. Phylogenetic analysis placed Koongol and Umbre viruses along with unclassified Witwatersrand virus (see below) into a monophyletic group regardless of the protein assayed for tree reconstruction. These viruses also share a last common ancestor with Gamboa serogroup viruses, which are exclusively distributed in North and South America.

Genomic sequence information for the Turlock serogroup was thus far limited to partial Umbre virus M segment sequences. We expanded the sequence space of this serogroup by determining the coding-complete Umbre virus genome sequence and confirmed the relationship of Umbre virus to Kongool and Witwatersrand viruses (Figure 1).

3.6. Tataguine and Witwatersrand Viruses (Unassigned Bunyaviruses)

We present sequence information for Tataguine virus, the closest relative of which appears to be Lukuni virus, with 47.5% aa identity for N, 35.5% aa identity for glycoprotein precursor polyprotein, and 53.5% aa identity for L. Anopheles A and Anopheles B group viruses, along with Tataguine virus, form a monophyletic group with Tataguine virus at its base. Our data indicate that Tataguine virus may have to be assigned to a new species in the genus *Orthobunyavirus*.

As mentioned above, based on the obtained phylogenies, Witwatersrand virus clusters together with Umbre (Turlock serogroup) and Koongol (Koongol serogroup) viruses in the same branching order independent of the analyzed protein. The closest relative of Witwatersrand virus is Koongol virus (48.5% to 59.6% aa identities for the three proteins), indicating that Witwatersrand virus should be classified as an orthobunyavirus.

3.7. Characteristics of S Segments: N and NSs proteins

The N proteins of bunyaviruses encapsidate viral RNA and are major CF determinants [3]. The lengths of the N proteins of the examined orthobunyaviruses range from 234 aa residues for Koongol virus (Kongool serogroup) to 258 aa for Bahig, Matruh, and Tete viruses (Tete serogroup) (Table 2). Our studies confirm that Tete serogroup viruses possess the longest N proteins in the genus *Orthobunyavirus* with unique extensions predominantly located at the amino termini [74].

Table 2. Protein lengths of examined orthobunyaviruses (in amino acid residues)*.

Orthobunyavirus Serogroup Virus Name (Virus Abbreviation)	N	NSs	Glycoprotein Precursor (# of Cysteines)	Gn	NSm	Gc	L
Anopheles A serogroup							
Lukuni virus (LUKV)	242	-	1408 (71)	286	168	940	2241
Capim serogroup							
Capim virus (CAPV)	235	-	1430 (74)	284	188	947	2252
Guajará virus (GJAV)	235	-	1435 (73)	286	188	948	2252
Guamá serogroup							
Bimiti virus (BIMV)	237	-	1443 (73)	284	191	954	2250
Catú virus (CATUV)	237	-	1440 (72)	284	191	952	2250
Guamá virus (GMAV)	237	-	1439 (78)	284	187	952	2250
Mahogany Hammock virus (MHV)	237	-	1436 (75)	284	189	946	2250
Moju virus (MOJUV)	237	-	1435 (75)	284	189	946	2250
Koongol serogroup							
Koongol virus (KOOV)	234	-	1105 (57)	284	38	777	2270
Mapputta serogroup							
Mapputta virus (MAPV)	236	-	1370 (77)	288	161	910	2241
Trubanaman virus (TRUV)	237	-	1371 (71)	286	164	908	2242
Tete serogroup							
Bahig virus (BAHV)	258	-	1433 (69)	286	179	955	2280
Matruh virus (MTRV)	258	-	1433 (67)	286	179	955	2280
Tete virus (TETEV)	258	-	1432 (68)	286	178	955	2281
Turlock serogroup							
Umbre virus (UMBV)	237	79	1466 (72)	284	176	991	2293
Unassigned							
Tataguine virus (TATV)	239	-	1446 (73)	287	171	976	2246
Witwatersrand virus (WITV)	245	111	1448 (70)	285	173	974	2288

* Gn, NSm, and Gc lengths were calculated by implying Gn-NSm cleavage at conserved residue R_{302} (Bunyamwera virus). NSm-Gc cleavage was predicted by SignalP 4.1 server.

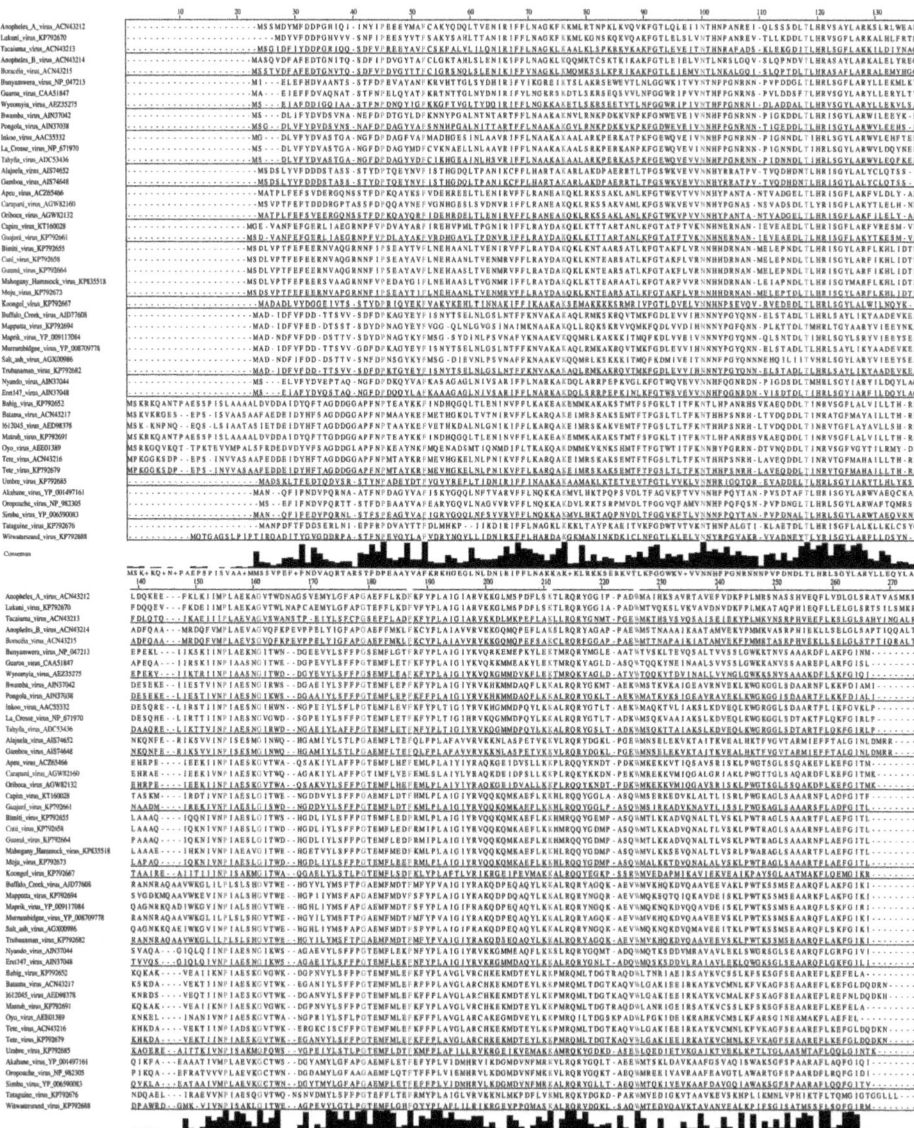

Figure 2. Amino-acid alignment of orthobunyaviral nucleocapsid protein sequences. Alignment was performed on 118 nucleocapsid sequences belonging to different orthobunyaviruses using the Clustal algorithm with default settings as implemented in Jalview 2.9 [76], followed by hiding all but selected representatives of the serogroups. Sites that are strictly conserved among all aligned sequences are depicted in red. Consensus histograms, calculated for all aligned sequences, represent proportions of sites matching corresponding positions of the consensus sequence.

Forty-six aa of the N protein are conserved among the previously determined 51 sequences of orthobunyaviruses from the Bunyamwera, California, Group C, and Simbu serogroups [77]. Our analyses reveal that only 11 aa of these 46 aa are strictly conserved among the N proteins of Anopheles A, Anopheles B, Bunyamwera, Bwamba, California, Capim, Gamboa, Guamá, Group C, Koongol,

Turlock, Nyando, Simbu, and Tete serogroup viruses (Figure 2). Five of these 11 aa (F26, P125, G131, K179, W193) are crucial for Bunyamwera virus mini-replicon rescue. Two of those aa, K179 and W193, are likely involved in RNA synthesis, and two other aa residues, P125 and G131, are thought to play a role in ribonucleoprotein packaging [77].

A number of researchers have evaluated whether NSs is involved in the immune response to orthobunyavirus infections. Despite not expressing NSs, Tacaiuma virus (Anopheles A serogroup) antagonizes host interferon (IFN) production through a yet unrecognized mechanism [74] and is associated with human febrile illness [35]. Similarly, Mapputta serogroup viruses, such as Maprik virus and Buffalo Creek virus, do not encode NSs, but are linked to human diseases [24]. Two out of 17 orthobunyaviruses studied in the present work, Umbre virus (Turlock serogroup) and Witwatersrand virus (ungrouped orthobunyavirus), encode NSs proteins of 79 and 111 aa, respectively (previously characterized orthobunyaviruses: 62 (Group C serogroup) to 130 aa (Gamboa serogroup)). Neither of the two viruses has been associated with human disease. In contrast, known human pathogens, such as the febrile disease-causing ungrouped Tataguine virus and Guamá and Catú viruses (Guamá serogroup), do not encode NSs. These findings suggest that the presence or absence of an NSs-encoding ORF alone does not predic human pathogenicity. Additionally, the presence of an NSs-encoding ORF is far less common for orthobunyaviruses than previously thought as only viruses from 8 out of 15 sequenced serogroups do encode this nonstructural protein.

3.8. Characterization of M segments: Glycoprotein Precursor Polyprotein and Gn and Gc Proteins Cleavage Products

Each orthobunyavirus M segment contains a single continuous ORF encoding a glycoprotein precursor polyprotein that is cotranslationally cleaved into glycoprotein Gn, the nonstructural protein NSm, and the glycoprotein Gc [78]. Among the polyprotein sequences derived from the sequenced M segments of our study, the Koongol virus glycoprotein precursor polyprotein (1105 aa) is notably smaller than all other glycoprotein precursors of orthobunyaviruses, which range in size from 1370 aa in the case of Mapputta virus to 1448 aa in the case of Witwatersrand virus, Table 1. A strictly conserved arginine residue of the glycoprotein precursor polyprotein sequences of all sequenced orthobunyaviruses (position 302 in the prototype Bunyamwera virus) is believed to mark the cleavage site between Gn and NSm proteins [78]. Therefore, Koongol virus Gn is comparable in size with Gn of other orthobunyaviruses whilst its Gc and NSm proteins are notably shorter.

Regions highly similar to the fusion peptide identified in La Crosse virus glycoprotein precursor polyprotein (positions 1066–1087) are present in the predicted polyproteins of the sequenced orthobunyaviruses of all serogroups. Ten out of twenty-two La Crosse virus fusion peptide aa are strictly conserved across the genus. This finding indicates that all orthobunyavirus Gc glycoproteins have analogous functions and thereby act as class II fusion proteins [79]. Supporting this hypothesis is the finding that the overall topology of orthobunyaviral glycoprotein precursor polyprotein appears to be conserved based on the number of conserved cysteine residues (ranging from 67 in the case of Matruh virus (Tete serogroup) to 78 in the case of Guamá virus (Guamá serogroup)). Once again, Koongol virus (Koongol serogroup) is the outlier with only 57 cysteine residues [80–82].

N-glycosylation of viral membrane proteins plays a crucial role in correct protein folding and functioning, including receptor binding, membrane fusion, and cell-penetration processes [83]. All three predicted *N*-glycosylation sites of the membrane glycoproteins of Bunyamwera virus (Bunyamwera serogroup) are indeed glycosylated. Glycosylation of Bunyamwera virus Gn's N60 site is essential for correct protein folding of both Gn and Gc [84]. Interestingly, this glycosylation site is highly conserved among almost all previously sequenced orthobunyaviruses, with the notable exception of Maprik virus (Mapputta serogroup) [24]. NetNGlyc 1.0 server predicted this glycosylation site to be present in the Gn proteins of all viruses sequenced in this study, with the exception of Lukuni virus (Anopheles A serogroup) and ungrouped Tataguine viruses. In general, glycosylation site locations were moderately conserved in orthobunyaviruses belonging to the same serogroup, but were not consistent among all viruses of the genus *Orthobunyavirus*. Finally, with the exception of Koongol virus, transmembrane prediction using hidden Markov models (TMHHM) 2.0 generally predicted the same distribution and type of transmembrane regions for the glycoprotein precursor polyprotein sequences of the analyzed viruses, which included two transmembrane domains in Gn, three in NSm, and one in Gc. The Koongol glycoprotein precursor polyprotein has four predicted transmembrane regions, lacking two domains usually located at the C-terminal half of NSm.

3.9. Characterization of L Segments: RNA-Dependent RNA Polymerase

The L sequence aa lengths of the studied orthobunyaviruses are comparable to those of previously studied viruses, ranging from 2241 aa in the case of Lukuni virus to 2293 aa in the case of Umbre virus (Table 2). Tete group (Bahig, Matruh, and Tete viruses) L (2280–2281 aa) possesses a serogroup-characteristic 24 aa insertion (E2185–E2208) at the C terminus. Umbre and Witwatersrand virus L possess several aa at the very end of the C terminus that are not conserved among other viruses. All analyzed L sequences have the same well-conserved topology, consisting of four distinct regions with readily distinguishable RNA-dependent RNA polymerase motifs pre-A to E inside the POL III block. The proposed site-active domain SDD1163–5 of Bunyamwera virus [85] is strictly conserved among all previously and newly characterized orthobunyaviruses.

4. Discussion

The family *Bunyaviridae* was originally established to group viruses that produce morphologically similar and often serologically cross-reactive virions [86]. Largely non-sequence-based efforts [9–14] further grouped the members of this family into various serogroups [87,88], which were later assembled in higher-order taxa, *i.e.*, the currently accepted genera *Hantavirus*, *Nairovirus*, *Orthobunyavirus*, *Phlebovirus*, and *Tospovirus* [2,89]. Recent large-scale efforts to sequence the genomes of bunyaviruses and other segmented negative- or ambisense-stranded RNA viruses revealed that numerous novel bunyaviruses cannot be assigned to the five existing genera and that a plethora of viruses that are clearly related to bunyaviruses do not fit their classical definition of being trisegmented, *i.e.*, arenaviruses, emaraviruses, tenuiviruses, and Mourilyan virus [1,5,23,28,90–109]. The International Committee on Taxonomy of Viruses (ICTV) *Bunyaviridae* Study Group has therefore recently initiated discussions on a taxonomic re-evaluation of the entire bunyavirus-like supergroup with the ultimate goal to establish a novel taxonomy that adequately reflects the phylogenetic relationships of all these viruses. Unfortunately, the genomic "sequence space" of the supergroup is still very limited, thereby impeding those efforts.

Within the classic orthobunyavirus group, genomic sequence information was limited to viruses of 10 of 18 serogroups [15–28]. Our work expands this sequence space by adding coding-complete genomic information on viruses of an additional five serogroups. Our efforts largely confirm the relationships of the studied viruses that had been established previously by non-sequence-based (serological) techniques. Despite the high genetic diversity of the orthobunyaviruses, reflecting their wide geographic distribution and variety of ecological features, the viruses of each serogroup are grouped together within the appropriate lineage on the three phylogenetic trees. Since the members of the *Bunyaviridae* family possess segmented genomes, the phenomenon of segment reassortment plays a significant role in their evolution [110]. Earlier, it was shown that many members of the genus *Orthobunyavirus* are intra-group reassortants. Our data show that all examined viruses of the Guamá serogroup are in all likelihood genome segment reassortants. Furthermore, viruses of the Guamá and Capim serogroups form a monophyletic lineage on the tree inferred for the glycoprotein polyprotein precursor protein (M segment), suggesting inter-group reassortment of M segments in their natural history. Another important observation that could be made is the classification of Witwatersrand virus and Tataguine virus as likely members of two new novel species in the genus *Orthobunyavirus*.

Finally, our data show that presence of an ORF encoding an NSs protein is not a universal feature for orthobunyaviruses. Among the viruses of 15 orthobunyavirus serogroups for which genomic data are now available, only viruses of eight serogroups along with Witwatersrand virus encode NSs proteins. Therefore, the presence or absence of NSs protein should not be considered as a taxonomic characteristic of the genus *Orthobunyavirus*, but may be important for differentiating pathogenic from nonpathogenic viruses.

Our analyses advance the overall resolution of orthobunyavirus phylogeny. Although sequence information for the 3′ and 5′ genomic segment termini could not obtained in this study, we are confident that the determined phylogenetic placement on the phylogenetic tree of all viruses studied here will hold. Complementation of our results by genomic sequences of viruses from the remaining unsequenced orthobunyavirus serogroups should allow official taxonomic re-organization of the genus *Orthobunyavirus*.

Acknowledgments: The authors thank Laura Bollinger (IRF-Frederick) for critically editing this paper. This work was funded by the Russian Foundation for Basic Research (RFBR) according to research project 13-04-01749a. The content of this publication does not necessarily reflect the views or policies of the US Department of Health and Human Services, or the institutions and companies affiliated with the authors. This work was supported in part through Battelle Memorial Institute's prime contract with the US National Institute of Allergy and Infectious Diseases (NIAID) under Contract No. HHSN272200700016I. A subcontractor to Battelle Memorial Institute who performed this work is: J.H.K., an employee of Tunnell Government Services, Inc. (Bethesda, MD, USA).

Author Contributions: D.K.L., P.G.D., A.M.S., and S.V.A. designed the study. A.M.S., A.K.G., A.G.B., and S.V.A. conducted experiments. A.M.S., J.H.K., and S.V.A. analyzed the results and wrote the paper. All authors have read and approved the final manuscript.

Conflicts of Interest: The authors declare no conflict of interest.

References

1. Whitehouse, C.A.; Kuhn, J.H.; Wada, J.; Ergunay, K. Family *Bunyaviridae*. In *Global Virology I—Identifying and Investigating Viral Diseases*; Shapshak, P., Sinnott, J., Somboonwit, C., Kuhn, J.H., Eds.; Springer: New York, USA, 2015; pp. 199–246.
2. Plyusnin, A.; Beaty, B.J.; Elliott, R.M.; Goldbach, R.; Kormelink, R.; Lundkvist, Å.; Schmaljohn, C.S.; Tesh, R.B. Family *Bunyaviridae*. In *Virus Taxonomy–Ninth Report of the International Committee on Taxonomy of Viruses*; King, A.M.Q., Adams, M.J., Carstens, E.B., Lefkowitz, E.J., Eds.; Elsevier/Academic Press: London, UK, 2011; pp. 725–741.
3. Elliott, R.M.; Schmaljohn, C.S. *Bunyaviridae*. In *Fields Virology*, 6th ed.; Knipe, D.M., Howley, P.M., Eds.; Lippincott/The Williams & Wilkins Company: Philadelphia, PA, USA, 2013; pp. 1244–1282.
4. Kuhn, J.H.; Peters, C.J. Arthropod-borne and rodent-borne virus infections. In *Harrison's Principles of Internal Medicine*, 19th ed.; Kasper, D.L., Fauci, A.S., Hauser, S.L., Longo, D.L., Jameson, J.L., Loscalzo, J., Eds.; McGraw-Hill Education: Columbus, OH, USA, 2015; Volume 2, pp. 1304–1323.
5. Marklewitz, M.; Zirkel, F.; Rwego, I.B.; Heidemann, H.; Trippner, P.; Kurth, A.; Kallies, R.; Briese, T.; Lipkin, W.I.; Drosten, C.; *et al.* Discovery of a unique novel clade of mosquito-associated bunyaviruses. *J. Virol.* **2013**, *87*, 12850–12865. [CrossRef] [PubMed]
6. Marklewitz, M.; Zirkel, F.; Kurth, A.; Drosten, C.; Junglen, S. Evolutionary and phenotypic analysis of live virus isolates suggests arthropod origin of a pathogenic rna virus family. *Proc. Natl. Acad. Sci. USA* **2015**, *112*, 7536–7541. [CrossRef] [PubMed]
7. Elliott, R.M. Orthobunyaviruses: Recent genetic and structural insights. *Nat. Rev. Microbiol.* **2014**, *12*, 673–685. [CrossRef] [PubMed]
8. Elliott, R.M.; Blakqori, G. Molecular Biology of Orthobunyaviruses. In *Bunyaviridae: Molecular and Cellular Biology*; Plyusnin, A., Elliott, R.M., Eds.; Caister Academic Press: Norfolk, UK, 2011; pp. 1–40.
9. Calisher, C.H. Taxonomy, classification, and geographic distribution of california serogroup bunyaviruses. *Prog. Clin. Biol. Res.* **1983**, *123*, 1–16. [PubMed]
10. Calisher, C.H.; Coimbra, T.L.; Lopez Ode, S.; Muth, D.J.; Sacchetta Lde, A.; Francy, D.B.; Lazuick, J.S.; Cropp, C.B. Identification of new guama and group c serogroup bunyaviruses and an ungrouped virus from southern brazil. *Am. J. Trop. Med. Hyg.* **1983**, *32*, 424–431. [PubMed]

11. Calisher, C.H.; Gutierrez, E.; Francy, D.B.; Alava, A.; Muth, D.J.; Lazuick, J.S. Identification of hitherto unrecognized arboviruses from ecuador: Members of serogroups b, c, bunyamwera, patois, and minatitlan. *Am. J. Trop. Med. Hyg.* **1983**, *32*, 877–885. [PubMed]
12. Calisher, C.H.; Lazuick, J.S.; Wolff, K.L.; Muth, D.J. Antigenic relationships among turlock serogroup bunyaviruses as determined by neutralization tests. *Acta Virol.* **1984**, *28*, 148–151. [PubMed]
13. Kinney, R.M.; Calisher, C.H. Antigenic relationships among simbu serogroup (bunyaviridae) viruses. *Am. J. Trop. Med. Hyg.* **1981**, *30*, 1307–1318. [PubMed]
14. Travassos da Rosa, A.P.; Tesh, R.B.; Pinheiro, F.P.; Travassos da Rosa, J.F.; Peterson, N.E. Characterization of eight new phlebotomus fever serogroup arboviruses (bunyaviridae: *Phlebovirus*) from the amazon region of brazil. *Am. J. Trop. Med. Hyg.* **1983**, *32*, 1164–1171. [PubMed]
15. Bowen, M.D.; Jackson, A.O.; Bruns, T.D.; Hacker, D.L.; Hardy, J.L. Determination and Comparative Analysis of the Small Rna Genomic Sequences of California Encephalitis, Jamestown Canyon, Jerry Slough, Melao, Keystone and Trivittatus Viruses (*Bunyaviridae*, Genus *Bunyavirus*, California Serogroup). *J. Gen. Virol.* **1995**, *76*, 559–572. [CrossRef] [PubMed]
16. Dunn, E.F.; Pritlove, D.C.; Elliott, R.M. The s rna genome segments of batai, cache valley, guaroa, kairi, lumbo, main drain and northway bunyaviruses: Sequence determination and analysis. *J. Gen. Virol.* **1994**, *75*, 597–608. [CrossRef] [PubMed]
17. Nunes, M.R.; Travassos da Rosa, A.P.; Weaver, S.C.; Tesh, R.B.; Vasconcelos, P.F. Molecular epidemiology of group c viruses (*Bunyaviridae, orthobunyavirus*) isolated in the americas. *J. Virol.* **2005**, *79*, 10561–10570. [CrossRef] [PubMed]
18. Saeed, M.F.; Wang, H.; Nunes, M.; Vasconcelos, P.F.; Weaver, S.C.; Shope, R.E.; Watts, D.M.; Tesh, R.B.; Barrett, A.D. Nucleotide sequences and phylogeny of the nucleocapsid gene of oropouche virus. *J. Gen. Virol.* **2000**, *81*, 743–748. [CrossRef] [PubMed]
19. Groseth, A.; Wollenberg, K.R.; Mampilli, V.; Shupert, T.; Weisend, C.; Guevara, C.; Kochel, T.J.; Tesh, R.B.; Ebihara, H. Spatiotemporal analysis of guaroa virus diversity, evolution, and spread in south america. *Emerg. Infect. Dis.* **2015**, *21*, 460–463. [CrossRef] [PubMed]
20. Hontz, R.D.; Guevara, C.; Halsey, E.S.; Silvas, J.; Santiago, F.W.; Widen, S.G.; Wood, T.G.; Casanova, W.; Vasilakis, N.; Watts, D.M.; *et al.* Itaya virus, a novel orthobunyavirus associated with human febrile illness, peru. *Emerg. Infect. Dis.* **2015**, *21*. [CrossRef] [PubMed]
21. Tilston-Lunel, N.L.; Hughes, J.; Acrani, G.O.; da Silva, D.E.; Azevedo, R.S.; Rodrigues, S.G.; Vasconcelos, P.F.; Nunes, M.R.; Elliott, R.M. Genetic analysis of members of the species Oropouche virus and identification of a novel M segment sequence. *J. Gen. Virol.* **2015**, *96*, 1636–1650. [CrossRef] [PubMed]
22. Van Eeden, C.; Harders, F.; Kortekaas, J.; Bossers, A.; Venter, M. Genomic and phylogenetic characterization of Shuni virus. *Arch. Virol.* **2014**, *159*, 2883–2892. [CrossRef] [PubMed]
23. Ladner, J.T.; Savji, N.; Lofts, L.; Travassos da Rosa, A.; Wiley, M.R.; Gestole, M.C.; Rosen, G.E.; Guzman, H.; Vasconcelos, P.F.; Nunes, M.R.; *et al.* Genomic and phylogenetic characterization of viruses included in the Manzanilla and Oropouche species complexes of the genus *Orthobunyavirus*, family *Bunyaviridae*. *J. Gen. Virol.* **2014**, *95*, 1055–1066. [CrossRef] [PubMed]
24. Gauci, P.J.; McAllister, J.; Mitchell, I.R.; Boyle, D.B.; Bulach, D.M.; Weir, R.P.; Melville, L.F.; Gubala, A.J. Genomic characterisation of three Mapputta group viruses, a serogroup of Australian and Papua New Guinean bunyaviruses associated with human disease. *PLoS One* **2015**, *10*, e0116561. [CrossRef] [PubMed]
25. Groseth, A.; Mampilli, V.; Weisend, C.; Dahlstrom, E.; Porcella, S.F.; Russell, B.J.; Tesh, R.B.; Ebihara, H. Molecular characterization of human pathogenic bunyaviruses of the Nyando and Bwamba/Pongola virus groups leads to the genetic identification of Mojui dos Campos and Kaeng Khoi virus. *PLoS Negl. Trop. Dis.* **2014**, *8*, e3147. [CrossRef] [PubMed]
26. Nunes, M.R.T.; Chiang, J.O.; de Lima, C.P.S.; Martins, L.C.; Aragão Dias, A.; Cardoso, J.F.; Silva, S.P.; da Silva, D.E.A.; Oliveira, L.F.; Vasconcelos, J.M.; *et al.* New genome sequences of Gamboa viruses (family *Bunyaviridae*, genus *Orthobunyavirus*) isolated in Panama and Argentina. *Genome Announc.* **2014**, *2*, e00940-14. [CrossRef] [PubMed]
27. Rodrigues, D.S.; Medeiros, D.B.; Rodrigues, S.G.; Martins, L.C.; de Lima, C.P.; de Oliveira, L.F.; de Vasconcelos, J.M.; da Silva, D.E.; Cardoso, J.F.; da Silva, S.P.; *et al.* Pacui Virus, Rio Preto da Eva Virus, and Tapirape Virus, three distinct viruses within the family *Bunyaviridae*. *Genome Announc.* **2014**, *2*, e00923-14. [CrossRef] [PubMed]

28. Liu, R.; Zhang, G.; Yang, Y.; Dang, R.; Zhao, T. Genome sequence of Abbey Lake virus, a novel orthobunyavirus isolated from china. *Genome Announc.* **2014**, *2*, e00433-14. [CrossRef] [PubMed]
29. Calisher, C.H.; Lazuick, J.S.; Muth, D.J.; de Souza Lopes, O.; Crane, G.T.; Elbel, R.E.; Shope, R.E. Antigenic relationships among Tacaiuma complex viruses of the Anopheles A serogroup (*Bunyaviridae*). *Bull. Pan Am. Health Organ.* **1980**, *14*, 386–391. [PubMed]
30. Da Rosa, J.F.S.; da Rosa, A.P.D.A.; Degallier, N.; Vasconcelos, P.F.D.C. Caracterização e relacionamento antigênico de três novos *Bunyavirus* no grupo Anopheles A (*Bunyaviridae*) dos arbovirus. *Rev. Saúd. Públ.* **1992**, *26*, 173–178, (In Portuguese). [CrossRef]
31. Calisher, C.H.; Sasso, D.R.; Maness, K.S.; Gheorghiu, V.N.; Shope, R.E. Relationships of anopheles a group arboviruses. *Proc. Soc. Exp. Biol. Med. Soc. Exp. Biol. Med.* **1973**, *143*, 465–468. [CrossRef]
32. Nichol, S.T.; Beaty, B.J.; Elliott, R.M.; Goldbach, R.; Plyusnin, A.; Schmaljohn, C.S.; Tesh, R.B. Family *Bunyaviridae*. In *Virus Taxonomy—Eighth Report of the International Committee on Taxonomy of Viruses*; Fauquet, C.M., Mayo, M.A., Maniloff, J., Desselberger, U., Eds.; Elsevier/Academic Press: London, UK, 2005; pp. 695–716.
33. Spence, L.; Anderson, C.R.; Aitken, T.H.; Downs, W.G. Bushbush, Ieri and Lukuni viruses, three unrelated new agents isolated from trinidadian forest mosquitoes. *Proc. Soc. Exp. Biol. Med. Soc. Exp. Biol. Med.* **1967**, *125*, 45–50. [CrossRef]
34. Woodall, J.P. Virus Research in Amazonia. In *Atas do Simpósio sôbre a Biota Amazônica*; Conselho Nacional de Pesquitas: Rio de Janeiro, Brazil, 1967; Volume 6, Patologia,f; pp. 31–63. (In Portuguese)
35. Karabatsos, N. *International Catalogue of Arboviruses, Including Certain Other Viruses of Vertebrates*, 3rd ed.; American Society of Tropical Medicine and Hygiene for the Subcommittee on Information Exchange of the American Committee on Arthropod-borne Viruses: San Antonio, TX, USA, 1985.
36. Spence, L.; Anderson, C.R.; Aitken, T.H.G.; Downs, W.G. *Bimiti virus*, a new agent isolated from Trinidadian mosquitoes. *Am. J. Trop. Med. Hyg.* **1962**, *11*, 414–417. [PubMed]
37. Whitman, L.; Casals, J. The guamá group: A new serological group of hitherto underscribed viruses. Immunological studies. *Am. J. Trop. Med. Hyg.* **1961**, *10*, 259–263. [PubMed]
38. Causey, O.R.; Causey, C.E.; Maroja, O.M.; Macedo, D.G. The isolation of arthropod-borne viruses, including members of two hitherto undescribed serological groups, in the Amazon region of Brazil. *Am. J. Trop. Med. Hyg.* **1961**, *10*, 227–249. [PubMed]
39. Coleman, P.H.; Ryder, S.; Work, T.H. Mahogany Hammock virus, a new guama group arbovirus from the florida everglades. *Am. J. Epidemiol.* **1969**, *89*, 217–221. [PubMed]
40. Doherty, R.L.; Carley, J.G.; Mackerras, M.J.; Marks, E.N. Studies of arthropod-borne virus infections in Queensland. III. Isolation and characterization of virus strains from wild-caught mosquitoes in North Queensland. *Aust. J. Exp. Biol. Med. Sci.* **1963**, *41*, 17–39. [CrossRef] [PubMed]
41. Doherty, R.L.; Whitehead, R.H.; Wetters, E.J.; Gorman, B.M. Studies of the epidemiology of arthropod-borne virus infections at Mitchell River Mission, Cape York Peninsula, North Queensland. II. Arbovirus infections of mosquitoes, man and domestic fowls, 1963–1966. *Trans. R. Soc. Trop. Med. Hyg.* **1968**, *62*, 430–438. [CrossRef]
42. Boughton, C.R.; Hawkes, R.A.; Naim, H.M. Arbovirus infection in humans in NSW: Seroprevalence and pathogenicity of certain Australian bunyaviruses. *Aust. N. Z. J. Med.* **1990**, *20*, 51–55. [CrossRef] [PubMed]
43. Williams, M.C.; Tukei, P.M.; Lule, M.; Mujomba, E.; Mukuye, A. Virology: Identification studies, An 1047–61. *East Afr. Virus Res. Inst. Rep.* **1966**, *16*, 24.
44. Taylor, R.M. *Catalogue of Arthropod-Borne Viruses of the World*; US Government Printing Office, Public Health Service Pub. No. 1760: Washington, DC, USA, 1967.
45. Fagbami, A.H.; Monath, T.P.; Murphy, F.A. Tataguine virus infection in mice and hamsters. *Microbios* **1980**, *27*, 89–96. [PubMed]
46. McIntosh, B.M.; Kokernot, R.H.; Paterson, H.E. Witwatersrand virus: An apparently new virus isolated from culicine mosquitoes. *S. Afr. J. Med. Sci.* **1960**, *23*, 33–37.
47. Shope, R.E.; Woodall, J.P.; Travassos da Rosa, A.P. The Epidemiology of Diseases Caused by Viruses in Groups C and Guama (*Bunyaviridae*). In *The Arboviruses: Epidemiology and Ecology*; Monath, T.P., Ed.; CRC Press: Boca Raton, Florida, 1988; Volume 3, pp. 38–51.
48. Doherty, R.L.; Wetters, E.J.; Gorman, B.M.; Whitehead, R.H. Arbovirus infection in Western Queensland: serological studies, 1963–1969. *Trans. R. Soc. Trop. Med. Hyg.* **1970**, *64*, 740–747. [CrossRef]

49. Doherty, R.L.; Whitehead, R.H.; Wetters, E.J.; Gorman, B.M.; Carley, J.G. A survey of antibody to 10 arboviruses (Koongol group, Mapputta group and ungrouped) isolated in Queensland. *Trans. R. Soc. Trop. Med. Hyg.* **1970**, *64*, 748–753. [CrossRef]
50. Sanderson, C.J. A serologic survey of Queensland cattle for evidence of arbovirus infections. *Am. J. Trop. Med. Hyg.* **1969**, *18*, 433–439. [PubMed]
51. Newton, S.E.; Short, N.J.; Irving, A.M.; Dalgarno, L. The Mapputta group of arboviruses: Ultrastructural and molecular studies which place the group in the *Bunyavirus* genus of the family Bunyaviridae. *Aust. J. Exp. Biol. Med. Sci.* **1983**, *61*, 201–217. [CrossRef] [PubMed]
52. Weir, R. Classification and Identification of Viruses Isolated from Mosquitoes in the Northern Territory, 1982–1992, Using a Range of Techniques. Ph.D. Thesis, University of Sydney, Sydney, Australia, 2002.
53. Coffey, L.L.; Page, B.L.; Greninger, A.L.; Herring, B.L.; Russell, R.C.; Doggett, S.L.; Haniotis, J.; Wang, C.; Deng, X.; Delwart, E.L. Enhanced arbovirus surveillance with deep sequencing: Identification of novel rhabdoviruses and bunyaviruses in Australian mosquitoes. *Virology* **2014**, *448*, 146–158. [CrossRef] [PubMed]
54. Balducci, M.; Verani, P.; Lopes, M.C.; Gregoric, B. Isolation in Italy of Bahig and Matruh viruses (Tete group) from migrating birds. *Ann. Microbiol. (Paris)* **1973**, *124B*, 231–237.
55. Converse, J.D.; Hoogstraal, H.; Moussa, M.I.; Stek, M., Jr.; Kaiser, M.N. Bahig virus (Tete group) in naturally- and transovarially-infected *Hyalomma marginatum* ticks from Egypt and Italy. *Arch Gesamte Virusforsch* **1974**, *46*, 29–35. [CrossRef] [PubMed]
56. Hubalek, Z.; Rudolf, I. Tick-borne viruses in Europe. *Parasitol. Res.* **2012**, *111*, 9–36. [CrossRef] [PubMed]
57. Brès, P.; Williams, M.C.; Chambon, L. Isolement au Sénégal d'un nouveau prototype d'arbovirus, la souche "Tataguine" (IPD/A 252). *Ann. Inst. Pasteur* **1966**, *111*, 585–591, (In French).
58. Fagbami, A.H. Studies on transmission of Tataguine virus by *Culex (pipens) fatigans* mosquitoes. *Afr. J. Med. Med. Sci.* **1979**, *8*, 31–33. [PubMed]
59. Fagbami, A.H.; Monath, T.P.; Tomori, O.; Lee, V.H.; Fabiyi, A. Studies on Tataguine infection in Nigeria. *Trop. Geogr. Med.* **1972**, *24*, 298–302. [PubMed]
60. Salaun, J.J.; Rickenbach, A.; Brès, P.; Germain, M.; Eouzan, J.P.; Ferrara, L. Isolement au Cameroun de trois souches de virus Tataguine. *Bull. Soc. Pathol. Exot. Ses Filiales* **1968**, *61*, 557–564, (In French).
61. Saluzzo, J.F.; Gonzalez, J.P.; Hervé, J.P.; Georges, A.J. Enquête sérologique sur la prévalence de certains arbovirus dans la population humaine du sud-est de la République Centrafricaine en 1979. *Bull. Soc. Pathol. Exot. Ses Filiales* **1981**, *74*, 490–499, (In French).
62. Causey, O.R.; Kemp, G.E.; Madbouly, M.H.; Lee, V.H. Arbovirus surveillance in Nigeria, 1964–1967. *Bull. Soc. Pathol. Exot. Ses Filiales* **1969**, *62*, 249–253.
63. Moore, D.L.; Causey, O.R.; Carey, D.E.; Reddy, S.; Cooke, A.R.; Akinkugbe, F.M.; David-West, T.S.; Kemp, G.E. Arthropod-borne viral infections of man in Nigeria, 1964–1970. *Ann. Trop. Med. Parasitol.* **1975**, *69*, 49–64. [PubMed]
64. Fagbami, A.H.; Tomori, O. Tataguine virus isolations from humans in Nigeria, 1971–1975. *Trans. R. Soc. Trop. Med. Hyg.* **1981**, *75*, 788. [CrossRef]
65. Digoutte, J.P.; Brès, P.; Nguyen Trung Luong, P.; Durand, B. Isolement du virus Tataguine à partir de deux fièvres exanthématiques. *Bull. Soc. Pathol. Exot. Ses Filiales* **1969**, *62*, 72–80, (In French).
66. Bishop, D.H.; Shope, R.E. *Bunyaviridae*. In *Comprehensive Virology*; Fraenkel-Conrat, H., Wagner, R.R., Eds.; Plenum Press: New York, NY, USA, 1979; Volume 14.
67. Center for Genomic Regulation. M-Coffee Aligns DNA, RNA or Proteins by Combining the Output of Popular Aligners. Available Online: http://tcoffee.crg.cat/apps/tcoffee/do:mcoffee (accessed on 11 September 2015).
68. Abascal, F.; Zardoya, R.; Posada, D. ProtTest: Selection of best-fit models of protein evolution. *Bioinformatics* **2005**, *21*, 2104–2105. [CrossRef] [PubMed]
69. Ronquist, F.; Teslenko, M.; van der Mark, P.; Ayres, D.L.; Darling, A.; Hohna, S.; Larget, B.; Liu, L.; Suchard, M.A.; Huelsenbeck, J.P. MrBayes 3.2: Efficient Bayesian phylogenetic inference and model choice across a large model space. *Syst. Biol.* **2012**, *61*, 539–542. [CrossRef] [PubMed]
70. Stöver, B.C.; Müller, K.F. TreeGraph 2: Combining and visualizing evidence from different phylogenetic analyses. *BMC Bioinform.* **2010**, *11*, 7. [CrossRef] [PubMed]
71. Petersen, T.N.; Brunak, S.; von Heijne, G.; Nielsen, H. SignalP 4.0: Discriminating signal peptides from transmembrane regions. *Nat. Methods* **2011**, *8*, 785–786. [CrossRef] [PubMed]

72. Center for Biological Sequence Analysis. TMHMM Server v. 2.0. Prediction of Transmembrane Helices. Available online: http://www.cbs.dtu.dk/services/TMHMM-2.0/ (accessed on 14 April 2015).
73. Center for Biological Sequence Analysis. NetNGlyc 1.0 Server. Available online: http://www.cbs.dtu.dk/services/NetNGlyc/ (accessed on 14 April 2015).
74. Mohamed, M.; McLees, A.; Elliott, R.M. Viruses in the Anopheles A, Anopheles B, and Tete serogroups in the *Orthobunyavirus* genus (family *Bunyaviridae*) do not encode an NSs protein. *J. Virol.* **2009**, *83*, 7612–7618. [CrossRef] [PubMed]
75. Yadav, P.D.; Mishra, A.C.; Mourya, D.T. Molecular characterization of Umbre virus (*Bunyaviridae*). *Virol. J.* **2008**, *5*, 115. [CrossRef] [PubMed]
76. Waterhouse, A.M.; Procter, J.B.; Martin, D.M.; Clamp, M.; Barton, G.J. Jalview Version 2—A multiple sequence alignment editor and analysis workbench. *Bioinformatics* **2009**, *25*, 1189–1191. [CrossRef] [PubMed]
77. Eifan, S.A.; Elliott, R.M. Mutational analysis of the Bunyamwera orthobunyavirus nucleocapsid protein gene. *J. Virol.* **2009**, *83*, 11307–11317. [CrossRef] [PubMed]
78. Fazakerley, J.K.; Gonzalez-Scarano, F.; Strickler, J.; Dietzschold, B.; Karush, F.; Nathanson, N. Organization of the middle RNA segment of snowshoe hare bunyavirus. *Virology* **1988**, *167*, 422–432. [CrossRef]
79. Plassmeyer, M.L.; Soldan, S.S.; Stachelek, K.M.; Roth, S.M.; Martin-Garcia, J.; Gonzalez-Scarano, F. Mutagenesis of the La Crosse Virus glycoprotein supports a role for Gc (1066–1087) as the fusion peptide. *Virology* **2007**, *358*, 273–282. [CrossRef] [PubMed]
80. Grady, L.J.; Sanders, M.L.; Campbell, W.P. The sequence of the M RNA of an isolate of La Crosse virus. *J. Gen. Virol.* **1987**, *68*, 3057–3071. [CrossRef] [PubMed]
81. Lees, J.F.; Pringle, C.R.; Elliott, R.M. Nucleotide sequence of the Bunyamwera virus M RNA segment: Conservation of structural features in the bunyavirus glycoprotein gene product. *Virology* **1986**, *148*, 1–14. [CrossRef]
82. Pardigon, N.; Vialat, P.; Gerbaud, S.; Girard, M.; Bouloy, M. Nucleotide-sequence of the M-segment of Germiston virus - comparison of the M-gene product of several bunyaviruses. *Virus Res.* **1988**, *11*, 73–85. [CrossRef]
83. Braakman, I.; van Anken, E. Folding of viral envelope glycoproteins in the endoplasmic reticulum. *Traffic* **2000**, *1*, 533–539. [CrossRef] [PubMed]
84. Shi, X.; Brauburger, K.; Elliott, R.M. Role of N-linked glycans on Bunyamwera virus glycoproteins in intracellular trafficking, protein folding, and virus infectivity. *J. Virol.* **2005**, *79*, 13725–13734. [CrossRef] [PubMed]
85. Jin, H.; Elliott, R.M. Mutagenesis of the L protein encoded by Bunyamwera virus and production of monospecific antibodies. *J. Gen. Virol.* **1992**, *73*, 2235–2244. [CrossRef] [PubMed]
86. Murphy, F.A.; Harrison, A.K.; Whitfield, S.G. Bunyaviridae: Morphologic and morphogenetic similarities of Bunyamwera serologic supergroup viruses and several other arthropod-borne viruses. *Intervirology* **1973**, *1*, 297–316. [CrossRef] [PubMed]
87. Porterfield, J.S.; Casals, J.; Chumakov, M.P.; Gaidamovich, S.Y.; Hannoun, C.; Holmes, I.H.; Horzinek, M.C.; Mussgay, M.; Oker-Blom, N.; Russell, P.K. Bunyaviruses and Bunyaviridae. *Intervirology* **1975**, *6*, 13–24. [CrossRef] [PubMed]
88. Porterfield, J.S.; Casals, J.; Chumakov, M.P.; Gaidamovich, S.Y.; Hannoun, C.; Holmes, I.H.; Horzinek, M.C.; Mussgay, M.; Russell, P.K. Bunyaviruses and Bunyaviridae. *Intervirology* **1974**, *2*, 270–272. [CrossRef] [PubMed]
89. Bishop, D.H.; Calisher, C.H.; Casals, J.; Chumakov, M.P.; Gaidamovich, S.Y.; Hannoun, C.; Lvov, D.K.; Marshall, I.D.; Oker-Blom, N.; Pettersson, R.F.; *et al.* Bunyaviridae. *Intervirology* **1980**, *14*, 125–143. [CrossRef] [PubMed]
90. Ballinger, M.J.; Bruenn, J.A.; Hay, J.; Czechowski, D.; Taylor, D.J. Discovery and evolution of bunyavirids in arctic phantom midges and ancient bunyavirid-like sequences in insect genomes. *J. Virol.* **2014**, *88*, 8783–8794. [CrossRef] [PubMed]
91. Bakonyi, T.; Kolodziejek, J.; Rudolf, I.; Bercic, R.; Nowotny, N.; Hubalek, Z. Partial genetic characterization of Sedlec virus (*Orthobunyavirus*, *Bunyaviridae*). *Infect. Genet. Evol.* **2013**, *19*, 244–249. [CrossRef] [PubMed]
92. Palacios, G.; da Rosa, A.T.; Savji, N.; Sze, W.; Wick, I.; Guzman, H.; Hutchison, S.; Tesh, R.; Lipkin, W.I. *Aguacate virus*, a new antigenic complex of the genus *Phlebovirus* (family *Bunyaviridae*). *J. Gen. Virol.* **2011**, *92*, 1445–1453. [CrossRef] [PubMed]

93. Palacios, G.; Savji, N.; Travassos da Rosa, A.; Desai, A.; Sanchez-Seco, M.P.; Guzman, H.; Lipkin, W.I.; Tesh, R. Characterization of the Salehabad virus species complex of the genus *Phlebovirus* (*Bunyaviridae*). *J. Gen. Virol.* **2013**, *94*, 837–842. [CrossRef] [PubMed]
94. Palacios, G.; Savji, N.; Travassos da Rosa, A.; Guzman, H.; Yu, X.; Desai, A.; Rosen, G.E.; Hutchison, S.; Lipkin, W.I.; Tesh, R. Characterization of the Uukuniemi virus group (*Phlebovirus*: *Bunyaviridae*): Evidence for seven distinct species. *J. Virol.* **2013**, *87*, 3187–3195. [CrossRef] [PubMed]
95. Palacios, G.; Tesh, R.; Travassos da Rosa, A.; Savji, N.; Sze, W.; Jain, K.; Serge, R.; Guzman, H.; Guevara, C.; Nunes, M.R.; *et al.* Characterization of the Candiru antigenic complex (*Bunyaviridae*: *Phlebovirus*), a highly diverse and reassorting group of viruses affecting humans in tropical America. *J. Virol.* **2011**, *85*, 3811–3820. [CrossRef] [PubMed]
96. Palacios, G.; Tesh, R.B.; Savji, N.; Travassos da Rosa, A.P.; Guzman, H.; Bussetti, A.V.; Desai, A.; Ladner, J.; Sanchez-Seco, M.; Lipkin, W.I. Characterization of the Sandfly fever Naples species complex and description of a new Karimabad species complex (genus *Phlebovirus*, family *Bunyaviridae*). *J. Gen. Virol.* **2014**, *95*, 292–300. [CrossRef] [PubMed]
97. Palacios, G.; Wiley, M.R.; Travassos da Rosa, A.P.; Guzman, H.; Quiroz, E.; Savji, N.; Carrera, J.P.; Bussetti, A.V.; Ladner, J.T.; Lipkin, W.I.; *et al.* Characterization of the Punta Toro species complex (genus *Phlebovirus*, family *Bunyaviridae*). *J. Gen. Virol.* **2015**, *96*, 2079–2085. [CrossRef] [PubMed]
98. Matsuno, K.; Weisend, C.; Kajihara, M.; Matysiak, C.; Williamson, B.N.; Simuunza, M.; Mweene, A.S.; Takada, A.; Tesh, R.B.; Ebihara, H. Comprehensive molecular detection of tick-borne phleboviruses leads to the retrospective identification of taxonomically unassigned bunyaviruses and the discovery of a novel member of the genus *phlebovirus*. *J. Virol.* **2015**, *89*, 594–604. [CrossRef] [PubMed]
99. Matsuno, K.; Weisend, C.; Travassos da Rosa, A.P.; Anzick, S.L.; Dahlstrom, E.; Porcella, S.F.; Dorward, D.W.; Yu, X.J.; Tesh, R.B.; Ebihara, H. Characterization of the Bhanja serogroup viruses (*Bunyaviridae*): A novel species of the genus *Phlebovirus* and its relationship with other emerging tick-borne phleboviruses. *J. Virol.* **2013**, *87*, 3719–3728. [CrossRef] [PubMed]
100. Tokarz, R.; Williams, S.H.; Sameroff, S.; Sanchez Leon, M.; Jain, K.; Lipkin, W.I. Virome Analysis of *Amblyomma americanum*, *Dermacentor variabilis*, and *Ixodes scapularis* Ticks Reveals Novel Highly Divergent Vertebrate and Invertebrate Viruses. *J. Virol.* **2014**, *88*, 11480–11492. [CrossRef] [PubMed]
101. Li, C.-X.; Shi, M.; Tian, J.-H.; Lin, X.-D.; Kang, Y.-J.; Chen, L.-J.; Qin, X.-C.; Xu, J.; Holmes, E.C.; Zhang, Y.-Z. Unprecedented genomic diversity of RNA viruses in arthropods reveals the ancestry of negative-sense RNA viruses. *Elife* **2015**, *4*, e05378. [CrossRef] [PubMed]
102. L'vov, D.K.; Al'khovskii, S.V.; Shchelkanov, M.Y.; Shchetinin, A.M.; Deriabin, P.G.; Gitel'man, A.K.; Aristova, V.A.; Botikov, A.G. [Taxonomic status of the Burana virus (BURV) (*Bunyaviridae*, *Nairovirus*, Tamdy group) isolated from the ticks *Haemaphysalis punctata* Canestrini et Fanzago, 1877 and *Haem. concinna* Koch, 1844 (*Ixodidae*, *Haemaphysalinae*) in Kyrgyzstan]. *Vopr. Virusol.* **2014**, *59*, 10–15, (In Russian). [PubMed]
103. L'vov, D.K.; Al'khovskii, S.V.; Shchelkanov, M.Y.; Shchetinin, A.M.; Deriabin, P.G.; Samokhvalov, E.I.; Gitel'man, A.K.; Botikov, A.G. [Genetic characterization of the Caspiy virus (CASV) (*Bunyaviridae*, *Nairovirus*) isolated from the Laridae (Vigors, 1825) and Sternidae (Bonaparte, 1838) birds and the Argasidae (Koch, 1844) ticks *Ornithodoros capensis* Neumann, 1901, in Western and Eastern coasts of the Caspian Sea]. *Vopr. Virusol.* **2014**, *59*, 24–29, (In Russian). [PubMed]
104. Al'khovskii, S.V.; L'Vov, D.K.; Shchelkanov, M.; Shchetinin, A.M.; Deriabin, P.G.; Gitel'man, A.K.; Botikov, A.G.; Samokhvalov, E.I.; Zakarian, V.A. [Taxonomic status of the Artashat virus (ARTSV) (*Bunyaviridae*, *Nairovirus*) isolated from the ticks *Ornithodoros alactagalis* Issaakjan, 1936 and O. verrucosus Olenev, Sassuchin et Fenuk, 1934 (*Argasidae* Koch, 1844) collected in Transcaucasia]. *Vopr. Virusol.* **2014**, *59*, 24–28, (In Russian). [PubMed]
105. Al'khovskii, S.V.; L'Vov, D.K.; Shchelkanov, M.; Shchetinin, A.M.; Deriabin, P.G.; Samokhvalov, E.I.; Gitel'man, A.K.; Botikov, A.G. [The taxonomy of the Issyk-Kul virus (ISKV, *Bunyaviridae*, *Nairovirus*), the etiologic agent of the Issyk-Kul fever isolated from bats (Vespertilionidae) and ticks *Argas* (*Carios*) *vespertilionis* (Latreille, 1796)]. *Vopr. Virusol.* **2013**, *58*, 11–15, (In Russian). [PubMed]
106. Ishii, A.; Ueno, K.; Orba, Y.; Sasaki, M.; Moonga, L.; Hang'ombe, B.M.; Mweene, A.S.; Umemura, T.; Ito, K.; Hall, W.W.; Sawa, H. A nairovirus isolated from African bats causes haemorrhagic gastroenteritis and severe hepatic disease in mice. *Nat. Commun.* **2014**, *5*, 5651. [CrossRef] [PubMed]

107. Cowley, J.A.; McCulloch, R.J.; Spann, K.M.; Cadogan, L.C.; Walker, P.J. Preliminary Molecular and Biological Characterization of Mourilyan Virus (MoV): A New Bunya-Related Virus of Penaeid Prawns. In *Diseases in Asian Aquaculture V. Proceedings of the 5th Symposium on Diseases in Asian Aquaculture*; Walker, P.J., Lester, R.G., Bondad-Reantaso, M.G., Eds.; Fish Health Section, Asian Fisheries Society: Manila, The Philippines, 2005; pp. 113–124.
108. Mielke-Ehret, N.; Mühlbach, H.-P. *Emaravirus*: A novel genus of multipartite, negative strand RNA plant viruses. *Viruses* **2012**, *4*, 1515–1536. [CrossRef] [PubMed]
109. Kormelink, R.; Garcia, M.L.; Goodin, M.; Sasaya, T.; Haenni, A.L. Negative-strand RNA viruses: The plant-infecting counterparts. *Virus Res.* **2011**, *162*, 184–202. [CrossRef] [PubMed]
110. Briese, T.; Calisher, C.H.; Higgs, S. Viruses of the family *Bunyaviridae*: Are all available isolates reassortants? *Virology* **2013**, *446*, 207–216. [CrossRef] [PubMed]

© 2015 by the authors. Licensee MDPI, Basel, Switzerland. This article is an open access article distributed under the terms and conditions of the Creative Commons Attribution (CC BY) license (http://creativecommons.org/licenses/by/4.0/).

Article

Mutational Analysis of the Rift Valley Fever Virus Glycoprotein Precursor Proteins for Gn Protein Expression

Inaia Phoenix [1], Nandadeva Lokugamage [1], Shoko Nishiyama [1] and Tetsuro Ikegami [1,2,3,*]

1 Department of Pathology, The University of Texas Medical Branch, Galveston, TX 77555, USA; inphoeni@UTMB.EDU (I.P.); nalokuga@UTMB.EDU (N.L.); shnishiy@UTMB.EDU (S.N.)
2 The Sealy Center for Vaccine Development, The University of Texas Medical Branch, Galveston, TX 77555, USA
3 The Center for Biodefense and Emerging Infectious Diseases, The University of Texas Medical Branch, Galveston, TX 77555, USA
* Correspondence: teikegam@utmb.edu; Tel.: +1-409-772-2563

Academic Editors: Jane Tao and Pierre-Yves Lozach
Received: 31 March 2016; Accepted: 19 May 2016; Published: 24 May 2016

Abstract: The Rift Valley fever virus (RVFV) M-segment encodes the 78 kD, NSm, Gn, and Gc proteins. The 1st AUG generates the 78 kD-Gc precursor, the 2nd AUG generates the NSm-Gn-Gc precursor, and the 3rd AUG makes the NSm′-Gn-Gc precursor. To understand biological changes due to abolishment of the precursors, we quantitatively measured Gn secretion using a reporter assay, in which a *Gaussia* luciferase (gLuc) protein is fused to the RVFV M-segment pre-Gn region. Using the reporter assay, the relative expression of Gn/gLuc fusion proteins was analyzed among various AUG mutants. The reporter assay showed efficient secretion of Gn/gLuc protein from the precursor made from the 2nd AUG, while the removal of the untranslated region upstream of the 2nd AUG (AUG2-M) increased the secretion of the Gn/gLuc protein. Subsequently, recombinant MP-12 strains encoding mutations in the pre-Gn region were rescued, and virological phenotypes were characterized. Recombinant MP-12 encoding the AUG2-M mutation replicated slightly less efficiently than the control, indicating that viral replication is further influenced by the biological processes occurring after Gn expression, rather than the Gn abundance. This study showed that, not only the abolishment of AUG, but also the truncation of viral UTR, affects the expression of Gn protein by the RVFV M-segment.

Keywords: Rift Valley fever virus; M-segment; Gn; 78 kD; NSm; precursor; expression strategy; reporter assay; reverse genetics

---

1. Introduction

Rift Valley fever (RVF) is a mosquito-borne zoonotic disease affecting humans and ruminants. The disease was originally endemic to sub-Saharan Africa, but it has since spread to Egypt, Madagascar, Saudi Arabia, and Yemen [1,2]. RVF causes a high-rate of abortion in sheep, cattle, and goats, and hemorrhagic fever, encephalitis, or blindness in humans [3]. The mortality rate of RVF patients is considered to be less than 0.5% to 1% [3,4]. However, RVF outbreaks have been known to involve a large number of patients: For example, 20,000 to 200,000 infections and 600 deaths in Egypt in 1977–1978. Floodwater *Aedes* mosquitoes can transovarially transmit RVFV [5]. These eggs are resistant to draught, and flooding, due to heavy rainfall, facilitates the hatching of infected eggs [1]. RVF is an important public health and agricultural concern, and vaccination of susceptible animals is important to minimize the spread of disease [6]. Though live-attenuated RVF vaccines are available for veterinary use in endemic countries, RVF outbreaks still occur in Africa and the surrounding countries, indicating the requirement of more effective control measures.

Rift Valley fever virus (RVFV: genus *Phlebovirus*, family *Bunyaviridae*) is comprised of three segmented, negative-stranded RNA: the Large (L), Medium (M), and Small (S)-segments. The L-segment encodes the RNA-dependent RNA polymerase (L protein). The S-segment encodes nucleocapsid (N) and nonstructural S (NSs) proteins in an ambi-sense manner, and the NSs protein is dispensable for viral replication. However, the NSs serves as the major virulence factor for RVF, and counteracts host antiviral responses by shutting-off host general transcription, including the interferon (IFN)-β gene, and promoting the posttranslational degradation of transcription factor (TF)IIH p62 subunits and dsRNA-dependent protein kinase (PKR) [7–13]. The M-segment encodes the envelope glycoproteins Gn and Gc, and two accessory proteins, NSm and 78 kD [14–16]. The Kozak consensus sequence (5′-**G**CC **R**CC AUG **G**-3′) affects translation initiation efficiency, through the nucleotides located at the −6, −3 and +4 positions in vertebrate cells [17]. In the RVFV M mRNA, the first AUG is surrounded by a weak Kozak context (5′-CAU UAA AUG U-3′), for vertebrate cells [18]. On the other hand, the 2nd AUG partially matches the Kozak context (5′-CCA **G**AG AUG A) via the guanosine at the −3 position. As a result, RVFV M-segment allows a leaky scanning of the ribosome at the 1st AUG [17]. Thus, two polypeptides can be generated using those two initiation codons starting at nt. 21 (1st AUG) and 135 (2nd AUG) (Figure 1A). The precursor protein from the 1st AUG is cleaved to produce 78 kD and Gc, while the precursor from the 2nd AUG is cleaved to produce NSm, Gn, and Gc. When the 2nd AUG is abolished, the precursor from the 3rd AUG generate NSm′, Gn, and Gc [18]. Since the 3rd, 4th, or 5th AUG are also able to generate Gn-Gc precursors, the 2nd AUG can be abolished without affecting the expression of Gn and Gc.

The live-attenuated RVF MP-12 vaccine is safe and efficacious in ruminants, and is conditionally licensed for veterinary use in the U.S. [19–24]. However, the MP-12 vaccine lacks a marker for the differentiation of infected from vaccinated animals (DIVA). The NSm and 78 kD proteins are dispensable for viral replication [25,26]. Based on the strong immunogenicity, a recombinant MP-12 vaccine encoding an in-frame truncation in the 78 kD/NSm region in the M-segment (rMP12-ΔNSm21/384) is considered as one of the next generation of MP-12 vaccines. The rMP12-ΔNSm21/384 lacking both 78 kD and NSm induces apoptosis earlier than parental rMP-12 in Vero E6, 293, and J774.1 cells [25]. In another study, the C-terminal (aa. 71–115) region of the NSm protein, which overlaps with NSm′, a truncated NSm generated from the 3rd AUG, was shown to be sufficient to suppress apoptosis [27]. NSm and NSm′ localize to the mitochondrial outer membrane, through the C-terminal transmembrane domain [18,27]. Thus, an rMP-12 strain lacking the 78 kD and NSm proteins would be a viable candidate vaccine as it would have good immunogenicity and a DIVA marker.

Since the 78 kD and NSm/NSm′ proteins are synthesized from distinct precursor polyproteins made from the 1st and 2nd AUG (*i.e.*, 78 kD-Gc and NSm-Gn-Gc, respectively), the alteration of the 1st or 2nd AUG may also affect the synthesis of Gn or Gc. Using recombinant vaccinia viruses, Suzich *et al.* analyzed the impact of specific AUG abolishment (*i.e.*, AUG to CUC substitution) in the pre-Gn region on Gn expression levels. The abolishment of the 1st AUG (Δ1) slightly increased the Gn expression, whereas the abolishment of the 2nd AUG (Δ2), the 2nd and 3rd AUGs (Δ2 + 3), the 2nd, 3rd, and 4th AUGs (Δ2 + 3 + 4), or the 2nd, 3rd, 4th and 5th AUGs (Δ2 + 3 + 4 + 5) decreased the accumulation of Gn, compared to that from the parental wild-type M-segment. Thus, the "default" NSm-Gn-Gc precursor protein from the 2nd AUG is apparently more efficient than the NSm′-Gn-Gc precursor (from the 3rd AUG) or the Gn-Gc precursors (from the 4th or 5th AUGs), for Gn expression. However, further quantitative analysis of Gn expression changes by AUG alterations will be required to correctly understand the impact of mutagenesis in the M-segment start codons.

(A) RVFV M-segment precursors

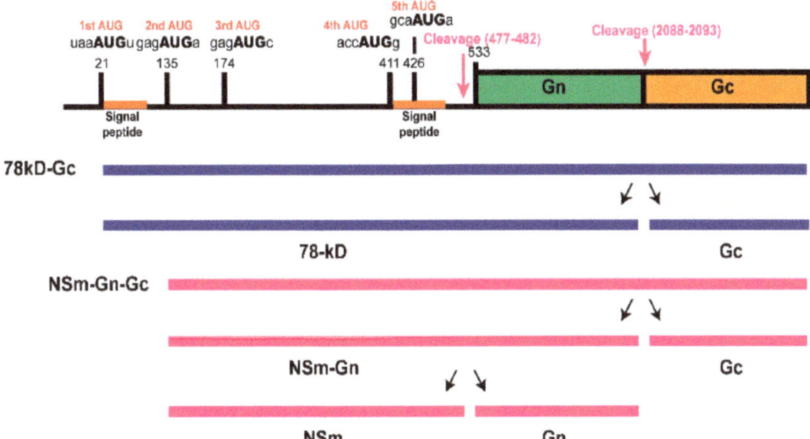

(B) pCAGGS-PreGn-gLuc-SF expression system

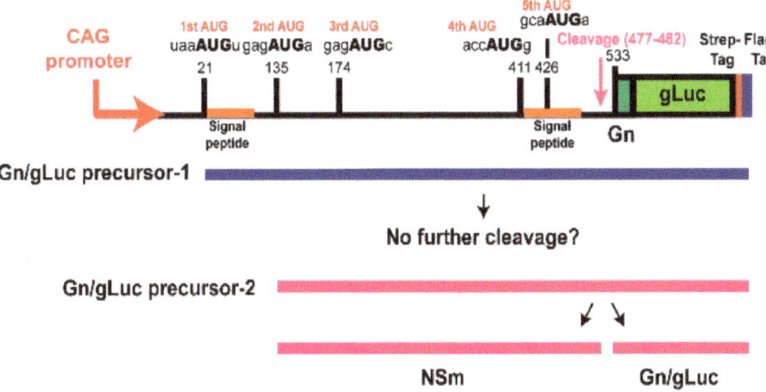

Figure 1. Gene expression of the RVFV M-segment and pCAGGS-PreGn-gLuc-SF. (**A**) The polypeptides synthesized from the 1st AUG (78 kD-Gc) or the 2nd AUG (NSm-Gn-Gc) are cleaved by signal peptidases [14,16,28]. The 78 kD protein and Gc are generated from the 78 kD-Gc precursor, while NSm, Gn, and Gc are made from the NSm-Gn-Gc precursor; (**B**) the pre-Gn region was fused to the gLuc ORF lacking the intrinsic signal peptide, which allows for secretion from the cell via Gn signal peptide. The Gn/gLuc precursor-1 makes a chimeric protein consisting of the pre-Gn region and gLuc, while the Gn/gLuc precursor-2 generates NSm and the Gn/gLuc fusion protein.

Little information is available about the consequence of pre-Gn region mutagenesis in terms of viral phenotypes, other than the expression of 78 kD or NSm/NSm′. We hypothesized that an optimization of the M-segment pre-Gn region increases the secretion of Gn, which in turn, will lead to an increase in viral titer. We aimed to analyze the consequences of the abolishment of the 1st, 2nd, 3rd, 4th, and/or 5th AUG, in terms of Gn secretion and viral phenotype. We established a reporter assay, in which *Gaussia* luciferase (gLuc), fused to the N-terminal region of Gn, is expressed from the pre-Gn region of the RVFV M-segment. In the reporter assay, the Gn/gLuc fusion protein is cleaved from the Gn/gLuc precursor proteins made from either the 1st or 2nd AUG, while the 3rd, 4th, or 5th AUG can serve as surrogates for precursor production, as shown in Figure 1B. Normally, gLuc encodes an intrinsic signal sequence, which allows it to be secreted from expressed cells [29]. In our reporter assay, we deleted the intrinsic signal sequence for gLuc so that the cleavage of Gn/gLuc fusion proteins occurs only through the Gn signal sequence, and subsequently, the Gn/gLuc fusion proteins are secreted from expressed cells. The Gn/gLuc fusion proteins produced from this construct do not accumulate in the Golgi, due to a lack of the Golgi retention signal at the C-terminus of Gn [30], and the secretion occurs through the endocytic recycling pathway from the Golgi to plasma membrane (e.g., Rab8-positive vesicles) [31]. Although the level of Gn/gLuc fusion protein secretion does not predict virion release efficiency, as RVFV Gn/Gc accumulates in the Golgi, we aimed to measure relative expression level of Gn protein from precursor proteins made from RVFV M-segment. We analyzed the relative secretion of the Gn/gLuc fusion proteins into the culture supernatant among different AUG (Met) to CUC (Leu) substitution mutants. Furthermore, recombinant MP-12 (rMP-12) encoding the AUG mutations, which can secrete distinct levels of Gn, were rescued by reverse genetics, and their phenotypes were characterized. Our study provides fundamental information for the consequences of mutagenesis in the pre-Gn region of the M-segment, and will support the understanding of current and future studies using NSm or 78 kD knockout mutants.

2. Materials and Methods

2.1. Media, Cells, and Viruses

Human embryonic kidney (293) cells, Vero cells, (ATCC CCL-81), and Vero E6 cells (ATCC C1008) were maintained in Dulbecco's Modified Eagle Medium (DMEM) supplemented with 10% FBS, penicillin (100 U/mL), and streptomycin (100 μg/mL). Minimum Essential Medium (MEM)-alpha supplemented with 10% fetal bovine serum (FBS) (Life Technologies, Carlsbad, CA, USA), penicillin (100 U/mL), streptomycin (100 μg/mL), and hygromycin B (600 μg/mL) was used to culture BHK/T7-9 cells that stably express T7 RNA polymerase [32]. Recombinant RVFV MP-12 strains encoding mutation(s) in the M-segment pre-Gn region were rescued by reverse genetics, as described previously [33]. rMP-12 mutants were titrated by plaque assay using Vero E6 cells [34].

2.2. Plasmids

The pCAGGS-PreGn-gLuc-SF plasmid was made as follows: Synthetic DNA was created (gBlocks, Integrated DNA Technologies, Coralville, IA, USA) for the RVFV MP-12 M-segment (nt. 1-533) fused with gLuc (lacking nt.1-51 to remove an intrinsic signal sequence for gLuc), and two tandem Strep-tags, and a Flag-tag. The DNA fragment was cloned into pCAGGS plasmid by Gibson assembly mastermix (New England BioLabs, Ipswich, MA, USA). To introduce mutations in pCAGGS-PreGn-gLuc-SF plasmid, the DNA fragment (nt. 1-830) was first transferred to a pProT7 plasmid. Then, site-directed mutagenesis was performed using the pProT7 plasmid back-bone, before cloning the insert sequence into a pCAGGS plasmid. Corresponding AUGs were replaced with CUC (Leu) to abolish the AUG codon.

2.3. Western Blotting

Cells were suspended in 2x SDS sample buffer, and boiled for 10 min. Samples were separated by SDS-polyacrylamide gel electrophoresis (SDS-PAGE) under reducing conditions. Western blot was performed as described previously [35]. Anti-Flag M2 antibody (Sigma-Aldrich, St. Louis, MO, USA), and anti-actin antibody (I-19, Santa Cruz Biotechnology, Inc., Dallas, TX, USA) were used.

2.4. The Gaussia and Cypridia Luciferase Assays

Sub-confluent 293 cells (1×10^6 cells) were mock-transfected or co-transfected with 0.1 µg of pSV40-CLuc (encoding *Cypridia* luciferase, cLuc, downstream of SV40 promoter) and 2.0 µg of pCAGGS-PreGn-gLuc-SF, or the mutants in the pre-Gn region. At 36 h post transfection, culture supernatants were harvested, and gLuc assays (BioLuc Gaussia Luciferase Assay Kit, New England BioLabs) and cLuc assays (BioLuc Cypridina Luciferase Assay Kit, New England BioLabs) were performed, according to manufacturer's instructions.

2.5. Measurement of Plaque Sizes

Plaque images of rMP-12, Δ2 + 3, or the AUG2-M mutants, formed in VeroE6 cells in 6-well plates, were incorporated by a scanner, and the diameters (mm) of small and large plaques (n = 10 each) were measured using Adobe Photoshop Element version 7.0 [36]. The average and standard errors were plotted onto the graph using GraphPad Prism version 6.05 [37].

2.6. Statistical Analysis

Statistical analysis was performed using GraphPad Prism version 6.05. For the gLuc/cLuc values normalized to parental construct value in Figures 2 and 3 or virus titers in Figure 4, arithmetic means of $\log_{10}$ values were analyzed by one-way ANOVA followed by Tukey's multiple comparisons test.

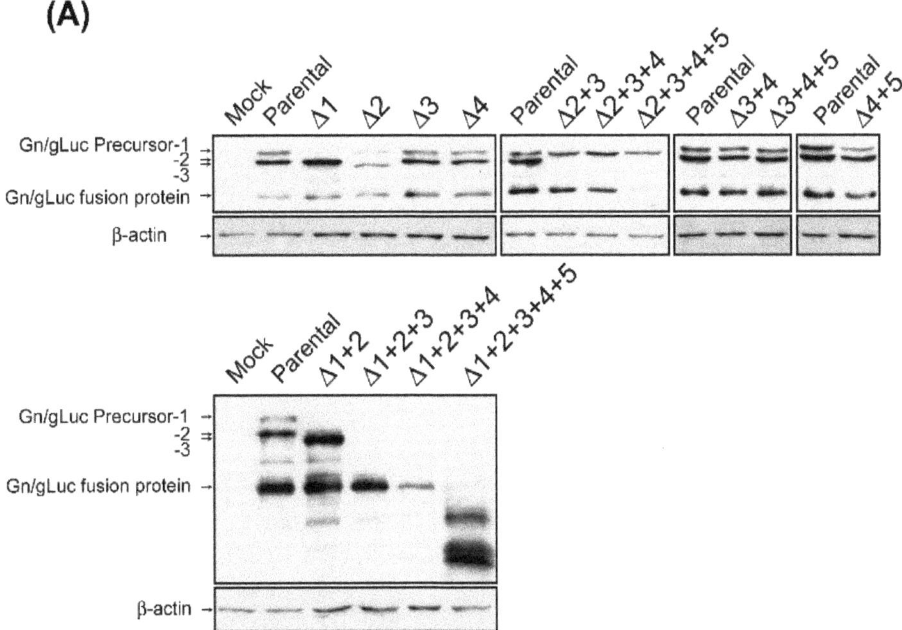

Figure 2. *Cont.*

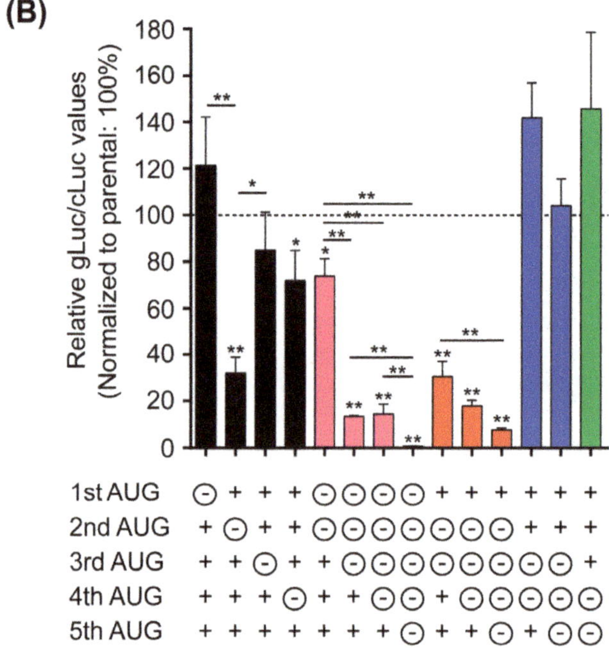

Figure 2. Relative expression of precursor proteins and cleaved Gn/gLuc fusion proteins after AUG abolishment. (**A**) Western blot of cell lysates. 293 cells were co-transfected with pSV40-CLuc (transfection control), and pCAGGS-PreGn-gLuc-SF or the AUG mutant plasmids. At 36 h post transfection, cell lysates were collected and analyzed by Western blot using anti-Flag antibody; (**B**) the extracellular Gn/gLuc fusion proteins were measured using the culture supernatant of transfected cells. The ratio of gLuc to cLuc (control plasmid) was normalized to that of parental pCAGGS-PreGn-gLuc-SF. The graph represents the mean + the standard error of three independent experiments. Asterisks represent statistically significant differences (one-way ANOVA followed by Tukey's multiple comparisons test, * $p < 0.05$, ** $p < 0.01$). Asterisks shown on error bars represent the comparison with compared to the Δ4 + 5 mutant.

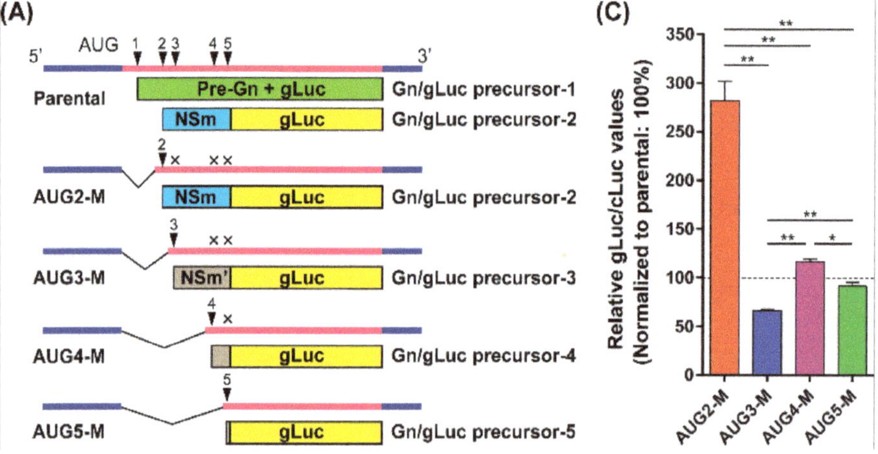

Figure 3. *Cont.*

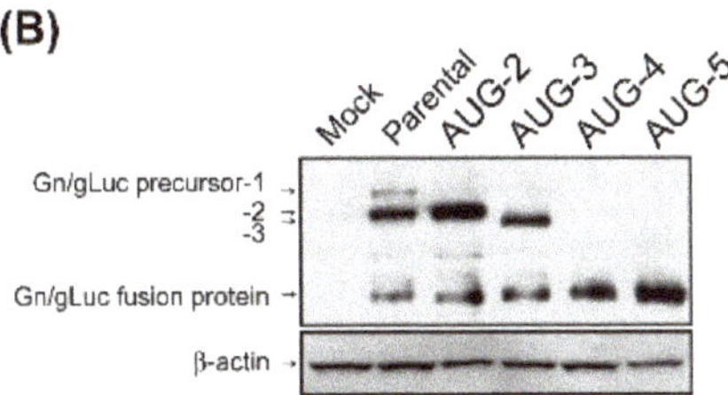

Figure 3. Relative expression of precursor proteins and cleaved gLuc after the truncation of the viral sequence upstream of AUG. (**A**) Schematic representation of AUG2-M, AUG3-M, AUG4-M, or AUG5-M, which express a single precursor from the 2nd, 3rd, 4th, or the 5th AUG, respectively. AUG2-M lacks the 3rd, 4th, and 5th AUGs, AUG3-M lacks the 4th and 5th AUGs, and AUG4-M lacks the 5th AUG. The mutant plasmids encode a common 20 nucleotide viral sequence upstream of the 1st AUG (5′-ACACAAAGACGGUGCACGAGAUG (initiation codon is underlined)); (**B**) 293 cells were co-transfected with pSV40-CLuc (transfection control), and pCAGGS-PreGn-gLuc-SF or the mutant plasmids. At 36 h post transfection, culture supernatants were collected, and the gLuc and cLuc activities were measured. Then, the ratio of gLuc to cLuc was normalized to that of parental pCAGGS-PreGn-gLuc-SF plasmid. The graph represents the mean + the standard error of three independent experiments. Asterisks represent statistically significant differences (one-way ANOVA followed by Tukey's multiple comparisons test, * $p < 0.05$, ** $p < 0.01$).

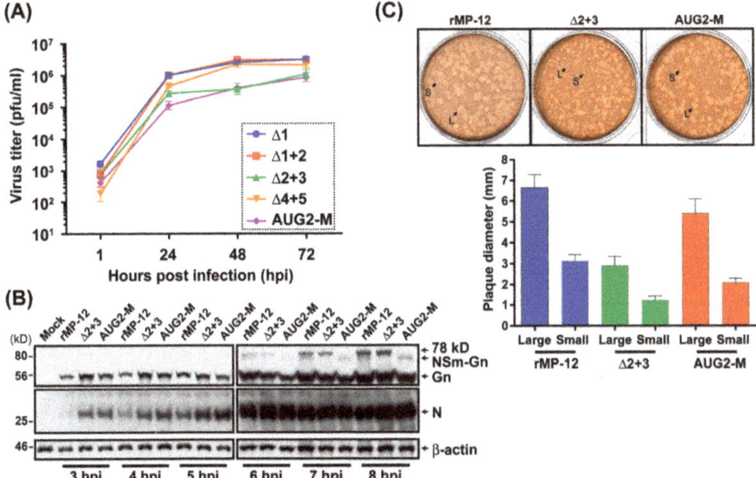

Figure 4. Characterization of recombinant MP-12 encoding mutations in the pre-Gn region. (**A**) Replication of rMP-12 mutants encoding the mutations either in the 1st AUG (Δ1), the 1st and 2nd AUG (Δ1 + 2), the 2nd and 3rd AUG (Δ2 + 3), the 4th and 5th AUG (Δ4 + 5), or the deletion of the UTR upstream of the 2nd AUG (AUG2-M: see Figure 3A). Vero cells were infected with each virus at a multiplicity of infection (MOI) of 0.15. The graph represents the antilog of the arithmetic mean of the $\log_{10}$-transformed virus titers + the standard deviation of three independent experiments; (**B**) Western blot using Vero cells infected with either rMP-12, Δ2 + 3, or the AUG2-M mutants (an MOI of 1). The 78 kD and Gn were detected by mouse anti-Gn monoclonal antibody (4D4). The N proteins were detected using mouse anti-RVFV polyclonal antibody. β-actin is shown as sample loading controls; (**C**) plaque phenotypes of rMP-12, Δ2 + 3, or the AUG2-M mutants in VeroE6 cells. Small (S) and large (L) plaques are shown with arrows. The diameters (mm) of small and large plaques were measured (n = 10 per sample), and the average and standard errors are shown in the graph.

3. Results

3.1. The Gn/gLuc Precursor-1 Does Not Efficiently Generate Gn/gLuc Fusion Proteins

293 cells were co-transfected with pSV40-CLuc (transfection control), and pCAGGS-PreGn-gLuc-SF or the AUG mutant plasmids. At 36 h post transfection, cell lysates were collected and the intracellular expression of Gn/gLuc proteins or precursor proteins was analyzed by Western blot using anti-Flag antibody. Culture supernatants were used to measure the level of secreted extracellular Gn/gLuc fusion proteins by reporter assay. The cLuc protein encodes an intrinsic signal peptide and is secreted into the secretory pathway after expression. Thus, secreted cLuc served as a control to measure the secretion of the Gn/gLuc fusion proteins from transfected cells. Western blot using anti-Flag antibody was performed to confirm the expression of precursor proteins and Gn/gLuc fusion proteins (Figure 2A). As expected, the Gn/gLuc precursor-1 disappeared after the abolishment of the 1st AUG. Similarly, the Gn/gLuc precursor-2 disappeared when the 2nd AUG was abolished. When the 2nd AUG was abolished, the Gn/gLuc precursor-3 appeared (*i.e.*, Δ2 or Δ1 + 2). The Gn/gLuc precursor-4 and -5 were indistinguishable from the cleaved Gn/gLuc fusion protein based on band migrations. When the 1st, 2nd, 3rd, 4th, and 5th AUGs were abolished, no Gn/gLuc fusion proteins were detected. On the other hand, when the 2nd, 3rd, 4th, and 5th AUGs were abolished, the band intensity of the Gn/gLuc fusion protein was largely decreased but still detectable in the Western blot. Next, both gLuc and cLuc activities in culture supernatants were measured. We calculated the ratio of gLuc to cLuc, and the value was normalized to that of parental pCAGGS-PreGn-gLuc-SF (Figure 2B). When the 1st, 2nd, 3rd, 4th, and 5th AUGs were all abolished, no gLuc activity (0.79% compared to parental 100%) could be measured in the supernatant. On the other hand, when the 2nd, 3rd, 4th, and 5th AUGs were abolished, a decreased level of gLuc activity (7.7% compared to parental 100%) was detected in the supernatant. These results indicated that the precursor protein synthesized from the 1st AUG can also generate Gn, though at a decreased level.

3.2. Precursor-2 Plays a Major Role in Gn/gLuc Expression

We further analyzed the relative secretion of the cleaved gLuc using various AUG mutants (Figure 2B). The 3rd and 4th AUGs (Δ3 + 4), the 3rd, 4th, and 5th AUGs (Δ3 + 4 + 5); or the 4th and 5th AUGs (Δ4 + 5) were abolished, the gLuc activity was 142%, 104%, or 146%, respectively, compared to the parental plasmid (100%). It indicates that abolishment of the 3rd and the 4th AUGs, that of the 4th and the 5th AUGs, and that of the 3rd, the 4th, and the 5th AUGs, do not result in a decrease of relative gLuc activity, in the presence of the 1st and the 2nd AUGs. The individual abolishment of the 1st (Δ1), 2nd (Δ2), 3rd (Δ3), or 4th AUG (Δ4), resulted in relative gLuc activities of 121%, 32%, 85%, or 72%, respectively, compared to the parental plasmid (100%) (Figure 2B). The Δ2 mutant generated significantly lower gLuc activity than the Δ4 + 5 mutant, the Δ1 mutant, or the Δ3 mutant. Thus, the abolishment of the 2nd AUG significantly affected the gLuc activity. When the 1st and 2nd AUGs (Δ1 + 2); the 1st, 2nd, and 3rd AUGs (Δ1 + 2 + 3); or the 1st, 2nd, 3rd, and 4th AUGs (Δ1 + 2 + 3 + 4) were concomitantly abolished, the gLuc activity was 74%, 13%, or 15%, respectively, compared to the parental plasmid (100%). The gLuc activities of the Δ1 + 2 + 3 mutant and the Δ1 + 2 + 3 + 4 mutant were significantly lower than that of the Δ1 + 2 mutant. Furthermore, when the 2nd and 3rd AUGs (Δ2 + 3); the 2nd, 3rd, and 4th AUGs (Δ2 + 3 + 4); or the 2nd, 3rd, 4th, and 5th AUGs (Δ2 + 3 + 4 + 5) were abolished, the gLuc activity was 31%, 18%, or 8%, respectively, compared to the parental plasmid, and all of which were significantly lower than that of the Δ4 + 5 mutant. Although it was not statistically significant, the differences of gLuc activity between the Δ1 + 2 + 3 (13%) and Δ2 + 3 mutants (31%), or between the Δ1 + 2 + 3 + 4 (15%) and Δ2 + 3 + 4 (18%), indicate that the precursor from the 1st AUG slightly increases gLuc activity. Taken together, these results indicate that the secretion of Gn/gLuc fusion proteins into the culture supernatant occurs efficiently through the Gn/gLuc precursor-2, while the Gn/gLuc precursor-3 can serve as a surrogate of Gn/gLuc precursor production in the absence of

AUG-2. However, it was unclear why the secretion of the Gn/gLuc fusion protein from the precursor-4 or 5 was lower than those of precursor-2 or 3.

3.3. The Viral Untranslated Region Sequence, Upstream of the 4th or 5th AUG, Affects Efficient Generation of Gn/gLuc Fusion Proteins

Spik *et al.* previously showed that the cloned open reading frame (ORF) of Gn-Gc (starting from the 4th AUG), without an upstream viral untranslated region (UTR), expresses slightly higher Gn/Gc than the cloned ORF of the NSm-Gn-Gc precursor (starting from the 2nd AUG) [38]. Thus, we suspected that the decrease in the gLuc activity from the Gn/gLuc precursor-4 or -5 occurred due to the presence of the UTR upstream of the 4th or 5th AUG. Therefore, we truncated the upstream viral UTR sequence from pCAGGS-PreGn-gLuc-SF (Figure 3A). The plasmids, AUG2-M, AUG3-M, AUG4-M, or AUG-5-M, generate the Gn/gLuc precursor-2, -3, -4, or -5, respectively, and share the common UTR sequence (5′-ACACAAAGACGGUGCACGAGAUG (initiation codon is underlined)). Each plasmid also lacks downstream initiation codons, to prevent the generation of additional precursor proteins. This abolishment allowed us to analyze the role of a single precursor protein in the production of the Gn/gLuc fusion protein. Using those plasmids, we analyzed the secretion of the Gn/gLuc fusion proteins into the culture supernatant. Surprisingly, AUG2-M had 282% gLuc activity, which was significantly higher than that of AUG3-M, AUG4-M, or AUG5-M. On the other hand, the AUG3-M had significantly lower gLuc activity (66%) than that of the AUG4-M (116%) or the AUG5-M (92%). The difference of gLuc activity between AUG4-M and AUG5-M was marginally significant. The results indicated that precursor-3, produced from the AUG3-M, is less efficient than precursor-4 from the AUM4-M in the secretion of the Gn/gLuc fusion protein. Since the gLuc activities of Δ1 + 2 + 3, and Δ1 + 2 + 3 + 4 mutants were 13%, and 15% (Figure 2B), respectively, the viral UTR upstream of the 4th or 5th AUG affects the translation efficiency of precursor-4 or 5. Furthermore, the AUG2-M mutant increased the secretion of the Gn/gLuc fusion protein, compared to the Δ1 mutant, indicating that viral UTR upstream of the 2nd AUG also affects the translation efficiency of precursor-2.

3.4. The rMP-12 Encoding the AUG2-M Mutation or the rMP-12 Encoding the Δ2 + 3 Mutation Replicate Less Efficiently Than Parental rMP-12

We, next, characterized the viral phenotypes caused by modification of the M-segment precursors. The reporter assay results indicated that the AUG2-M mutant plasmid generates Gn proteins efficiently, due to a lack of upstream viral UTR sequence. Thus, we aimed to test whether the recombinant MP-12 encoding the AUG2-M mutations in the preglycoprotein coding region can replicate more efficiently than other mutants. However, the modification of preglycoprotein region also affects the expression of the 78 kD, NSm, or NSm′. The AUG2-M mutant does not encode the 78 kD and NSm′ proteins, but encodes the NSm protein with 3 mutations (Met to Leu, at the 3rd, 4th, and 5th AUGs). For comparison, we also analyzed the Δ1 mutant (lacking the 78 kD, but still encoding NSm and NSm′), Δ1 + 2 mutant (lacking 78 kD and NSm, but encoding NSm′), and the Δ2 + 3 mutant (encoding 78 kD, but lacking NSm and NSm′). Those constructs were predicted to express Gn protein less efficiently than the AUG2-M mutant construct, based on reporter assay result. As a control, the Δ4 + 5 mutant was analyzed, which produces the default precursors from the 1st and 2nd AUGs.

We, first, analyzed the replication of rMP-12 mutants (the AUG2-M, Δ1, Δ1 + 2, Δ2 + 3, or Δ4 + 5) at a multiplicity of infection (MOI) of 0.15. All the rMP-12 mutants replicated efficiently in Vero cells, while the rMP-12 encoding the Δ2 + 3 mutation, or that encoding the AUG2-M mutation replicated slightly more slowly than the others (Figure 4A). The arithmetic means of $\log_{10}$ titers of AUG2-M mutant at 48 and 72 hpi were 7 and 2 times lower than those of the Δ4 + 5 mutant at 48 and 72 hpi, respectively ($p < 0.05$). On the other hand, the arithmetic means of $\log_{10}$ titers of Δ2 + 3 mutant were 6 and 2 times lower than those of the Δ4 + 5 mutant at 48 and 72 hpi, respectively, and the differences were not statistically significant. These two mutants, AUG2-M and Δ2 + 3, were further analyzed at a higher MOI infection (1 MOI). After virus infection at 37 °C for 1 h, Vero cells were washed 6 times with media, and the cell lysates were collected at 3, 4, 5, 6, 7, and 8 hpi. Western blot analysis showed that parental rMP-12 generated a detectable level of 78 kD at 6, 7, and 8 hpi, and NSm-Gn at 7 and 8 hpi (Figure 4B). On the other hand, 78 kD was not synthesized from the AUG2-M mutant, and the NSm-Gn was not made from the Δ2 + 3 mutant. Viral RNA replication most likely started between 5 and 6 hpi, as there was an increase of all viral proteins at 6 hpi. The parental rMP-12 made plaques with heterogeneous sizes (2.6 to 5.7 mm in diameter). The Δ2 + 3 mutant made smaller plaques than parental rMP-12 ranging from 0.9 to 2.3 mm in diameter (Figure 4C). On the other hand, the AUG2-M made the intermediate sized plaques between the Δ2 + 3 and parental rMP-12 (1.7 to 4.5 mm in diameter). Taken together, these results indicate that rMP-12 encoding the AUG2-M mutation replicates less efficiently than Δ4 + 5 mutant, though the NSm-Gn-Gc precursor made from the AUG2-M mutant was predicted to generate more Gn than Δ4 + 5 mutant. Thus, the discrepancy of reporter assay and virus replication efficiency suggested a role of 78 kD, NSm, or NSm′ in the downstream expression of Gn.

4. Discussion

A live-attenuated RVFV vaccine, MP-12, is conditionally licensed for veterinary use in the U.S. Though the MP-12 vaccine is safe and efficacious [19–22,39], the vaccine lacks a marker for DIVA (differentiation of infected from vaccinated animals). Reverse genetics is a useful tool to generate RVFV lacking either the 78 kD or NSm proteins. The Δ78 kD or ΔNSm mutant can be made by the abolishment of the 1st or the 2nd AUG, respectively. An introduction of a DIVA marker in the M-segment can be made by truncating the 78 kD/NSm coding region ranging from the 1st AUG to the downstream of the 3rd AUG, while leaving a short UTR upstream of the 4th AUG. The rMP12-ΔNSm21/384 (similar to our AUG4-M mutant plasmid), which lacks both 78 kD and NSm expression, showed a similar immunogenicity and efficacy with parental MP-12 [40,41]. However, without knowing the role of each AUG or UTR in the Gn/Gc expression or virion productions, the impact of alterations of the preglycoprotein coding region on the virological phenotype cannot be predicted. In this study, we analyzed the effect of AUG abolishment or an in-frame deletion of viral UTR sequence upstream of the AUG on the Gn expression levels using a quantitative reporter assay system. Subsequently, we also characterized the virological phenotypes of representative AUG mutants. Initially, we hypothesized that increased expression levels of Gn/Gc would increase progeny virus titer. However, that assumption was not correct, and the results showed an unexpectedly complex regulation of viral progeny production through Gn and Gc. As increased production of Gn did not lead to an increase in virus production, regulation of viral production may be regulated at a later step in the viral life cycle.

The reporter assay expressing the Gn/gLuc fusion protein downstream of the RVFV preglycoprotein coding region is useful to measure the level of cleaved Gn/gLuc from the precursor proteins translated from the 1st, 2nd, 3rd, 4th, or 5th AUG. Since the Gn/gLuc fusion protein does not encode the Golgi retention signal, the protein is secreted out from transfected cells without accumulation in the Golgi. Thus, the relative reporter activities in the culture supernatants indicate the efficiency of precursor expression, and subsequent cleavage of the precursor protein. We confirmed that the plasmid lacking the 2nd, 3rd, 4th, and 5th AUGs (Δ2 + 3 + 4 + 5) still generates a small amount of the Gn/gLuc fusion protein and a detectable gLuc activity in the culture supernatant. The 78 kD protein encodes a signal sequence at the N-terminus [28], and the 78 kD-Gc precursor is synthesized in the ER membrane. The second signal sequence for 78 kD protein, which is located between the NSm and Gn coding region, may not be efficiently recognized by signal peptidase, which may be hindered by the folding of ectodomain in the ER lumen. Indeed, we could not rescue the rMP-12 encoding the Δ2 + 3 + 4 + 5 mutation in the M-segment, most likely due to such low expression of Gn from the 78 kD-Gc precursor. The Δ2 showed decreased secretion of the Gn/gLuc fusion proteins (32%) in culture supernatants. However, the Δ1, Δ3, or Δ4 mutant still efficiently secreted the Gn/gLuc fusion proteins (121%, 85%, or 72%, respectively). In addition, in the presence of the 1st and 2nd AUG, the abolishment of the 3rd, 4th, or 5th AUG (Δ3 + 4 + 5) did not affect the gLuc activity (104%). Thus, the precursor-2, produced by AUG 2, plays a role in the production of the Gn/gLuc fusion protein. Since the Δ1 + 2 mutant still expresses relatively high gLuc activity (74%), compared to the Δ1 + 2 + 3 mutant (13%), the precursor-3, but not precursor-4, or 5, serves as an efficient surrogate of precursor-2 in the production of the Gn/gLuc fusion protein. Thus, our results indicated that the NSm-Gn-Gc precursor plays a default role, and the NSm′-Gn-Gc precursor plays a surrogate role, in the expression of the Gn protein.

Although we introduced mutations to abolish specific AUGs, this approach did not address the effect of long viral UTR upstream of AUG in the translation efficiency of Gn. Relative gLuc activities of the Δ1 + 2 + 3 or Δ1 + 2 + 3 + 4 mutants were low. To address this concern, we generated additional reporter constructs encoding the 2nd, 3rd, 4th, or 5th AUG, without upstream viral UTR sequences (AUG2-M, AUG3-M, AUG4-M, or AUG5-M, respectively). We also abolished downstream AUGs, to prevent the expression of more than one precursor protein. The AUG4-M (116%), and AUG5-M (92%) constructs showed increased gLuc activity, compared to Δ1 + 2 + 3 (13%), and Δ1 + 2 + 3 + 4 (15%) mutant plasmids. Thus, the deletion of the UTR upstream of the 4th or 5th AUG improved the secretion of the Gn/gLuc fusion protein from precursor-4, or 5, respectively. Those results indicated that an in-frame deletion of UTR sequence increases the expression of Gn/Gc from the precursor made from the 4th or 5th AUG.

A limitation of this reporter assay is the lack of natural viral assembly and budding from the Golgi. The results from reporter assay predict the expression levels of Gn proteins from mRNA. However, RVFV Gn encodes a Golgi retention signal at the C-terminus, and co-localizes with Gc to form heterodimers [30,42,43]. Gn and Gc are assembled with the viral ribonucleocapsid, bud from the Golgi, and out of the cell. On the other hand, little is known about the functions of 78 kD, NSm, or NSm′ in the assembly process. We generated recombinant rMP-12 encoding mutations in the preglycoprotein coding region (Δ1, Δ1 + 2, Δ2 + 3, Δ4 + 5, or AUG2-M). Our study showed that the rMP-12 encoding the AUG2-M mutations replicated slightly less efficiently than parental rMP-12 in Vero cells, despite having increased Gn expression. The AUG2-M does not make 78 kD and NSm′, but encodes NSm having Met-to-Leu substitutions at the 3rd, 4th, and 5th AUGs. The Met-to-Ala substitution at the 3rd AUG is known to decrease the migration of NSm protein [18], indicating the occurrence of posttranslational modification of NSm. We assume that NSm and NSm′ play a role at a later step of viral protein synthesis: For example, viral assembly process. The Δ2 + 3 mutant encode neither NSm nor NSm′ but still generates the 78 kD protein [18]. In our study, this mutant also showed relatively inefficient replication kinetics in Vero cells, although it was not statistically significant. The NSm of Bunyamwera virus (BUNV: genus *Orthobunyavirus*) serves as a scaffold to form a "viral tube" structure to facilitate the assembly of the ribonucleocapsid with the Gn/Gc complex at the peripheral Golgi, and a lack of NSm reduce virus production by 10 to 100-fold [44]. The 78 kD protein may also play a role in viral replication. Kreher *et al.* showed that the AUG (Met) to GCG (Ala) mutation at the 1st AUG (Δ1), at the 1st and 2nd AUGs (Δ1 + 2), or at the 1st, 2nd, and the 3rd AUGs (Δ1 + 2 + 3) led to the emergence of RVFV mutant encoding a new AUG upstream of the original 1st AUG, during 5 serial passages in mammalian cells [18]. This new AUG can generate a 78 kD-Gc-like precursor protein, in addition to the NSm-Gn-Gc precursor. Previous studies indicated that the 78 kD plays a major role in viral dissemination in mosquito vectors, while NSm and NSm′ contribute to the RVFV propagation in vertebrate cells [18]. However, little is known about the role of 78 kD, NSm, or NSm′ proteins in viral replication, and further studies will be required to elucidate the mechanisms.

5. Conclusions

This study showed that, not only the abolishment of AUG, but also the truncation of viral UTR, affects the expression of Gn protein by the RVFV M-segment. Increased production of Gn did not lead to an increase in virus production, and thus, regulation of viral production may be further regulated at a later step in the viral life cycle.

Acknowledgments: We thank Robert B. Tesh at the University of Texas Medical Branch at Galveston for the mouse anti-RVFV antibody, and David. A. Norwood at the United States Army Medical Research Institute for Infectious Diseases for the mouse 4D4 monoclonal antibody. This work was supported by the National Institute of Health (grant number R01 AI087643-01A1) (Tetsuro Ikegami); the James W. McLaughlin predoctoral fellowship (Inaia Phoenix); the funding from the Sealy Center for Vaccine Development at The University of Texas Medical Branch at Galveston (Tetsuro Ikegami).

Author Contributions: Tetsuro Ikegami conceived and designed the experiments; Inaia Phoenix, Shoko Nishiyama, and Nandadeva Lokugamage performed the experiments; Inaia Phoenix, Shoko Nishiyama, and Tetsuro Ikegami analyzed the data; Inaia Phoenix and Tetsuro Ikegami wrote the paper.

Conflicts of Interest: The authors declare no conflict of interest.

Abbreviations

The following abbreviations are used in this manuscript:

RVF	Rift Valley fever
RVFV	Rift Valley fever virus
L-segment	Large-segment
M-segment	Medium-segment
S-segment	Small-segment
DIVA	Differentiation of infected from Vaccinated animals
gLuc	*Gaussia* luciferase
cLuc	*Cypridia* luciferase
PKR	dsRNA-dependent protein kinase
IFN	Interferon
FBS	Fetal bovine serum
BHK	Baby hamster kidney
MOI	Multiplicity of infection
SDS-PAGE	Sodium dodecyl sulfate-polyacrylamide gel electrophoresis
CAG promoter	The cytomegalovirus early enhancer/chicken β actin promoter
UTR	Untranslated region

References

1. Pepin, M.; Bouloy, M.; Bird, B.H.; Kemp, A.; Paweska, J. Rift Valley fever virus (*Bunyaviridae: Phlebovirus*): An update on pathogenesis, molecular epidemiology, vectors, diagnostics and prevention. *Vet. Res.* **2010**, *41*. [CrossRef]
2. Ikegami, T. Molecular biology and genetic diversity of Rift Valley fever virus. *Antivir. Res.* **2012**, *95*, 293–310. [CrossRef] [PubMed]
3. Ikegami, T.; Makino, S. The pathogenesis of Rift Valley fever. *Viruses* **2011**, *3*, 493–519. [CrossRef] [PubMed]
4. Bird, B.H.; Ksiazek, T.G.; Nichol, S.T.; Maclachlan, N.J. Rift Valley fever virus. *J. Am. Vet. Med. Assoc.* **2009**, *234*, 883–893. [CrossRef] [PubMed]
5. Linthicum, K.J.; Davies, F.G.; Kairo, A.; Bailey, C.L. Rift Valley fever virus (family Bunyaviridae, genus *Phlebovirus*). Isolations from Diptera collected during an inter-epizootic period in Kenya. *J. Hyg.* **1985**, *95*, 197–209. [CrossRef] [PubMed]
6. Ikegami, T.; Makino, S. Rift valley fever vaccines. *Vaccine* **2009**, *27*, D69–D72. [CrossRef] [PubMed]
7. Billecocq, A.; Spiegel, M.; Vialat, P.; Kohl, A.; Weber, F.; Bouloy, M.; Haller, O. NSs protein of Rift Valley fever virus blocks interferon production by inhibiting host gene transcription. *J. Virol.* **2004**, *78*, 9798–9806. [CrossRef] [PubMed]
8. Bouloy, M.; Janzen, C.; Vialat, P.; Khun, H.; Pavlovic, J.; Huerre, M.; Haller, O. Genetic evidence for an interferon-antagonistic function of Rift Valley fever virus nonstructural protein NSs. *J. Virol.* **2001**, *75*, 1371–1377. [CrossRef] [PubMed]
9. Le May, N.; Dubaele, S.; Proietti de Santis, L.; Billecocq, A.; Bouloy, M.; Egly, J.M. TFIIH transcription factor, a target for the Rift Valley hemorrhagic fever virus. *Cell* **2004**, *116*, 541–550. [CrossRef]
10. Le May, N.; Mansuroglu, Z.; Leger, P.; Josse, T.; Blot, G.; Billecocq, A.; Flick, R.; Jacob, Y.; Bonnefoy, E.; Bouloy, M. A SAP30 complex inhibits IFN-beta expression in Rift Valley fever virus infected cells. *PLoS Pathog.* **2008**, *4*, e13. [CrossRef] [PubMed]

11. Ikegami, T.; Narayanan, K.; Won, S.; Kamitani, W.; Peters, C.J.; Makino, S. Rift Valley fever virus NSs protein promotes post-transcriptional downregulation of protein kinase PKR and inhibits eIF2α phosphorylation. *PLoS Pathog.* **2009**, *5*, e1000287. [CrossRef] [PubMed]
12. Kalveram, B.; Lihoradova, O.; Ikegami, T. NSs protein of Rift Valley fever virus promotes post-translational downregulation of the TFIIH subunit p62. *J. Virol.* **2011**, *85*, 6234–6243. [CrossRef] [PubMed]
13. Habjan, M.; Pichlmair, A.; Elliott, R.M.; Overby, A.K.; Glatter, T.; Gstaiger, M.; Superti-Furga, G.; Unger, H.; Weber, F. NSs protein of Rift Valley fever virus induces the specific degradation of the double-stranded RNA-dependent protein kinase. *J. Virol.* **2009**, *83*, 4365–4375. [CrossRef] [PubMed]
14. Kakach, L.T.; Suzich, J.A.; Collett, M.S. Rift Valley fever virus M segment: Phlebovirus expression strategy and protein glycosylation. *Virology* **1989**, *170*, 505–510. [CrossRef]
15. Suzich, J.A.; Collett, M.S. Rift Valley fever virus M segment: Cell-free transcription and translation of virus-complementary RNA. *Virology* **1988**, *164*, 478–486. [CrossRef]
16. Suzich, J.A.; Kakach, L.T.; Collett, M.S. Expression strategy of a phlebovirus: Biogenesis of proteins from the Rift Valley fever virus M segment. *J. Virol.* **1990**, *64*, 1549–1555. [PubMed]
17. Kozak, M. An analysis of vertebrate mRNA sequences: Intimations of translational control. *J. Cell Biol.* **1991**, *115*, 887–903. [CrossRef] [PubMed]
18. Kreher, F.; Tamietti, C.; Gommet, C.; Guillemot, L.; Ermonval, M.; Failloux, A.B.; Panthier, J.J.; Bouloy, M.; Flamand, M. The Rift Valley fever accessory proteins NSm and P78/NSm-Gn are determinants of virus propagation in vertebrate and invertebrate hosts. *Emerg. Microb. Infect.* **2014**, *3*. [CrossRef] [PubMed]
19. Morrill, J.C.; Carpenter, L.; Taylor, D.; Ramsburg, H.H.; Quance, J.; Peters, C.J. Further evaluation of a mutagen-attenuated Rift Valley fever vaccine in sheep. *Vaccine* **1991**, *9*, 35–41. [CrossRef]
20. Morrill, J.C.; Jennings, G.B.; Caplen, H.; Turell, M.J.; Johnson, A.J.; Peters, C.J. Pathogenicity and immunogenicity of a mutagen-attenuated Rift Valley fever virus immunogen in pregnant ewes. *Am. J. Vet. Res.* **1987**, *48*, 1042–1047. [PubMed]
21. Morrill, J.C.; Mebus, C.A.; Peters, C.J. Safety of a mutagen-attenuated Rift Valley fever virus vaccine in fetal and neonatal bovids. *Am. J. Vet. Res.* **1997**, *58*, 1110–1114. [PubMed]
22. Ikegami, T.; Hill, T.E.; Smith, J.K.; Zhang, L.; Juelich, T.L.; Gong, B.; Slack, O.A.; Ly, H.J.; Lokugamage, N.; Freiberg, A.N. Rift Valley fever virus MP-12 vaccine is fully attenuated by a combination of partial attenuations in the S, M, and L segments. *J. Virol.* **2015**, *89*, 7262–7276. [CrossRef] [PubMed]
23. Nishiyama, S.; Lokugamage, N.; Ikegami, T. The L-, M- and S-segments of Rift Valley fever virus MP-12 vaccine independently contribute to a temperature-sensitive phenotype. *J. Virol.* **2016**, *90*, 3735–3744. [CrossRef] [PubMed]
24. Miller, M.M.; Bennett, K.E.; Drolet, B.S.; Lindsay, R.; Mecham, J.O.; Reeves, W.K.; Weingartl, H.M.; Wilson, W.C. Evaluation of the efficacy, potential for vector transmission, and duration of immunity of MP-12, an attenuated Rift Valley fever virus vaccine candidate, in sheep. *Clin. Vaccine Immunol.* **2015**, *22*, 930–937. [CrossRef] [PubMed]
25. Won, S.; Ikegami, T.; Peters, C.J.; Makino, S. NSm and 78-kilodalton proteins of Rift Valley fever virus are nonessential for viral replication in cell culture. *J. Virol.* **2006**, *80*, 8274–8278. [CrossRef] [PubMed]
26. Gerrard, S.R.; Bird, B.H.; Albarino, C.G.; Nichol, S.T. The NSm proteins of Rift Valley fever virus are dispensable for maturation, replication and infection. *Virology* **2007**, *359*, 459–465. [CrossRef] [PubMed]
27. Terasaki, K.; Won, S.; Makino, S. The C-terminal region of Rift Valley fever virus NSm protein targets the protein to the mitochondrial outer membrane and exerts anti-apoptotic function. *J. Virol.* **2013**, *87*, 676–682. [CrossRef] [PubMed]
28. Gerrard, S.R.; Nichol, S.T. Synthesis, proteolytic processing and complex formation of N-terminally nested precursor proteins of the Rift Valley fever virus glycoproteins. *Virology* **2007**, *357*, 124–133. [CrossRef] [PubMed]
29. Tannous, B.A.; Kim, D.E.; Fernandez, J.L.; Weissleder, R.; Breakefield, X.O. Codon-optimized Gaussia luciferase cDNA for mammalian gene expression in culture and *in vivo*. *Mol. Ther.* **2005**, *11*, 435–443. [CrossRef] [PubMed]
30. Gerrard, S.R.; Nichol, S.T. Characterization of the Golgi retention motif of Rift Valley fever virus G(N) glycoprotein. *J. Virol.* **2002**, *76*, 12200–12210. [CrossRef] [PubMed]
31. Bar, S.; Rommelaere, J.; Nuesch, J.P. Vesicular transport of progeny parvovirus particles through ER and Golgi regulates maturation and cytolysis. *PLoS Pathog.* **2013**, *9*, e1003605. [CrossRef] [PubMed]

32. Ito, N.; Takayama-Ito, M.; Yamada, K.; Hosokawa, J.; Sugiyama, M.; Minamoto, N. Improved recovery of rabies virus from cloned cDNA using a vaccinia virus-free reverse genetics system. *Microb. Immunol.* **2003**, *47*, 613–617. [CrossRef]
33. Ikegami, T.; Won, S.; Peters, C.J.; Makino, S. Rescue of infectious Rift Valley fever virus entirely from cDNA, analysis of virus lacking the NSs gene, and expression of a foreign gene. *J. Virol.* **2006**, *80*, 2933–2940. [CrossRef] [PubMed]
34. Kalveram, B.; Lihoradova, O.; Indran, S.V.; Ikegami, T. Using reverse genetics to manipulate the NSs gene of the Rift Valley fever virus MP-12 strain to improve vaccine safety and efficacy. *J. Vis. Exp.* **2011**. [CrossRef] [PubMed]
35. Kalveram, B.; Lihoradova, O.; Indran, S.V.; Lokugamage, N.; Head, J.A.; Ikegami, T. Rift Valley fever virus NSs inhibits host transcription independently of the degradation of dsRNA-dependent protein kinase PKR. *Virology* **2013**, *435*, 415–424. [CrossRef] [PubMed]
36. *Adobe Photoshop Elements*; version 7.0 for Windows; Adobe Systems Incorporated: San Jose, CA, USA, 2008.
37. *GraphPad Prism*; version 6.05 for Windows; GraphPad Software, Inc.: La Jolla, CA, USA, 2014.
38. Spik, K.; Shurtleff, A.; McElroy, A.K.; Guttieri, M.C.; Hooper, J.W.; SchmalJohn, C. Immunogenicity of combination DNA vaccines for Rift Valley fever virus, tick-borne encephalitis virus, Hantaan virus, and Crimean Congo hemorrhagic fever virus. *Vaccine* **2006**, *24*, 4657–4666. [CrossRef] [PubMed]
39. Morrill, J.C.; Mebus, C.A.; Peters, C.J. Safety and efficacy of a mutagen-attenuated Rift Valley fever virus vaccine in cattle. *Am. J. Vet. Res.* **1997**, *58*, 1104–1109. [PubMed]
40. Morrill, J.C.; Laughlin, R.C.; Lokugamage, N.; Pugh, R.; Sbrana, E.; Weise, W.J.; Adams, L.G.; Makino, S.; Peters, C.J. Safety and immunogenicity of recombinant Rift Valley fever MP-12 vaccine candidates in sheep. *Vaccine* **2013**, *31*, 559–565. [CrossRef] [PubMed]
41. Morrill, J.C.; Laughlin, R.C.; Lokugamage, N.; Wu, J.; Pugh, R.; Kanani, P.; Adams, L.G.; Makino, S.; Peters, C.J. Immunogenicity of a recombinant Rift Valley fever MP-12-NSm deletion vaccine candidate in calves. *Vaccine* **2013**, *31*, 4988–4994. [CrossRef] [PubMed]
42. Huiskonen, J.T.; Overby, A.K.; Weber, F.; Grunewald, K. Electron cryo-microscopy and single-particle averaging of Rift Valley fever virus: Evidence for GN-GC glycoprotein heterodimers. *J. Virol.* **2009**, *83*, 3762–3769. [CrossRef] [PubMed]
43. Rusu, M.; Bonneau, R.; Holbrook, M.R.; Watowich, S.J.; Birmanns, S.; Wriggers, W.; Freiberg, A.N. An assembly model of Rift Valley fever virus. *Front Microb.* **2012**, *3*. [CrossRef] [PubMed]
44. Fontana, J.; Lopez-Montero, N.; Elliott, R.M.; Fernandez, J.J.; Risco, C. The unique architecture of Bunyamwera virus factories around the Golgi complex. *Genom. Comput. Sci. Virus Res.* **2008**, *10*, 2012–2028. [CrossRef] [PubMed]

© 2016 by the authors. Licensee MDPI, Basel, Switzerland. This article is an open access article distributed under the terms and conditions of the Creative Commons Attribution (CC BY) license (http://creativecommons.org/licenses/by/4.0/).

Article

N-Glycans on the Rift Valley Fever Virus Envelope Glycoproteins Gn and Gc Redundantly Support Viral Infection via DC-SIGN

Inaia Phoenix [1], Shoko Nishiyama [1], Nandadeva Lokugamage [1], Terence E. Hill [1,†], Matthew B. Huante [2], Olga A.L. Slack [1,‡], Victor H. Carpio [2], Alexander N. Freiberg [1,3,4,5] and Tetsuro Ikegami [1,3,4,*]

1 Department of Pathology, The University of Texas Medical Branch, Galveston, TX 77555, USA; inphoeni@utmb.edu (I.P.); shnishiy@utmb.edu (S.N.); nalokuga@utmb.edu (N.L.); tehill@utmb.edu (T.E.H.); olga.slack@novartis.com (O.A.L.S.); anfreibe@utmb.edu (A.N.F.)

2 Department of Microbiology and Immunology, The University of Texas Medical Branch, Galveston, TX 77555, USA; mbhuante@utmb.edu (M.B.H.); vhcarpio@utmb.edu (V.H.C.)

3 The Sealy Center for Vaccine Development, The University of Texas Medical Branch, Galveston, TX 77555, USA

4 The Center for Biodefense and Emerging Infectious Diseases, The University of Texas Medical Branch, Galveston, TX 77555, USA

5 Galveston National Laboratory, The University of Texas Medical Branch, Galveston, TX 77555, USA

* Correspondence: teikegam@utmb.edu; Tel.: +1-409-772-2563

† Present address: Applied Research Associates, Alexandria, VA 22314-4576, USA.

‡ Present address: Novartis Institute for Biomedical Research, Cambridge, MA 02139, USA.

Academic Editors: Jane Tao and Pierre-Yves Lozach
Received: 8 March 2016; Accepted: 20 May 2016; Published: 23 May 2016

Abstract: Rift Valley fever is a mosquito-transmitted, zoonotic disease that infects humans and ruminants. Dendritic cell specific intercellular adhesion molecule 3 (ICAM-3) grabbing non-integrin (DC-SIGN) acts as a receptor for members of the phlebovirus genus. The Rift Valley fever virus (RVFV) glycoproteins (Gn/Gc) encode five putative N-glycan sequons (asparagine (N)–any amino acid (X)–serine (S)/threonine (T)) at positions: N438 (Gn), and N794, N829, N1035, and N1077 (Gc). The *N*-glycosylation profile and significance in viral infection via DC-SIGN have not been elucidated. Gc *N*-glycosylation was first evaluated by using Gc asparagine (N) to glutamine (Q) mutants. Subsequently, we generated a series of recombinant RVFV MP-12 strain mutants, which encode N-to-Q mutations, and the infectivity of each mutant in Jurkat cells stably expressing DC-SIGN was evaluated. Results showed that Gc N794, N1035, and N1077 were *N*-glycosylated but N829 was not. Gc N1077 was heterogeneously *N*-glycosylated. RVFV Gc made two distinct *N*-glycoforms: "Gc-large" and "Gc-small", and N1077 was responsible for "Gc-large" band. RVFV showed increased infection of cells expressing DC-SIGN compared to cells lacking DC-SIGN. Infection via DC-SIGN was increased in the presence of either Gn N438 or Gc N1077. Our study showed that *N*-glycans on the Gc and Gn surface glycoproteins redundantly support RVFV infection via DC-SIGN.

Keywords: Rift Valley fever virus; *N*-glycosylation; Gn; Gc; sequon; DC-SIGN; L-SIGN

---

1. Introduction

Rift Valley fever (RVF) is a mosquito-borne, viral disease endemic to Africa and is characterized by high rates of abortion, fetal deformities, and high rates of newborn mortality, particularly in sheep, goats, and cattle [1]. Humans can be infected through close contact with the body fluids of infected animals or from the bites of infected mosquitoes [2]. Human RVF is typically characterized by a self-limiting febrile illness (e.g., biphasic fever, severe headaches, muscle pain, or nausea). However, some patients may develop more severe disease, such as lethal hemorrhagic fever, neurologic disorders, or blindness [3]. RVF is caused by Rift Valley fever virus (RVFV), which belongs to the genus *Phlebovirus* of the family *Bunyaviridae*. Because of the major social and economic impacts of the disease in both public health and agriculture, RVFV is classified as a Category A Priority Pathogen by National Institutes of Health (NIH)/National Institute of Allergy and Infectious Diseases (NIAID) and also listed as an overlap select agent by the United States Department of Health and Human Services (HHS) and the United States Department of Agriculture (USDA), which could pose a severe threat to public health and agriculture [4]. Currently, there is no approved, effective treatment for patients with RVFV infection, and the development of an antiviral treatment is of great importance to improve the prognosis of patients.

RVFV is a three-segmented, negative-stranded RNA virus, and the genome consists of the Large (L), Medium (M), and Small (S) segments [5]. The L-segment encodes the L protein, which is a RNA-dependent RNA polymerase. The S-segment encodes nucleoprotein (N) and a nonstructural protein (NSs), in an ambi-sense manner. The M-segment encodes the 78-kD protein, NSm, Gn (537 amino acids (aa), and Gc (507 aa) [6]. The 78-kD and NSm are accessory proteins and dispensable for viral replication [7,8]. The NSm protein is a minor virulence factor that delays apoptosis in infected cells [8–10]. The 78-kD protein plays a role in viral dissemination in mosquito vectors [11], and is incorporated into virions from infected mosquito C6/36 cells but not from Vero cells [12].

As shown in Figure 1A, six N-X-S or N-X-T sequons are encoded by the RVFV M-segment, where N is asparagine, X is any amino acid, S is serine, and T is threonine: *i.e.*, N88 (78-kD), N438 (78-kD and Gn), N794 (Gc), N829 (Gc), N1035 (Gc), and N1077 (Gc). The sites N88 and N438 are *N*-glycosylated in the 78-kD protein, and N438 is *N*-glycosylated in the Gn protein [13,14]; however, NSm is not *N*-glycosylated [14]. A previous study indicated that the Gc protein is *N*-glycosylated at three sites [13]. The recently solved crystal structure of Gc revealed architectural similarity with the envelope proteins of flaviviruses and alphaviruses, and it is categorized as a class II fusion protein. Similar to flaviviruses and alphaviruses [15], RVFV Gc consists of three domains: domain I (aa. 691–759, 852–901 and 981–1024), domain II (759–852, 901–981), and domain III (1024–1120) [16]. The fusion loop is located within domain II (aa. 820–830), which includes the N829 sequon. In the crystal, Gc forms head-to-tail homodimers, in which the fusion loop is buried by the domain III. An assembly model of Gn and Gc indicated that domain II, along with the fusion loop, may potentially be covered by the Gn surface protein [17].

The C-type lectin, dendritic cell specific intercellular adhesion molecule 3 (ICAM-3) grabbing non-integrin (DC-SIGN), was identified as a receptor for RVFV [18]. On the other hand, liver/lymph node-specific ICAM-3-grabbing non-integrin (L-SIGN or DC-SIGNR), a homologue of DC-SIGN, does not support the infectivity of RVFV or La Crosse virus (genus *Orthobunyavirus*) [19]. L-SIGN and DC-SIGN share 77% of their amino acid identity, but they are expressed on different cell types [20]. DC-SIGN is expressed on dendritic cells and some macrophages, whereas L-SIGN is expressed on the endothelial cells of liver and lymph node sinuses, as well as the endothelial cells lining capillaries of the placenta [20]. Because the *N*-glycans on the envelope glycoproteins (Gn and Gc) are ligands for the C-type lectin, they are expected to play an important role in RVFV infection via DC-SIGN.

The usage of each Gc sequon for *N*-glycosylation has not been determined. Furthermore, the requirement of RVFV Gn/Gc *N*-glycosylation for viral infection via DC-SIGN is not known. Using mutagenesis and reverse genetics, we aimed to identify the *N*-glycosylation sites utilized by RVFV Gc among the four potential sites (N794, N829, N1035, and N1077), and determine their individual role in viral infection via DC-SIGN.

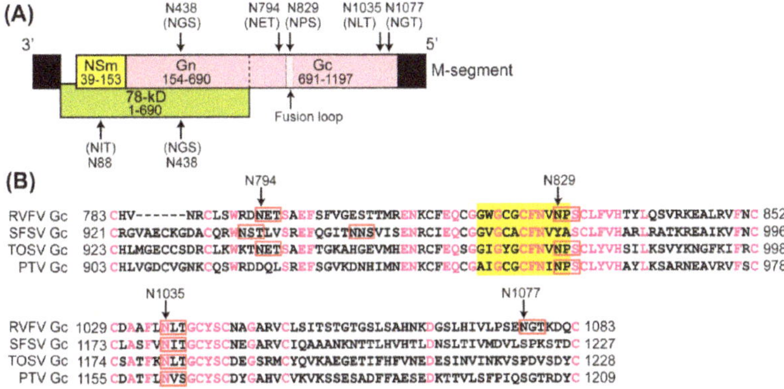

Figure 1. The asparagine (N)-any amino acid (X)-serine (S)/threonine (T) sequons of Rift Valley fever virus (RVFV) and other related phleboviruses. (**A**) Schematic diagram of the M-segment RNA and protein expressions. The RVFV medium (M)-segment encodes a single M mRNA, and co-translational cleavage and leaky ribosomal scanning of the initiation codons produce at least four proteins: 78-kD, nonstructural protein m (NSm) and glycoproteins Gn, and Gc. The Gn, and Gc are structural proteins, and the 78-kD protein is structural when virions are made in mosquito cells. There are six N-X-S/T sequons, which are potentially utilized for *N*-glycosylation: N88 (78-kD), N438 (Gn and 78-kD), N794 (Gc), N829 (Gc), N1035 (Gc), and N1077 (Gc); (**B**) Partial alignment of Gc amino acid sequences among RVFV, Sandfly fever Sicilian virus (SFSV: Genbank Accession No. U30500), Toscana virus (TOSV: Genbank Accession No. X89628), and Punta Toro virus (PTV: Genbank Accession No. DQ363407). Conserved amino acids are shown in pink, while the N-X-S/T sequons are shown in red squares. Positions of amino acids (aa.) at the amino (N)- and carboxyl (C)-termini of sequences are also shown. The fusion loop at aa. 820 to 830 is shown in yellow.

2. Materials and Methods

2.1. Media, Cells, and Viruses

Minimum Essential Medium (MEM)-alpha supplemented with 10% fetal bovine serum (FBS) (Life Technologies, Carlsbad, CA, USA), penicillin (100 U/mL), streptomycin (100 µg/mL), and hygromycin B (600 µg/mL) was used to maintain BHK/T7-9 cells that express the T7 RNA polymerase [21]. Vero E6 cells (ATCC C1008) were grown in Dulbecco's Modified Eagle Medium (DMEM) supplemented with 10% FBS, penicillin (100 U/mL), and streptomycin (100 µg/mL). A Jurkat parental cell line and Jurkat cells stably expressing human DC-SIGN (Jurkat-DC-SIGN) or L-SIGN (Jurkat-L-SIGN) were kindly provided by Dr. Rafael Delgado (Molecular Microbiology Laboratory, Hospital Universitario 12 de Octubre, Madrid, Spain) [22]. Roswell Park Memorial Institute (RPMI) media supplemented with 10% FBS, penicillin (100 U/mL), and streptomycin (100 µg/mL) was used to grow the Jurkat cells, Jurkat-DC-SIGN, and Jurkat-L-SIGN cells. The RVFV vaccine strain MP-12 was derived from two passages of MP-12 lot 7 vaccine [23] in Vero cells. Recombinant RVFV MP-12 (rMP-12) and the corresponding *N*-glycosylation mutants (N438Q, N794Q, N1035Q, and N1077Q) were recovered by reverse genetics [24]. Plaque assay using Vero E6 cells was performed to determine viral titers [25].

2.2. Plasmids

To generate rMP-12 mutants encoding either N438Q, N794Q, N1035Q, N1077Q, N438Q/N729Q, N438Q/N1035Q or N438Q/N1077Q, we constructed seven plasmids: pProT7-vM(+)N438Q, pProT7-vM(+)N794Q, pProT7-vM(+)N1035Q, pProT7-vM(+)N1077Q, pProT7-vM(+)N438Q/N794Q, pProT7-vM(+)N438Q/N1035Q or pProT7-vM(+)N438Q/N1077Q by site-directed mutagenesis using PfuUltra High-Fidelity DNA polymerase (Agilent Technologies, La Jolla, CA, USA) according to the manufacturer's instructions. The plasmids encode a point non-synonymous mutation (N to Q) at the indicated asparagine position (aa.1 represents the first methionine of M-segment open reading frame). The presence of individual mutations was confirmed by sequencing. To analyze the *N*-glycosylation of Gc, we modified pCAGGS-vG to encode all of N438Q, N729Q, N829Q, N1035Q, and N1077Q mutations (pCAGGS-vG-Gly-null). Then, one of those sequons was encoded in the following plasmids: pCAGGS-vG-N438(+), pCAGGS-vG-N794(+), pCAGGS-vG-N829(+), pCAGGS-vG-N1035(+), and pCAGGS-vG-N1077(+).

2.3. Precipitation of Gn/Gc by Concanavalin A Beads

Human embryonic kidney 293 cells (5×10^6 cells) were transfected with pCAGGS-vG or the mutants (2 μg) by TransIT-293 (Mirus Bio LLC., Madison, WI, USA) according to the manufacturer's instructions. At 48 h post transfection, cells were washed once with PBS, and harvested in RIPA buffer (150 mM NaCl, 50 mM Tris-HCl, 1% NP-40, 0.5% sodium deoxycholate, 0.1% sodium dodecyl sulfate) containing 1 mM of $CaCl_2$, $MgCl_2$, and $MnCl_2$. Samples were mixed with 20 μL of agarose bound concanavalin A (AL-1003, Vector Laboratories, Burlingame, CA, USA), and further incubated at 4 °C for 16 h. After washing 3 times with phosphate buffered saline (PBS) with $CaCl_2$ and $MgCl_2$, each sample was re-suspended in denaturing buffer, and heated at 95 °C for 5 min. Samples were either not treated, or treated with 1000 units of PNGase F or Endo H (New England BioLabs, Ipswich, MA, USA) for 24 h at 37 °C. Then, samples were boiled in 2× sodium dodecyl sulfate (SDS) sample buffer and separated in 12% SDS-polyacrylamide gel electrophoresis (PAGE).

2.4. Western Blotting

Samples were analyzed by SDS-PAGE under reducing conditions. Western blot analysis was performed, as described previously [26]. Anti-RVFV Gn mouse monoclonal antibody (4D4) or anti-RVFV Gc rabbit polyclonal antibody (CAT#4521, ProSci, Inc., Poway, CA, USA) were used for the detection of RVFV Gn or Gc, respectively.

2.5. Radiolabeling of Virus Particles

Vero E6 cells were infected with rMP-12 or the mutants at a multiplicity of infection (MOI) of 0.1 to 3 at 37 °C. Cells were incubated for 30 min at 37 °C with MEM lacking methionine/cysteine and L-glutamine, which was complemented with 1% dialyzed FBS, 20 mM L-glutamine, penicillin (100 U/mL) and streptomycin (100 μg/mL) before the [^{35}S] labeling of polypeptides. Trans [^{35}S] label metabolic reagent (MP Biomedicals, Santa Ana, CA, USA) was added to infected cells at 1 hour post infection (hpi) (single asparagine mutant experiment) or 4 hpi (double asparagine mutant experiment) to radiolabel virus particles. At 16 hpi, supernatant was harvested and clarified by low-speed centrifugation [4800 revolutions per minute (rpm) for 5 min]. Virus particles were then immunoprecipitated using anti-RVFV antibody, and washed four times in PBS. Samples were then re-suspended in 2× sample buffer containing 5% mercaptoethanol, and boiled for 10 min at 100 °C. Then, Gn and Gc mobility was analyzed by SDS-PAGE and subsequent autoradiography.

2.6. Infectivity of rMP-12 or the N-Glycan Mutants in Jurkat-DC-SIGN or Jurkat-L-SIGN Cells

Jurkat, or Jurkat-DC-SIGN cells (1×10^6 cells) were mock-infected or infected with 6.3×10^6 RNA copies of rMP-12 or the mutants (MOI = 3.6). After infection, cells were incubated at 37 °C for 6 h. Cells were then fixed with 4% paraformaldehyde for 30 min at 4 °C, followed by washing with PBS. Then, cells were permeabilized with permeabilization buffer (Affimetrix eBioScience, San Diego, CA, USA) at 4 °C for 25 min. Then, cells were incubated with anti-RVFV mouse ascite and Alexa Fluor 488-conjugated anti-green fluorescent protein (GFP) rabbit antibody (Life Technologies, Carlsbad, CA, USA) diluted in permeabilization buffer at 4 °C for 40 min. As an antibody control for GFP detection, Alexa Fluor 488-conjugated normal rabbit immunoglobulin (Ig)G (EMD Millipore, Billerica, MA, USA) was used. Cells were washed twice with permeabilization buffer, and the cells were incubated at 4 °C for 40 min with Alexa Fluor 647-conjugated goat anti-mouse IgG (Life Technologies). Permeabilization buffer was used to wash cells three times, and then cells were resuspended in fluorescence-activated cell sorting (FACS) buffer. Cells were analyzed by flow cytometry on the Canto or LSRII Fortessa (BD Biosciences, San Jose, CA, USA) in the UTMB Flow Cytometry and Cell Sorting Core Facility using FACSDiva software (version 8.0.1, BD Biosciences) and analyzed in FlowJo version 9.7 (TreeStar, Ashland, OR, USA).

2.7. Statistical Analysis

The statistical analyses were performed using the GraphPad Prism version 6.05 for Windows (GraphPad Software Inc., La Jolla, CA, USA). The unpaired Student's *t* test was used for the comparison of two groups.

2.8. Ethics Statement

All the recombinant DNA and RVFV were created upon the approval of the Notification of Use by the Institutional Biosafety Committee at UTMB.

3. Results

3.1. RVFV Gc N829 Is N-P-S Sequon and Is Located at Fusion Loop

As shown in Figure 1A, RVFV encodes six distinct N-X-S or N-X-T sequons in the M-segment: N88 (78-kD: N-I-T), N438 (78-kD/Gn: N-G-S), N794 (Gc: N-E-T), N829 (Gc: N-P-S), N1035 (Gc: N-L-T), and N1077 (Gc: N-G-T). The proline (P) at the X-site does not grant access of the oligosaccharyltransferase (OST) to the asparagine, and thus, N-P-S/T sequons cannot be *N*-glycosylated [27–29]. The N-P-S sequon at RVFV Gc N829 is located in the fusion loop, and an alignment with closely related phleboviruses revealed the N-P-S sequon was also found in TOSV and PTV (Figure 1B). On the other hand, the RVFV Gc N1035 could be aligned with SFSV (N1180), TOSV (N1180), and PTV (N1161), while the sequon corresponding to RVFV N794 was present in SFSV, and TOSV Gc, but not in PTV Gc. The sequon corresponding to N1077 was found in neither SFSV, TOSV, nor PTV. These results indicate the evolutional conservation of N794 and N1035 among RVFV, Sicilian, Naples serocomplexes, and the N1077 sequon was uniquely found in RVFV.

3.2. RVFV Gc N794, N1035, and N1077, but Not N829, Are N-Glycosylated

To determine the *N*-glycosylation status of each sequon encoded by RVFV Gc, we generated plasmids encoding RVFV M-segment open reading frame (ORF) lacking all sequons (pCAGGS-vG-Gly-null), or encoding only one sequon (Figure 2A). Human embryonic kidney 293 cells were transfected with parental pCAGGS-vG, or the mutant: pCAGGS-vG-Gly-null, pCAGGS-vG-N438(+), pCAGGS-vG-N794(+), pCAGGS-vG-N829(+), pCAGGS-vG-N1035(+), or pCAGGS-vG-N1077(+). At 48 h post transfection, cell lysates were harvested, and incubated with concanavalin A agarose beads. After washing, precipitates were subjected to Western blot analysis using anti-Gn or anti-Gc antibody (Figure 2B). *N*-glycans are classified into three types, based on the sugar structure attached to the common core sugar structure of three mannoses (Man) and two *N*-acetyl-glucosamines (GlucNAc) linked to the asparagine residue: (a) high mannose: only mannoses are attached to the core; (b) hybrid: the Manα1-6 arm is attached to only mannoses, and the Manα1-3 arm is attached to "antennae" (GlucNAc); and (c) complex: two "antennae" (GlucNAc) are attached via the Manα1-3 and Manα1-6 arms. To evaluate the presence of *N*-glycosylation, paired precipitated samples were treated with either PNGase F (cleaves between the innermost GlcNAc and the asparagine of high mannose-type, hybrid-type, and complex-type *N*-glycans) or Endo H (cleaves between two GlucNAc in the core of high mannose-type and some hybrid-type *N*-glycans) [30]. Parental Gn and Gc showed increased migration after PNGase F or Endo H enzyme treatments, confirming that RVFV Gn and Gc are *N*-glycosylated. As expected, the Gn and Gc lacking all sequons (*N*-Gly null) did not show migration changes after either enzyme treatment. N438(+) Gn reacted to both PNGase F and Endo H treatment, and the bands showed increased migration. N794(+) and N1035(+) Gc also reacted to both PNGase F and Endo H treatments, and bands showed increased migration compared to non-treated samples. N829(+) showed a similar migration pattern to *N*-Gly null, and no change in band migration was observed after either enzyme treatment. Interestingly, N1077(+) showed doublet bands, and the slower migrating band was not present after either PNGase F or Endo H treatment. Our results demonstrated that Gn N438, and Gc N794, N1035, and N1077 are *N*-glycosylated, and all are high mannose-type (or hybrid-type) *N*-glycans.

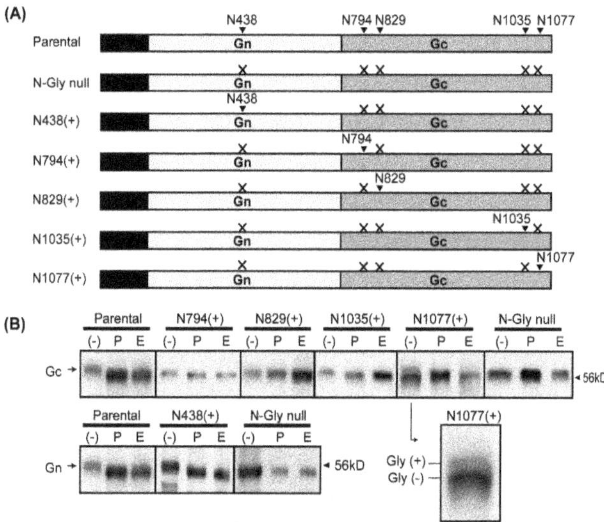

Figure 2. Mapping of *N*-glycosylation sites for RVFV Gc. (**A**) Schematics of the Medium (M)-segment open reading frame (ORF) encoded in the following plasmids: Parental (pCAGGS-vG) , N-Gly null (pCAGGS-vG-Gly-null), N438(+) (pCAGGS-vG-N438(+)), N794(+) (pCAGGS-vG-N794(+)), N829(+) (pCAGGS-vG-N829(+)), N1035(+) (pCAGGS-vG-N1035(+)), or N1077(+) (pCAGGS-vG-N1077(+)). N-to-Q mutations were made in the X-sites, and the remaining single asparagine site is shown with an arrowhead; (**B**) 293 cells were transfected with plasmids shown in A. Then, Gn/Gc were precipitated with concanavalin agarose beads, and subsequently detected by anti-Gc antibody (top panels) or anti-Gn antibody (bottom panels). Samples were either not treated (-), or treated with Peptide-N-Glycosidase F (PNGase F) (P) or Endoglycosidase H (E). An image of untreated N1077(+) lane is enlarged to show both glycosylated (Gly(+)) and unglycosylated (Gly(−)) bands.

3.3. Generation of Recombinant RVFV Encoding N-to-Q Substitutions at One or Two N-Glycan Sequons

To understand the role of the individual *N*-glycans of RVFV Gn/Gc, we replaced the asparagine (N) of each sequon with a glutamine (Q), and generated recombinant RVFV MP-12 strains, encoding N794Q, N1035Q, N1077Q, N438Q/N794Q, N438Q/N1035Q, or N438Q/N1077Q (Figure 3A). The N-to-Q substitution was used to minimize any structural changes due to mutagenesis because glutamine, H2N-CO-(CH2)2-CH(NH2)-COOH, is a polar and neutral amino acid, which is similar to asparagine, H2N-CO-(CH2)-CH(NH2)-COOH. Next, we analyzed migrations of Gn and Gc bands using immunoprecipitated radiolabeled virions. As expected, all rMP-12 mutants encoding N438Q (e.g., N438Q, N438Q/N794Q, N438Q/N1035Q, and N438Q/N1077Q) showed a fast migrating Gn band (Figure 3A,B). Gc displayed as doublet bands (Gc-large and Gc-small), while Gc-large was not present in N1077Q, N438Q/N1035Q, or N438Q/N1077Q. Since heterogeneous N-glycosylation occurs at N1077 (Figure 2B), Gc-large for N1035Q could be *N*-glycosylated at N794 and N1077, and Gc-large for N794Q could be N-glycosylated at N1035 and N1077. Though the N438Q/N1035Q mutant could be also N-glycosylated at N794 and N1077, we could not detect a distinct Gc-large band (Figure 1B). All single or double mutants replicated efficiently in Vero cells and showed similar growth kinetics to parental rMP-12 (MOI = 0.01) (Figure 3C). We also tried to rescue rMP-12 encoding N1035Q/N1077Q, but no viable viruses could be recovered after three independent attempts. The rMP-12 encoding N829Q was viable and replicated efficiently in Vero cells (data not shown), but was not further studied because N829 was not utilized for N-glycosylation (Figure 2B).

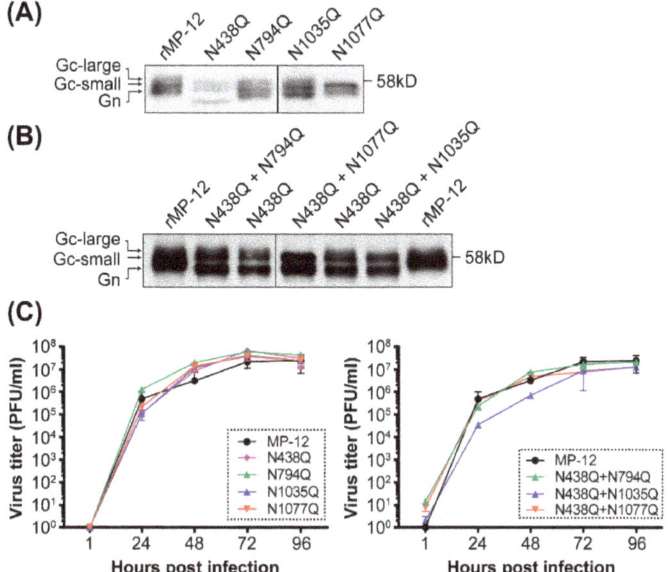

Figure 3. Generation of recombinant MP-12 encoding asparagine (N)-to-glutamine (Q) mutation(s) in Gn/Gc sequons. The Gn/Gc migration patterns of rMP-12, or that encoding N-to-Q mutation either in Gn or Gc (N438Q, N794Q, N1035Q, and N1077Q mutants) (**A**) or that encoding N-to-Q mutations both in Gn and Gc (N438Q/N794Q, N438Q/N1035Q, and N438Q/N1077Q mutants) (**B**). Vero E6 cells were infected with rMP-12 or the mutants at a multiplicity of infection (MOI) of 0.1 to 3, and metabolically labeled with [^{35}S] methionine/cysteine from 1 to 16 hours post infection (hpi). The cleared culture supernatants were subjected to immunoprecipitation using anti-Rift Valley fever virus (RVFV) antibody. Precipitated virions were analyzed by 7.5% sodium dodecyl sulfate- polyacrylamide gel electrophoresis (SDS-PAGE) and autoradiography. Gc-large: slow migrating Gc; Gc-small: fast migrating Gc; (**C**) Virus growth kinetics in Vero cells. Vero cells were infected with indicated rMP-12 mutants at an MOI of 0.01. Virus titers were determined at 1, 24, 48, 72 and 96 hpi. Means +/− standard deviations of three independent experiments are shown.

3.4. RVFV Gn and Gc N-Glycans Redundantly Support Viral Infection via DC-SIGN

Next, we analyzed the function of Gn and Gc N-glycans during infection of host target cells. A previous study indicated that a C-type lectin, DC-SIGN (CD209), which is expressed on dendritic cells and macrophages, functions as a receptor for phleboviruses [18]. Jurkat cells stably expressing human DC-SIGN [22] that also co-express GFP in up to 17% of the population (Figures S1 and S2) were used to determine infectivity via the lectin [22]. Since GFP and DC-SIGN were overall co-expressed, we analyzed cell populations expressing GFP to analyze RVFV infection via DC-SIGN. Though Jurkat-DC-SIGN cells expressed intrinsic GFP signals, the signal was very weak in the flow cytometry data analysis (Figure 4A, left top panel). Therefore, anti-GFP antibody was used to detect a population expressing GFP (Figure 4A, center and right top panels). To determine the relative infection efficiency of rMP-12, Jurkat-DC-SIGN cells were either mock-infected or infected with rMP-12 at an MOI of 3.6 (measured in Vero E6 cells). At 6 hpi, cells were fixed and permeabilized, and we analyzed the infectivity of virus in GFP-positive and in GFP-negative cells.

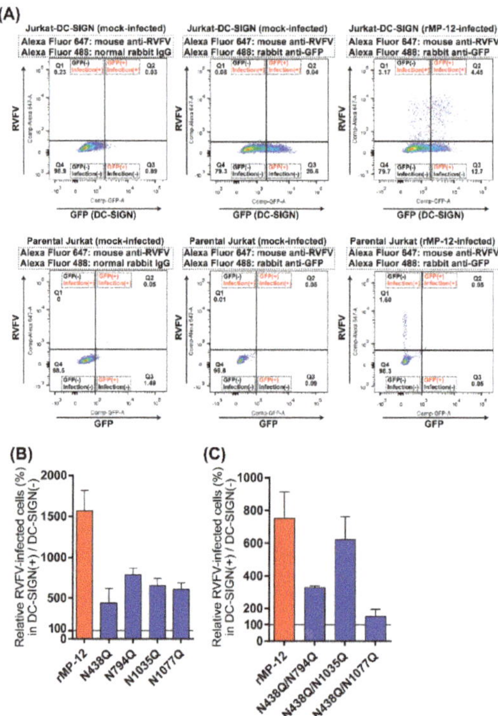

Figure 4. Infectivity of recombinant MP-12 encoding a glutamine (Q) in place of an asparagine (N) at N-X-S/T sequon(s) in Gn or Gc. (**A**) Jurkat-DC-SIGN cells stably co-expressing green fluorescent protein (GFP) and human DC-SIGN (~17% in cell population) (top panels), or parental Jurkat cells, which express neither DC-SIGN nor GFP (bottom panels), were mock-infected (left and center panels) or infected with rMP-12 (right panels) at a multiplicity of infection (MOI) of 3.6. At 6 hpi, cells were fixed, permeabilized, and then stained with a cocktail of mouse anti-RVFV antibody and Alexa Fluor 488-conjugated rabbit anti-GFP antibody, or a cocktail of mouse anti-RVFV antibody and Alexa Fluor 488-conjugated normal rabbit IgG. Subsequently, cells were stained with Alexa Fluor 647-conjugated goat anti-mouse IgG, and analyzed by flow cytometry. Since permeabilized Jurkat-DC-SIGN cells showed poor intrinsic GFP signals, as shown with Alexa Fluor 488-conjugated normal rabbit IgG (left top panel), rabbit anti-GFP antibody (center and right top panels) was used to detect GFP signals of permeabilized Jurkat-DC-SIGN cells. Q1, GFP-negative (DC-SIGN-negative) and RVFV-infected cell population; Q2, GFP-positive (DC-SIGN-positive) and RVFV-infected cell population; Q3, GFP-positive (DC-SIGN-positive) and uninfected cell population; Q4, GFP-negative (DC-SIGN-negative) and uninfected cell population; (**B**,**C**) Jurkat-DC-SIGN cells were infected with rMP-12 or that lacking one (**B**) or two sequons (**C**). Relative number of RVFV-infected cells in the GFP-positive cell population (Q2/(Q2+Q3)) normalized to that of RVFV-infected cells in the GFP-negative cell population (Q1/(Q1+Q4)) are shown. Graphs represent mean + standard deviations for three independent experiments.

Both Jurkat-DC-SIGN and parental Jurkat cells could be infected with rMP-12 (Figure 4A, right top and right bottom panels). Thus, it was important to analyze viral infection via DC-SIGN, over the nonspecific viral infection in Jurkat cells. To analyze viral infection via DC-SIGN, relative number of RVFV-infected cells in the GFP-positive cell population (Q2/(Q2+Q3)) was normalized to that of RVFV-infected cells in the GFP-negative cell population (Q1/(Q1+Q4)) (background level infectivity in Jurkat cells: 100%). This value most probably represents the infectivity of rMP-12 via DC-SIGN, over the nonspecific infectivity independent of DC-SIGN. Using this approach, we measured the infectivity of rMP-12 lacking one or two sequons, as shown in Figure 4B,C.

The rMP-12-positive cells increased more in DC-SIGN-positive cells (7.5 to 15-fold) compared to cells not expressing DC-SIGN (Figure 4B,C). All single asparagine mutants (N438Q, N794Q, N1035Q, or N1077Q) still showed an increased infection in DC-SIGN-positive cells (N438Q: 4.4-fold, N794Q: 7.9-fold, N1035Q: 6.5-fold, and N1077Q: 6.1-fold). Double asparagine mutants, N438Q/N794Q (3.3-fold) and N438Q/N1035Q (6.2-fold) still had an increased infection in DC-SIGN-positive cells, while N438Q/N1077Q (1.5-fold) no longer showed enhanced infectivity in DC-SIGN-positive cells. Taken together, it is indicated that either Gn N438, Gc N794 + N1077, or Gn N1035 + N1077 can increase viral infection via DC-SIGN. Thus, Gn and Gc N-glycans redundantly support virus infection via DC-SIGN.

4. Discussion

Modification of proteins by *N*-glycosylation occurs co-translationally in the rough endoplasmic reticulum (ER) by the en bloc transfer of an oligosaccharide *N*-glycan precursor onto the N-X-S/T sequons in the nascent polypeptide by the OST complex. Subsequently, the *N*-glycans are processed (e.g., trimming or addition of sugar residues) in the ER and the Golgi by glycosidases and glycosyltransferases. In this study, we determined that three (N794, N1035, and N1077) out of four Gc sequons are *N*-glycosylated. The presence of proline at the X-site (N-X-S/T) of the sequons inhibits *N*-glycosylation [27–29], and, the N-P-S sequon at RVFV Gc (N829) was not *N*-glycosylated. Our results also indicated that RVFV Gn and Gc *N*-glycans are high mannose or hybrid-type *N*-glycans, susceptible to Endo H treatment. *N*-glycosylation is catalyzed by either the STT3A or STT3B subunit of the OST complex of mammalian cells [31]. In general, *N*-glycosylation of nascent polypeptides occurs co-translationally by the STT3A-OST-complex, while posttranslational *N*-glycosylation is less common [32]. STT3B-mediated posttranslational glycosylation occurs near the C-terminal of unfolded proteins when the sequon is skipped by the STT3A-OST complex [31]. We observed that *N*-glycosylation of RVFV Gc at N1077 occurs only partially, which leads to the generation of Gc-large and Gc-small. Heterogeneous *N*-glycosylation of Gc suggests that N-glycosylation at N1077, which is located at 121 amino acids upstream of the C-terminus, occurs post-translationally.

We analyzed viral infection via DC-SIGN, using Jurkat-DC-SIGN cells. The Jurkat-DC-SIGN cells expressed GFP in up to 17% of cell population, and GFP-positive cells were overall co-expressed with DC-SIGN (Figure S2). In the samples prepared at an early stage of infection (6 hpi), we compared the number of infected cells in GFP-positive population, and GFP-negative population. MP-12 infection occurred preferably in GFP-positive population, indicating that DC-SIGN acts as a receptor for RVFV [18]. The N438Q/N1077Q mutant no longer retained enhanced infectivity via DC-SIGN, while the N438Q, N794Q, N1035Q, N1077Q, N438Q/N794Q, and N438Q/N1035Q mutants showed an increased viral infection via DC-SIGN. Thus, our results indicated that Gn N438 or Gc N1077 play an important role in viral infection via DC-SIGN, and that Gn and Gc N-glycans redundantly support virus infection via DC-SIGN.

It should be noted that recombinant RVFV lacking one or more sequons may have unpredictable effects on the usage of other sequons. We observed reduced infectivity of the N438Q or N1035Q mutants in Jurkat-DC-SIGN cells, while the N438Q/N1035Q double mutation had little impact on viral infection in those cells. In addition, the N438Q/N1035Q mutant did not show a distinct Gc-large band, unlike the N438Q/N794Q mutant. Since the Gc-large for the N438Q/N1035Q mutant could be *N*-glycosylated at N794 and N1077, the *N*-glycosylation status either at N794 or N1077 may be altered in the N438Q/N1035Q mutant. Effects of N-terminal *N*-glycosylation on C-terminal sites have been studied in the rabies virus glycoprotein (GP) [33]. The rabies virus GP is type I membrane glycoprotein and encodes three sequons: N37, N247, and N319. The GP lacking N37 and N319 was largely susceptible to Endo H, while intact GP or GP lacking only N37 were resistant to Endo H. Thus, the presence of an *N*-glycan at one site can affect the processing of an *N*-glycan at another site. Another study showed that the insertion of a new sequon (N58) in human plasminogen activator influences the processing of the *N*-glycosylation at N117 [34]. While this is beyond the scope of the current study, determination of carbohydrate chains at each asparagine residue will elucidate this mechanism of *N*-glycosylation modification during Gn/Gc maturation of each mutant.

We also analyzed the infectivity of MP-12 and the mutants in Jurkat cells expressing L-SIGN (Jurkat-L-SIGN cells) (Figure S3) [22]. However, we observed only a 1.2-fold augmentation of MP-12 infection via L-SIGN, in contrast to the 7.5–15-fold augmentation of infectivity via DC-SIGN. This result is consistent with the previous study by Hofmann *et al.* that demonstrated mosquito-borne RVFV or La Crosse virus (genus *Orthobunyavirus*) specifically utilize DC-SIGN, but not L-SIGN, while tick-borne severe fever with thrombocytopenia syndrome virus (SFTSV; genus *Phlebovirus*) uses both DC-SIGN and L-SIGN for entry [19]. Both DC-SIGN and L-SIGN are homotetrameric type II membrane proteins and retain 77% amino acid identity [20]. L-SIGN selectively binds to the trisaccharide Manα1-3(Manα1-6)Manα1 on high mannose glycans, while DC-SIGN binds to high mannose glycans (preferably with eight or nine mannoses) or fucose-containing structures including the Lewis-X trisaccharide: *i.e.*, Galβ1-4(Fucα1-3)GlcNAc [35–37]. Though both DC-SIGN and L-SIGN bind to high mannose-type *N*-glycans, the pH-dependent release of the oligosaccharide ligand by L-SIGN is not as efficient as DC-SIGN [37], which might explain the poor infectivity of RVFV via L-SIGN.

Though the infectivity was not high, parental Jurkat cells, which do not express those C-type lectins, could be also infected with MP-12. It was shown that RVFV entry is inhibited in Chinese hamster ovary (CHO) cells pgs-745 mutant (deficient in glycosaminoglycan synthesis) and the pgsD-677 mutant (deficient in synthesis of heparin sulfate: HS), or in CHO cells pretreated with heparinases [38]. Thus, HS also plays a role in RVFV entry. Since Jurkat cells synthesize HS [39], MP-12 infection of parental Jurkat cells is most likely mediated by HS. Indeed, in another study where DC-SIGN was expressed in Raji cells, a B-cell lymphoma cell line deficient in HS synthesis [40], RVFV infection was supported [18], indicating that RVFV entry via DC-SIGN does not require HS. In our study, an increased MP-12 infection occurred in Jurkat-DC-SIGN cells in the presence of both DC-SIGN and HS. However, further study is required to understand if the co-expression of DC-SIGN and HS synergically facilitates the entry of RVFV.

We also noted that RVFV Gn/Gc lacking all *N*-glycans could be still expressed without showing unstable characteristics. The N-to-Q mutation of Bunyamwera virus (genus *Orthobunyavirus*) Gn N60 resulted in the loss of immunoreactivity with an anti-Gc monoclonal antibody [41]. Further, the N-to-Q mutation of Hantaan virus (genus *Hantavirus*) Gn N134 resulted in poor accumulation of Gn and poor immunoreactivity to anti-Gc monoclonal antibodies [42]. Thus, RVFV *N*-glycans might be dispensable for protein stability. On the other hand, rMP-12 encoding N1035Q/N1077Q, N438Q/N794Q/N829Q/N1035Q/N1077Q, or N794Q/N829Q/N1035Q/N1077Q were not rescued successfully. Thus, *N*-glycans may play a role in combination to form a functional Gn/Gc complex for viral assembly.

In addition to Gn and Gc, RVFV also encodes 78 kD proteins, which are incorporated into virions matured from mosquito cells, but not those from mammalian cells [12]. Though the 78 kD protein shares the amino acid sequence with Gn, including the N438 sequon, it makes a distinct structure from the Gn and does not function as a precursor for Gn production [43,44]. The N-terminus encodes the N88 sequon, which is unique to 78 kD protein. A lack of 78 kD affects viral dissemination in mosquitoes [11,45,46], and it may have a distinct role from Gn and Gc in viral entry mechanism. Future studies involving the N-glycosylation of 78 kD and its potential role in viral entry will prove valuable in further elucidating the function of this protein.

5. Conclusions

We demonstrated the presence of *N*-glycans in Gn (N438) and Gc (N794, N1035, and N1077). RVFV Gc consists of two distinct *N*-glycoforms (Gc-large and Gc-small), due to heterogeneous *N*-glycosylation at N1077. We found that RVFV infection via DC-SIGN occurs in a redundant manner through Gn and Gc, and that *N*-glycans at Gn N438 and Gc N1077 play an important role in viral infection via DC-SIGN. Our study will support a better understanding of the post-translational *N*-glycan modification of Gn/Gc and its role in progeny infection.

Supplementary Materials: The following are available online at www.mdpi.com/1999-4915/8/5/149/s1, Figure S1: Co-expression of GFP and DC-SIGN in Jurkat-DC-SIGN cells, Figure S2: Populations of cells co-expressing GFP and DC-SIGN in Jurkat-DC-SIGN cells, Figure S3: Infectivity of recombinant MP-12 encoding a glutamine (Q) in place of an asparagine (N) at N-X-S/T sequon(s) in Gn or GC in Jurkat-L-SIGN cells.

Acknowledgments: We thank Robert. B. Tesh at the University of Texas Medical Branch at Galveston (UTMB) for the mouse anti-RVFV antibody, David. A. Norwood at the United States Army Medical Research Institute for Infectious Diseases for the 4D4 monoclonal antibody for RVFV Gn, and Rafael Delgado (Molecular Microbiology Laboratory, Hospital Universitario 12 de Octubre, Madrid, Spain) for the Jurkat, Jurkat-DC-SIGN, and Jurkat-L-SIGN cells. We also thank Birte Kalveram for technical suggestions of FACS analysis, and Mark Griffin (the UTMB Flow Cytometry Core) for all his help in analyzing FACS samples. This work was supported by the National Institute of Health (grant number R01 AI087643-01A1) (Tetsuro Ikegami and Alexander N. Freiberg); the James W. McLaughlin predoctoral fellowship (Inaia Phoenix); the funding from the Sealy Center for Vaccine Development at the UTMB (Tetsuro Ikegami); Training in Emerging Infectious Diseases and Biodefense grant from the National Institute of Health (grant number T32 AI007536) (Terence E. Hill); and a postdoctoral fellowship through the James W. McLaughlin Fellowship Fund at the UTMB (Terence E. Hill).

Author Contributions: Tetsuro Ikegami and Alexander N. Freiberg conceived and designed the experiments; Inaia Phoenix, Shoko Nishiyama, Nandadeva Lokugamage, Terence E. Hill, Olga A. L. Slack, and Matthew B. Huante performed the experiments; Inaia Phoenix, Victor H. Carpio, and Tetsuro Ikegami analyzed the data; Inaia Phoenix and Tetsuro Ikegami wrote the paper.

Conflicts of Interest: The authors declare no conflict of interest.

Abbreviations

The following abbreviations are used in this manuscript:

BHK	Baby hamster kidney
DC-SIGN	Dendritic cell specific ICAM-3 grabbing non-integrin
EDTA	Ethylenediaminetetraacetic acid
FACS	Fluorescence-activating cell sorting
FBS	Fetal bovine serum
G	Glycine
GFP	Green fluorescent protein
HHS	U.S. Department of Health and Human Services
I	Isoleucine
L	Leucine
L-segment	Large-segment
L-SIGN	Liver/lymph node-specific ICAM-3-grabbing non-integrin
M-segment	Medium-segment
MOI	Multiplicity of infection
N	Asparagine
NIAID	National Institute of Allergy and Infectious Diseases
NIH	National Institutes of Health
OST	Oligosaccharyltransferase
P	Proline
PBS	Phosphate buffered saline
PTV	Punta Toro virus

Q	Glutamine
RIPA	Radioimmunoprecipitation assay buffer
RVF	Rift Valley fever
RVFV	Rift Valley fever virus
S	Serine
SDS-PAGE	Sodium dodecyl sulfate-polyacrylamide gel electrophoresis
SFSV	Sandfly fever Sicilian virus
SFTSV	Severe Fever with Thrombocytopenia Syndrome virus
S-segment	Small-segment
STT	Subunit of the oligosaccharyltransferase complex
T	Threonine
TOSV	Toscana virus
USDA	U.S. Department of Agriculture

References

1. Swanepoel, R.; Coetzer, J.A.W. Rift Valley fever. In *Infectious Diseases of Livestock with Special Reference to Southern Africa*, 2nd ed.; Coetzer, J.A.W., Tustin, R.C., Eds.; Oxford University Press: Cape Town, South Africa, 2004; pp. 1037–1070.
2. Pepin, M.; Bouloy, M.; Bird, B.H.; Kemp, A.; Paweska, J. Rift Valley fever virus (Bunyaviridae: Phlebovirus): An update on pathogenesis, molecular epidemiology, vectors, diagnostics and prevention. *Vet. Res.* **2010**, *41*, 61. [CrossRef] [PubMed]
3. Ikegami, T.; Makino, S. The Pathogenesis of Rift Valley Fever. *Viruses* **2011**, *3*, 493–519. [CrossRef] [PubMed]
4. Bird, B.H.; Ksiazek, T.G.; Nichol, S.T.; Maclachlan, N.J. Rift Valley fever virus. *J. Am. Vet. Med. Assoc.* **2009**, *234*, 883–893. [CrossRef] [PubMed]
5. Schmaljohn, C.; Nichol, S.T. Bunyaviridae. In *Fields Virology*, 5th ed.; Knipe, D.M., Howley, P.M., Griffin, D.E., Lamb, R.A., Martin, M.A., Roizman, B., Straus, S.E., Eds.; Lippincott, Williams & Wilkins: Philadelphia, PA, USA, 2007; pp. 1741–1789.
6. Suzich, J.A.; Collett, M.S. Rift Valley fever virus M segment: Cell-free transcription and translation of virus-complementary RNA. *Virology* **1988**, *164*, 478–486. [CrossRef]
7. Won, S.; Ikegami, T.; Peters, C.J.; Makino, S. NSm and 78-kilodalton proteins of Rift Valley fever virus are nonessential for viral replication in cell culture. *J. Virol.* **2006**, *80*, 8274–8278. [CrossRef] [PubMed]
8. Gerrard, S.R.; Bird, B.H.; Albarino, C.G.; Nichol, S.T. The NSm proteins of Rift Valley fever virus are dispensable for maturation, replication and infection. *Virology* **2007**, *359*, 459–465. [CrossRef] [PubMed]
9. Won, S.; Ikegami, T.; Peters, C.J.; Makino, S. NSm protein of Rift Valley fever virus suppresses virus-induced apoptosis. *J. Virol.* **2007**, *81*, 13335–13345. [CrossRef] [PubMed]
10. Bird, B.H.; Albarino, C.G.; Nichol, S.T. Rift Valley fever virus lacking NSm proteins retains high virulence *in vivo* and may provide a model of human delayed onset neurologic disease. *Virology* **2007**, *362*, 10–15. [CrossRef] [PubMed]
11. Kreher, F.; Tamietti, C.; Gommet, C.; Guillemot, L.; Ermonval, M.; Failloux, A.B.; Panthier, J.J.; Bouloy, M.; Flamand, M. The Rift Valley fever accessory proteins NSm and P78/NSm-Gn are determinants of virus propagation in vertebrate and invertebrate hosts. *Emerg. Microbe Infect.* **2014**, *3*, e71. [CrossRef] [PubMed]
12. Weingartl, H.M.; Zhang, S.; Marszal, P.; McGreevy, A.; Burton, L.; Wilson, W.C. Rift Valley fever virus incorporates the 78 kDa glycoprotein into virions matured in mosquito C6/36 cells. *PLoS ONE* **2014**, *9*, e87385. [CrossRef] [PubMed]
13. Kakach, L.T.; Suzich, J.A.; Collett, M.S. Rift Valley fever virus M segment: Phlebovirus expression strategy and protein glycosylation. *Virology* **1989**, *170*, 505–510. [CrossRef]
14. Kakach, L.T.; Wasmoen, T.L.; Collett, M.S. Rift Valley fever virus M segment: Use of recombinant vaccinia viruses to study Phlebovirus gene expression. *J. Virol.* **1988**, *62*, 826–833. [PubMed]
15. Kielian, M. Class II virus membrane fusion proteins. *Virology* **2006**, *344*, 38–47. [CrossRef] [PubMed]
16. Dessau, M.; Modis, Y. Crystal structure of glycoprotein C from Rift Valley fever virus. *Proc. Natl. Acad. Sci. USA* **2013**, *110*, 1696–1701. [CrossRef] [PubMed]

17. Rusu, M.; Bonneau, R.; Holbrook, M.R.; Watowich, S.J.; Birmanns, S.; Wriggers, W.; Freiberg, A.N. An assembly model of rift valley Fever virus. *Front. Microbiol.* **2012**, *3*, 254. [CrossRef] [PubMed]
18. Lozach, P.Y.; Kuhbacher, A.; Meier, R.; Mancini, R.; Bitto, D.; Bouloy, M.; Helenius, A. DC-SIGN as a Receptor for Phleboviruses. *Cell Host Microbe* **2011**, *10*, 75–88. [CrossRef] [PubMed]
19. Hofmann, H.; Li, X.; Zhang, X.; Liu, W.; Kuhl, A.; Kaup, F.; Soldan, S.S.; Gonzalez-Scarano, F.; Weber, F.; He, Y.; *et al.* Severe fever with thrombocytopenia virus glycoproteins are targeted by neutralizing antibodies and can use DC-SIGN as a receptor for pH-dependent entry into human and animal cell lines. *J. Virol.* **2013**, *87*, 4384–4394. [CrossRef] [PubMed]
20. Soilleux, E.J.; Barten, R.; Trowsdale, J. DC-SIGN; a related gene, DC-SIGNR; and CD23 form a cluster on 19p13. *J. Immunol.* **2000**, *165*, 2937–2942. [CrossRef] [PubMed]
21. Ito, N.; Takayama-Ito, M.; Yamada, K.; Hosokawa, J.; Sugiyama, M.; Minamoto, N. Improved recovery of rabies virus from cloned cDNA using a vaccinia virus-free reverse genetics system. *Microbiol. Immunol.* **2003**, *47*, 613–617. [CrossRef] [PubMed]
22. Alvarez, C.P.; Lasala, F.; Carrillo, J.; Muniz, O.; Corbi, A.L.; Delgado, R. C-type lectins DC-SIGN and L-SIGN mediate cellular entry by Ebola virus in cis and in trans. *J. Virol.* **2002**, *76*, 6841–6844. [CrossRef] [PubMed]
23. Lokugamage, N.; Freiberg, A.N.; Morrill, J.C.; Ikegami, T. Genetic Subpopulations of Rift Valley Fever ZH548, MP-12 and Recombinant MP-12 Strains. *J. Virol.* **2012**, *86*, 13566–13575. [CrossRef] [PubMed]
24. Ikegami, T.; Won, S.; Peters, C.J.; Makino, S. Rescue of infectious rift valley fever virus entirely from cDNA, analysis of virus lacking the NSs gene, and expression of a foreign gene. *J. Virol.* **2006**, *80*, 2933–2940. [CrossRef] [PubMed]
25. Kalveram, B.; Lihoradova, O.; Indran, S.V.; Ikegami, T. Using reverse genetics to manipulate the NSs gene of the Rift Valley fever virus MP-12 strain to improve vaccine safety and efficacy. *J. Vis. Exp.* **2011**, e3400. [CrossRef] [PubMed]
26. Kalveram, B.; Lihoradova, O.; Indran, S.V.; Lokugamage, N.; Head, J.A.; Ikegami, T. Rift Valley fever virus NSs inhibits host transcription independently of the degradation of dsRNA-dependent protein kinase PKR. *Virology* **2013**, *435*, 415–424. [CrossRef] [PubMed]
27. Atkinson, P.H.; Lee, J.T. Co-translational excision of alpha-glucose and alpha-mannose in nascent vesicular stomatitis virus G protein. *J. Cell Biol.* **1984**, *98*, 2245–2249. [CrossRef] [PubMed]
28. Kornfeld, R.; Kornfeld, S. Assembly of asparagine-linked oligosaccharides. *Annu. Rev. Biochem.* **1985**, *54*, 631–664. [CrossRef] [PubMed]
29. Ben-Dor, S.; Esterman, N.; Rubin, E.; Sharon, N. Biases and complex patterns in the residues flanking protein N-glycosylation sites. *Glycobiology* **2004**, *14*, 95–101. [CrossRef] [PubMed]
30. Trimble, R.B.; Tarentino, A.L. Identification of distinct endoglycosidase (endo) activities in Flavobacterium meningosepticum: Endo F1, endo F2, and endo F3. Endo F1 and endo H hydrolyze only high mannose and hybrid glycans. *J. Biol. Chem.* **1991**, *266*, 1646–1651. [PubMed]
31. Ruiz-Canada, C.; Kelleher, D.J.; Gilmore, R. Cotranslational and posttranslational N-glycosylation of polypeptides by distinct mammalian OST isoforms. *Cell* **2009**, *136*, 272–283. [CrossRef] [PubMed]
32. Bolt, G.; Kristensen, C.; Steenstrup, T.D. Posttranslational N-glycosylation takes place during the normal processing of human coagulation factor VII. *Glycobiology* **2005**, *15*, 541–547. [CrossRef] [PubMed]
33. Wojczyk, B.S.; Takahashi, N.; Levy, M.T.; Andrews, D.W.; Abrams, W.R.; Wunner, W.H.; Spitalnik, S.L. *N*-glycosylation at one rabies virus glycoprotein sequon influences N-glycan processing at a distant sequon on the same molecule. *Glycobiology* **2005**, *15*, 655–666. [CrossRef] [PubMed]
34. Pfeiffer, G.; Strube, K.H.; Schmidt, M.; Geyer, R. Glycosylation of two recombinant human uterine tissue plasminogen activator variants carrying an additional N-glycosylation site in the epidermal-growth-factor-like domain. *Eur. J. Biochem.* **1994**, *219*, 331–348. [CrossRef] [PubMed]
35. Van Liempt, E.; Bank, C.M.; Mehta, P.; Garcia-Vallejo, J.J.; Kawar, Z.S.; Geyer, R.; Alvarez, R.A.; Cummings, R.D.; Kooyk, Y.; van Die, I. Specificity of DC-SIGN for mannose- and fucose-containing glycans. *FEBS Lett.* **2006**, *580*, 6123–6131. [CrossRef] [PubMed]
36. Davis, C.W.; Nguyen, H.Y.; Hanna, S.L.; Sanchez, M.D.; Doms, R.W.; Pierson, T.C. West Nile virus discriminates between DC-SIGN and DC-SIGNR for cellular attachment and infection. *J. Virol.* **2006**, *80*, 1290–1301. [CrossRef] [PubMed]

37. Guo, Y.; Feinberg, H.; Conroy, E.; Mitchell, D.A.; Alvarez, R.; Blixt, O.; Taylor, M.E.; Weis, W.I.; Drickamer, K. Structural basis for distinct ligand-binding and targeting properties of the receptors DC-SIGN and DC-SIGNR. *Nat. Struct. Mol. Biol.* **2004**, *11*, 591–598. [CrossRef] [PubMed]
38. De Boer, S.M.; Kortekaas, J.; de Haan, C.A.; Rottier, P.J.; Moormann, R.J.; Bosch, B.J. Heparan sulfate facilitates Rift Valley fever virus entry into the cell. *J. Virol.* **2012**, *86*, 13767–13771. [CrossRef] [PubMed]
39. Kaur, S.; Kuznetsova, S.A.; Pendrak, M.L.; Sipes, J.M.; Romeo, M.J.; Li, Z.; Zhang, L.; Roberts, D.D. Heparan sulfate modification of the transmembrane receptor CD47 is necessary for inhibition of T cell receptor signaling by thrombospondin-1. *J. Biol. Chem.* **2011**, *286*, 14991–15002. [CrossRef] [PubMed]
40. De Jong, M.A.; de Witte, L.; Bolmstedt, A.; van Kooyk, Y.; Geijtenbeek, T.B. Dendritic cells mediate herpes simplex virus infection and transmission through the C-type lectin DC-SIGN. *J. Gen. Virol.* **2008**, *89*, 2398–2409. [CrossRef] [PubMed]
41. Shi, X.; Brauburger, K.; Elliott, R.M. Role of N-linked glycans on bunyamwera virus glycoproteins in intracellular trafficking, protein folding, and virus infectivity. *J. Virol.* **2005**, *79*, 13725–13734. [CrossRef] [PubMed]
42. Shi, X.; Elliott, R.M. Analysis of N-linked glycosylation of hantaan virus glycoproteins and the role of oligosaccharide side chains in protein folding and intracellular trafficking. *J. Virol.* **2004**, *78*, 5414–5422. [CrossRef] [PubMed]
43. Suzich, J.A.; Kakach, L.T.; Collett, M.S. Expression strategy of a phlebovirus: Biogenesis of proteins from the Rift Valley fever virus M segment. *J. Virol.* **1990**, *64*, 1549–1555. [PubMed]
44. Gerrard, S.R.; Nichol, S.T. Synthesis, proteolytic processing and complex formation of N-terminally nested precursor proteins of the Rift Valley fever virus glycoproteins. *Virology* **2007**, *357*, 124–133. [CrossRef] [PubMed]
45. Crabtree, M.B.; Kent Crockett, R.J.; Bird, B.H.; Nichol, S.T.; Erickson, B.R.; Biggerstaff, B.J.; Horiuchi, K.; Miller, B.R. Infection and transmission of Rift Valley fever viruses lacking the NSs and/or NSm genes in mosquitoes: Potential role for NSm in mosquito infection. *PLoS Negl. Trop. Dis.* **2012**, *6*, e1639. [CrossRef] [PubMed]
46. Kading, R.C.; Crabtree, M.B.; Bird, B.H.; Nichol, S.T.; Erickson, B.R.; Horiuchi, K.; Biggerstaff, B.J.; Miller, B.R. Deletion of the NSm virulence gene of Rift Valley fever virus inhibits virus replication in and dissemination from the midgut of Aedes aegypti mosquitoes. *PLoS Negl. Trop. Dis.* **2014**, *8*, e2670. [CrossRef] [PubMed]

© 2016 by the authors. Licensee MDPI, Basel, Switzerland. This article is an open access article distributed under the terms and conditions of the Creative Commons Attribution (CC BY) license (http://creativecommons.org/licenses/by/4.0/).

Review

The Role of Phlebovirus Glycoproteins in Viral Entry, Assembly and Release

Martin Spiegel [1,2,*], Teresa Plegge [1] and Stefan Pöhlmann [1,*]

1 Infection Biology Unit, Deutsches Primatenzentrum, Kellnerweg 4, Göttingen 37077, Germany; TPlegge@dpz.eu
2 Institute for Microbiology and Virology, Medizinische Hochschule Brandenburg Theodor Fontane, Grossenhainer Str. 57, Senftenberg 01968, Germany
* Correspondence: MSpiegel@dpz.eu (M.S.); SPoehlmann@dpz.eu (S.P.); Tel.: +49-551-3851-415 (M.S.); +49-551-3851-150 (S.P.)

Academic Editors: Jane Tao and Pierre-Yves Lozach
Received: 20 April 2016; Accepted: 14 July 2016; Published: 21 July 2016

Abstract: Bunyaviruses are enveloped viruses with a tripartite RNA genome that can pose a serious threat to animal and human health. Members of the *Phlebovirus* genus of the family *Bunyaviridae* are transmitted by mosquitos and ticks to humans and include highly pathogenic agents like Rift Valley fever virus (RVFV) and severe fever with thrombocytopenia syndrome virus (SFTSV) as well as viruses that do not cause disease in humans, like *Uukuniemi* virus (UUKV). Phleboviruses and other bunyaviruses use their envelope proteins, Gn and Gc, for entry into target cells and for assembly of progeny particles in infected cells. Thus, binding of Gn and Gc to cell surface factors promotes viral attachment and uptake into cells and exposure to endosomal low pH induces Gc-driven fusion of the viral and the vesicle membranes. Moreover, Gn and Gc facilitate virion incorporation of the viral genome via their intracellular domains and Gn and Gc interactions allow the formation of a highly ordered glycoprotein lattice on the virion surface. Studies conducted in the last decade provided important insights into the configuration of phlebovirus Gn and Gc proteins in the viral membrane, the cellular factors used by phleboviruses for entry and the mechanisms employed by phlebovirus Gc proteins for membrane fusion. Here, we will review our knowledge on the glycoprotein biogenesis and the role of Gn and Gc proteins in the phlebovirus replication cycle.

Keywords: Bunyaviridae; phlebovirus; glycoproteins; virus attachment; entry; membrane fusion; signal peptidase; assembly

---

1. Introduction

The family *Bunyaviridae* comprises over 350 viruses, which infect diverse animals, insects, and plants. Five *Bunyavirus* genera have been identified: *Orthobunyavirus*, *Hantavirus*, *Nairovirus*, *Phlebovirus* and *Tospovirus* based on serologic, morphologic and biochemical criteria [1]. Viruses within the *Orthobunyavirus*, *Nairovirus* and *Phlebovirus* genera are transmitted to animal hosts by arthropod vectors, such as ticks, mosquitoes, midges, and flies during blood meals [2]. Tospoviruses also employ arthropods and thrips for spread but infect plants [3]. In contrast, hantaviruses infect rodents, bats, shrews, and moles [4–10] and are transmitted to humans upon exposure to aerosolized rodent excreta [2,11]. Several bunyaviruses cause severe disease, including hemorrhagic fevers in humans, and are teratogenic in animals. In addition, many bunyaviruses are "emerging", since disease incidence and geographical distribution are increasing. Thus, bunyaviruses can pose a significant threat to human health and understanding how these viruses replicate, spread, and cause disease is required to identify targets for intervention.

Bunyaviruses are enveloped viruses which harbor a tripartite, single stranded RNA genome with negative polarity. The L-segment of the genome encodes for the viral polymerase (L), the M-segment for the viral glycoproteins, Gn and Gc, and the S-segment for the nucleocapsid (N) protein [12]. In addition, non-structural proteins can be encoded by the S- and M-segment, employing either an ambisense coding strategy, overlapping open reading frames or an open reading frame (ORF) encoding a polyprotein. The glycoproteins mediate the first step in the bunyavirus replication cycle—viral entry into host cells— and are the only targets for neutralizing antibodies. Gn and Gc are synthesized as a precursor protein, Gn/Gc, in the secretory pathway of infected cells. Gn and Gc are separated by proteolytic cleavage but may remain non-covalently associated [13,14]. The cleavage step is executed by a cellular enzyme, signal peptidase [15–17], during import of the Gn/Gc precursor into the endoplasmic reticulum (ER). In the ER, Gn and Gc are decorated with *N*-linked glycans [18,19] of the high-mannose type, which can be processed into hybrid and complex forms upon import of Gn and Gc into the Golgi apparatus [18–21]. The Golgi apparatus is the site of bunyavirus budding [22–26] and this process is facilitated by Gn and Gc, which play a key role in particle morphogenesis and genome incorporation [27–31]. Finally, infectious particles decorated with Gn and Gc are released from the infected cell by exocytosis.

Despite their important role in bunyavirus entry and release, biogenesis and biological activities of bunyavirus Gn and Gc proteins are incompletely understood. In the present manuscript, we will review our knowledge on phlebovirus glycoproteins. The genus *Phlebovirus* (*Phlebotominae*, sandflies) currently contains 10 species, with *Rift Valley fever virus* (*RVFV*) being the type species, and the viruses grouped into the sandfly fever virus (SFV) and the Uukuniemi virus (UUKV) groups, depending on their vector species. Several phleboviruses are important human pathogens: RVFV causes severe diseases in ruminants and humans in Africa and the Middle East [32] while severe fever with thrombocytopenia syndrome virus (SFTSV) was discovered as a novel agent responsible for cases of severe fever in Asia, which may take a fatal course particularly in elderly patients [33,34]. In contrast, UUKV is not pathogenic in humans. For further information on phlebovirus biology and disease in general, the reader is referred to recent reviews [35,36]. Here, we will discuss how the glycoproteins of phleboviruses are generated and how they promote virus entry and release. For this, we will describe the role of the glycoproteins at different stages of the viral replication cycle, starting with their configuration in the envelope of infectious particles, followed by their function during viral entry, their biogenesis in infected cells and finally their roles during assembly and release of progeny phlebovirus particles (see Figure 1).

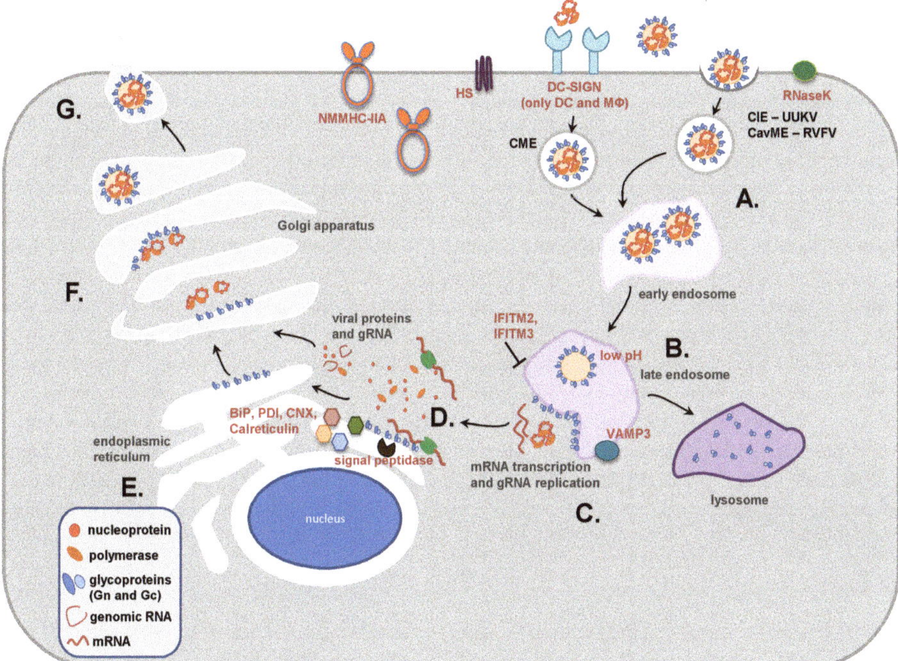

Figure 1. Replication cycle of phleboviruses. (**A**) Cellular attachment of phleboviruses is driven by glycoprotein interactions with host cell factors such as dendritic cell-specific intercellular adhesion molecule-3-grabbing non-integrin (DC-SIGN), heparan sulfate (HS), or non-muscle myosin heavy chain IIA (NMMHC-IIA). The binding to DC-SIGN and so far unknown entry factors induces uptake via caveolin-1-mediated endocytosis (CavME) (as for Rift Valley fever virus, RVFV) or incompletely defined clathrin-independent endocytic (CIE) mechanisms (as for Uukuniemi virus, UUKV). Ribonuclease kappa (RNaseK) promotes the internalization of virions by a yet unknown mechanism; (**B**) In late endosomes, the low pH induces the membrane fusion activity of the Gc protein. Expression of vesicle-associated membrane protein 3 (VAMP3) promotes UUKV penetration, while interferon-induced transmembrane protein (IFITM) 2 and IFITM3 inhibit the fusion of RVFV in late endosomes; (**C**) The fusion of viral and endosomal membranes allows release of the viral ribonucleoprotein complexes into the cytoplasm, the site of viral transcription and replication; (**D**) The viral glycoproteins Gn and Gc are translated at the rough endoplasmic reticulum (ER) as a precursor protein, Gn/Gc, which is cleaved by signal peptidase. The viral nucleoprotein and the viral polymerase are synthesized in the cytoplasm where they form together with newly produced genomic RNA (gRNA) ribonucleoprotein (RNP) complexes; (**E**) Binding immunoglobulin protein (BiP) and calnexin, two ER chaperones, are required for appropriate folding of Gn and Gc. Similarly, protein-disulfide-isomerase catalyzes Gn and Gc folding by promoting the formation of disulfide bonds, while calreticulin prevents misfolded Gn and Gc from being exported from the ER to the Golgi; (**F**) Correctly folded Gn/Gc heterodimers are transported into the Golgi apparatus where they associate with RNPs via the cytoplasmic tails of Gn during the budding process; (**G**) After budding of new virus particles into the Golgi is complete, virus-containing vesicles are transported to the plasma membrane where the virions are released by exocytosis. DC: dendritic cell; Mϕ: macrophage; CME: clathrin-mediated endocytosis; PDI: protein disulfide isomerase; CNX: calnexin.

2. Role of Gn and Gc in Phlebovirus Entry

2.1. Configuration of Gn and Gc Proteins in the Viral Envelope

Initial studies provided evidence that bunyavirus particles are pleomorphic [37]. It was therefore surprising that electron cryotomography revealed that both UUKV [38] and RVFV [39,40] particles display a spherical, highly ordered structure. The order is imposed by the configuration of Gn and Gc proteins in the viral envelope, which form an icosahedral lattice with a triangulation number of 12 [38–40]. The lattice is composed of 110 hexameric and 12 pentameric capsomers, and for RVFV it was proposed that the capsomers accommodate in total 720 Gn/Gc heterodimers [41,42], with Gn forming the capsomer spikes while Gc lies partially underneath, closer to the lipid membrane. The shape of the capsomers depends on the pH of the surrounding medium [38], since protonation triggers major conformational changes in Gc, which are associated with membrane fusion, as discussed below. Since the RVFV Gc ectodomain crystallizes as a dimer, an assembly model has been proposed for the RVFV envelope in which Gc dimers are oriented horizontally respective to the viral membrane [43]. In contrast, the virion interior does not display a particular organization, in keeping with the absence of a matrix protein in all bunyaviruses. Thus, the Gn and Gc proteins are presented in a highly ordered fashion on the virion surface. In the following paragraphs it will be discussed how these proteins mediate viral entry into target cells and cause assembly and budding of progeny particles in infected cells.

2.2. Attachment Factors and Receptors

Phlebovirus entry into cells commences with binding of particles to components of the plasma membrane. For the purpose of this discussion, we will define attachment factors as such plasma membrane components, which interact with viral glycoproteins and modulate entry efficiency but are ultimately dispensable for infectious entry. In contrast, cellular factors that bind to viral glycoproteins and are essential for entry will be termed receptors. For hantaviruses, a role for β1-3 integrins in host cell entry has been reported and integrin choice was found to correlate with viral pathogenicity [44–46]. These observations suggest that protein-protein interactions may orchestrate cellular entry of hantaviruses, although direct binding of Gn and/or Gc to integrins remains to be demonstrated [31]. In contrast, glycan-protein interactions seem to play a prominent role in phlebovirus entry, as discussed below.

2.3. Dendritic Cell-Specific Intercellular Adhesion Molecule-3-Grabbing Non-Integrin (DC-SIGN) Facilitates Phlebovirus Entry into DCs

DC-SIGN is a calcium-dependent lectin expressed on DCs, certain tissue macrophages, megakaryocytes, a subset of B-cells, and platelets [47]. DC-SIGN recognizes mannose and fucose residues on cellular ligands and several pathogens [48], including human immunodeficiency virus (HIV) [49] and mycobacterium [50,51], and tetramerization of DC-SIGN is required for avid ligand binding. In the context of phleboviruses, it was shown that DC-SIGN facilitates entry of UUKV, RVFV, Punta Toro virus (PTV) and Toscana virus (TOSV) [52]. Subsequent work showed that DC-SIGN also facilitates entry of vectors pseudotyped with SFTSV and La Crosse virus (LACV) glycoproteins [53]. Phleboviruses are transmitted by arthropod bites, hence, skin and tissue DCs are amongst the first cells encountered by these viruses, suggesting that DC-SIGN could be important for viral transmission. Indeed, initial studies on dengue virus (DENV), another arbovirus, showed that DC-SIGN promotes DENV infection [54,55], although a subsequent report demonstrated that DENV infection of human skin cells is DC-SIGN-independent [56]. Interactions of phleboviruses with DC-SIGN depend on *N*-glycans located on Gn and/or Gc and DC-SIGN expression was shown to be required for DC infection by UUKV [52] and for SFTSV Gn/Gc-mediated transduction of these cells [53]. In addition, DC-SIGN expression was sufficient to render cell lines susceptible to phlebovirus entry [52,53]. Thus, DC-SIGN is a bona fide phlebovirus receptor, which, due to endocytosis signals in its cytoplasmic tail [57], promotes uptake of phleboviruses into cells [52]. In early endosomes, the virions dissociate from DC-SIGN and continue the degradative pathway to late endosomes [52], the location of membrane fusion. It is noteworthy that many cell lines susceptible to phlebovirus infection do not express DC-SIGN [52], indicating that these viruses most likely also use other receptors for infectious entry. Other lectins like the DC-SIGN-related protein DC-SIGNR (L-SIGN) and LSECtin were shown to promote entry of several viruses [58–60] and might also augment phlebovirus infection of certain cells. Indeed, a recent report demonstrated that DC-SIGNR, which shares 77% amino acid sequence identity with DC-SIGN but is expressed on different cells (endothelial cells of liver and lymph nodes), can markedly augment phlebovirus entry into cell lines which are otherwise barely susceptible [61]. In contrast to DC-SIGN, L-SIGN mainly promotes viral attachment but not uptake into cells [61] in keeping with the established concept that DC-SIGN but not DC-SIGNR functions as an endocytic receptor [62]. Thus, it is conceivable that DC-SIGNR promotes phlebovirus entry by concentrating virions onto the cell surface, thereby increasing interactions with a so far unidentified receptor. The presence of such receptor(s) is strongly suggested by the broad cell tropism of several phleboviruses and the relatively narrow cell and tissue expression of the lectins discussed above. Finally, it is noteworthy that DC-SIGN might promote SFTSV pathogenesis independent of its function as a viral receptor: SFTSV was shown to associate with platelets [63], which are known to express DC-SIGN and to capture HIV and potentially other viruses in a DC-SIGN-dependent fashion [64]. Moreover, SFTSV-platelet complexes were found to be taken up into macrophages [63], suggesting that DC-SIGN-dependent SFTSV interactions with platelets could contribute to removal of platelets from the circulation and thus to thrombocytopenia, a hallmark of SFTS.

2.4. Heparan Sulfate (HS) Proteoglycans Promote Phlebovirus Attachment

HS is a glycosaminoglycan (GAG), an unbranched polysaccharide composed of disaccharide repeats, which can be linked to a protein via *O*-glycosylation, resulting in the formation of a proteoglycan. Several viruses engage HS for entry into target cells. Analysis of cell lines with defined glycosylation defects revealed that HS, but not complex *N*-glycans, is required for efficient cellular entry of RVFV [65]. This observation was confirmed by enzymatic removal of HS and competition experiments with heparin. Moreover, evidence was obtained that *O*-sulfation of HS is essential for RVFV entry [65]. Viral interactions with HS are frequently charge-dependent and sequence analysis revealed clusters of basic amino acids on the P78 protein, which might interact with negatively charged sulfate groups on HS [65]. In contrast, potential HS binding sites on Gn or Gc were not identified. The P78 protein is one out of four translation products of the M genomic segment of RVFV and its translation efficiency seems to be cell line-dependent. While P78 is quite abundant in RVFV-infected insect cells, mammalian cells produce only small amounts of P78 [66]. As a consequence, purified RVFV virions derived from Vero E6 cells did not contain detectable amounts of P78 [66]. Moreover, P78 is dispensable for RVFV virulence in mice [67]. Therefore, the HS binding sites on RVFV produced in mammalian cells await further investigation. However, it is noteworthy that P78 protein is efficiently incorporated into RVFV produced in mosquito cells [66] and is required for viral dissemination in mosquitos [67]. Whether HS binding accounts for the important role of P78 in viral spread in mosquitos remains to be elucidated. A role of HS in RVFV entry was also identified within a screen of haploid cells for factors required for RVFV spread [68]. This study reported that HS-dependence for entry of RVFV did not result from cell culture adaptation since primary isolates were found to depend on HS for entry [68]. Moreover, the role of HS in RVFV entry was shown to be cell type-dependent and evidence was provided that Crimean Congo hemorrhagic fever virus (CCHFV) and Hantaan virus (HNTV) rely on HS for efficient entry while Andes virus (ANDV) does not [68]. Additionally, separate work showed that TOSV uses GAGs for efficient cell entry [69]. Whether GAGs serve as attachment factors or as receptors is unknown. However, the cell line dependence of the role of HS in RVFV entry in combination with the detection of residual infection in the absence of HS suggest that GAGs might serve as attachment factors rather than receptors.

2.5. Non-Muscle Myosin Heavy Chain IIA (NMMHC-IIA) Promotes SFTSV Entry

NMMHC-IIA is an actin binding motor protein that induces actin crosslinking and contraction and plays a role in cell migration, adhesion, and polarization [70]. Sun and colleagues showed that recombinant SFTSV-Gn bound to susceptible cell lines and identified NMMHC-IIA as a cellular binding partner of Gn [71]. Moreover, evidence was obtained that inhibition of NMMHC-IIA expression or blockade by antibodies reduces viral entry while directed expression can increase entry efficiency [71]. However, formal proof that directed expression of NMMHC-IIA renders otherwise entirely refractory cells susceptible to infectious SFTSV entry remains to be provided. Binding of SFTSV to cells augmented total expression of NMMHC-IIA and increased surface levels within minutes (the protein is normally localized in the cytoplasm), indicating that SFTSV manipulates NMMHC-IIA trafficking to ensure efficient entry [71]. Moreover, SFTSV might parasitize the documented role of NMMHC-IIA in endocytosis and phagocytosis to ensure its uptake into the cells [71]. Finally, it has been suggested that SFTSV interactions with NMMHC-IIA might directly contribute to viral pathogenesis [71]. Thus, point mutations in NMMHC-IIA were found to be associated with thrombocytopenia [72] and obstruction of normal NMMHC-IIA function by SFTSV might have similar effects. In addition, NMMHC-IIA, like DC-SIGN, might promote viral attachment to platelets followed by uptake and destruction of virus-platelet complexes by macrophages [71]. Collectively, NMMHC-IIA could play an important role in SFTSV entry, although evidence for a bona fide receptor function is still missing. It is noteworthy that other viruses also exploit NMMHC-IIA for cellular entry: NMMHC-IIA was identified as a receptor for herpes simplex virus type 1 (HSV-1) and HSV-1 attachment to cells was shown to increase surface levels of NMMHC-IIA [73].

2.6. Phlebovirus Uptake: Clathrin-Dependent and -Independent Mechanisms

A seminal study by Lozach and colleagues examined the steps ensuing receptor binding, uptake of virions into cells and membrane fusion. They could show that UUKV, upon attachment to plasma membrane indentations and filopodia, is taken up into the cell within minutes by a mechanism that is mainly independent of clathrin coats (clathrin-independent endocytosis, CIE) [74]. Internalized UUKV is then transported into early and late endosomes, where low pH triggers membrane fusion [74], as discussed below. Expression of vesicle-associated membrane protein 3 (VAMP3), which belongs to the vesicle synaptosome-associated protein receptor (v-SNARE) family of membrane proteins, was required for UUKV infection and virus particles were found in VAMP3-positive late endosomal compartments [75]. Thus, adequate intracellular transport of UUKV seems to depend on the documented role of VAMP3 in late vesicular trafficking events [76,77]. In addition, expression of histone deacetylase (HDAC) 8 was shown to be required for UUKV entry [78], likely because of its role in microtubule organization and endosomal maturation. These studies point towards an essential role of UUKV transport into late endosomes for infectious entry while the uptake mechanism requires further analysis. In this context, it should be noted that cell entry of RVFV depends on ribonuclease kappa (RNaseK) [79], which is also essential for the uptake of other pH-dependent viruses [79], and on caveolin-1-mediated endocytosis (CavME), while macropinocytosis and clathrin-mediated endocytosis (CME) do not play a role [80]. In contrast, vesicular stomatitis virus (VSV)-particles pseudotyped with SFTSV-Gn/Gc employ a clathrin-dependent mechanism for viral uptake, [53] and orthobunyaviruses also enter cells in a clathrin-dependent fashion [81,82]. Thus, different phleboviruses might use different uptake mechanisms for entry and potential strain and cell line-dependent differences remain to be investigated.

2.7. Virus-Cell Fusion and Its Inhibition

2.7.1. Characteristics of Viral Membrane Fusion Proteins

Successful transport of virions into host cell endosomes and exposure to endosomal low pH initiates the last sequence of the phlebovirus entry cascade: the fusion of the viral envelope with an endosomal membrane. Three classes of viral proteins that can fuse viruses with cells have been identified. Class I membrane fusion proteins are usually oriented perpendicular to the viral membrane and α-helices are their predominant structural elements. In contrast, class II membrane fusion proteins frequently exhibit a parallel orientation relative to the viral membrane and a high content of β-sheets. Finally, class III membrane fusion proteins unite characteristics of both class I and II membrane fusion proteins [83,84]. All viral membrane fusion proteins have in common that a trigger, usually low pH or receptor binding (or a combination thereof), induces the membrane fusion reaction, which is facilitated by marked conformational changes in the glycoproteins. First, a fusion peptide or an internal fusion loop is propelled towards the target cell membrane and inserted into the bilayer. Then, a back-folding reaction brings the N- and C-termini of the glycoproteins and thus viral and cellular membranes into close contact and ultimately facilitates membrane fusion, allowing delivery of the viral nucleic acid into the host cell cytoplasm [83,84].

2.7.2. RVFV Gc is a Class II Membrane Fusion Protein

A computational study examining bunyavirus glycoprotein sequences provided the first evidence that phlebovirus Gc proteins might be class II membrane fusion proteins. Thus, similarities were noted between the sequences of SFV Gc and the E1 protein of Sindbis virus (SINV) [85], a bona fide class II membrane fusion protein. Moreover, SFV Gc sequences potentially involved in membrane fusion were found to be conserved among bunyavirus Gc proteins [85]. Formal proof that phlebovirus Gc proteins are indeed class II membrane fusion proteins was provided by the elucidation of the structure of the ectodomain of RVFV Gc in the pre-fusion state. Gc was found to be organized into three domains with a fold characteristic of class II membrane fusion proteins [43]. An internal fusion loop was identified, a feature of all class II membrane fusion proteins, and the location of certain histidines in Gc suggested a role in pH sensing [43], as expected. Thus, protonation of histidines is known to trigger the membrane fusion reaction of many glycoproteins and histidine 1087 in RVFV Gc, which is required for infectivity [86], was located at the same site as histidines critical for triggering of other class II membrane fusion proteins by low pH [43]. Despite the apparent structural similarities between RVFV Gc and class II membrane fusion proteins, differences were noted. For instance, the interface between domains I and II in RVFV Gc is more extensive and potentially more rigid than that of other viral class II membrane fusion proteins. Moreover, RVFV Gc exhibits an increased number and altered localization of disulfide bridges as compared to other class II membrane fusion proteins [43]. These results suggest that phlebovirus Gc proteins might employ similar strategies as flavivirus E proteins and alphavirus E1 protein to facilitate membrane fusion, although subtle differences might exist.

2.7.3. Low pH Triggers Membrane Fusion

The results discussed above suggest that the membrane fusion activity of phlebovirus Gc proteins is triggered by low pH upon transport of virions into endolysosomes. Indeed, treatment of target cells with lysosomotropic agents, which elevate intravesicular pH, blocks phlebovirus entry [53,86]. Moreover, exposure of Gn/Gc-expressing cells [87] or virions to low pH is sufficient to trigger Gc [74], and the ensuing conformational changes are irreversible, since triggering in the absence of target cells abrogates virus infectivity [86]. At present, no evidence has been reported that Gc proteins must first bind to a receptor or undergo proteolytic activation for subsequent triggering by low pH, although one report suggested that the activity of serine proteases in target cells is required for efficient SFTSV Gn/Gc-driven entry [53]. However, it is noteworthy that a trypsin-sensitive structure on target cell membranes might be required to support RVFV Gc-driven membrane fusion [87] and phospholipids with negatively charged headgroups were found to promote UUKV Gc-driven fusion in a liposome-based assay [88], indicating that specific components of the target cell membrane can impact fusion efficiency.

2.7.4. Inhibition of Membrane Fusion by Interferon-Induced Transmembrane (IFITM) Proteins

The alteration of the biological properties of endolysosomal membranes is an innate defense against viral invasion. Thus, the IFITM 1–3 proteins are synthesized in response to viral invasion and block entry of several viral agents by modifying target cell membranes [89,90]. IFITM1 localizes at or close to the cell surface and blocks viruses from entering at these sites while IFITM2 and IFITM3 are found in endolysosomal compartments and inhibit viruses entering via these compartments [89]. In accordance with RVFV entry being dependent on endolysosomal low pH, expression of IFITM2 and IFITM3 was shown to block RVFV entry and more than half of the antiviral activity associated with IFNα treatment of target cells was found to be due to expression of these proteins [91]. How exactly IFITM proteins modulate membrane properties to inhibit viral entry is not clear, but alteration of membrane curvature and/or fluidity due to IFITM insertion and IFITM-IFITM interactions as possible mechanisms has been proposed [92,93].

3. Role of Gn and Gc in Phlebovirus Assembly

3.1. M Segment Coding Strategy and Expression of the Glycoproteins Gn and Gc

After fusion of viral and endosomal membranes the three viral genomic segments (L, M, and S) which are associated with the viral polymerase are released into the cytoplasm and primary transcription of negative-sense genomic RNA (gRNA) into mRNA is initiated [94]. Transcription and translation are tightly coupled, i.e., the translation of the viral proteins starts before the transcription of the mRNA is completed [95].

The two phlebovirus glycoproteins (like the glycoproteins of members of other *Bunyavirus* genera) are encoded on the M-segment in a single ORF [96–99]. They are synthesized as a precursor which is cotranslationally processed into the glycoproteins Gn and Gc [19,100–103]. The Gn/Gc precursor protein cannot be detected in phlebovirus-infected cells. Only after expression of M-segment-based plasmid constructs followed by pulse-chase immunoprecipitations, or after in vitro translation in the absence of microsomal membranes, does the precursor become visible [18,103,104]. In the presence of microsomal membranes, the precursor is rapidly cleaved, indicating cotranslational cleavage by a host factor during viral protein synthesis [103,104]. The host factor responsible for precursor cleavage is the signal peptidase complex located in the ER membrane [16,17].

Due to a signal sequence preceding Gn, the nascent precursor polypeptide chain is translocated from the cytoplasm into the ER. The Gn signal peptide is cleaved off by signal peptidase and the growing polypeptide chain is translocated into the ER lumen [19,105,106]. Two hydrophobic domains in the Gn/Gc precursor located in the C-terminal parts of Gn and Gc are inserted into the ER membrane and serve as transmembrane domains of Gn and Gc [97,106]. Additionally, Gn and Gc are separated by a third hydrophobic domain acting as internal signal peptide for Gc which is also cleaved by signal peptidase thus separating Gn from Gc [96–99].

Currently, the signal peptidase is the only host enzyme known to be required for the cleavage of the phlebovirus glycoprotein precursor [16,17]. This implies that the Gc signal peptide remains connected to the cytoplasmic C-terminal end of Gn, thereby acting as a second transmembrane domain for Gn. Indeed, for UUKV it has been shown that the Gc signal peptide is not removed from the cytoplasmic tail of Gn—at least not during glycoprotein synthesis and maturation [17]. However, it is not known if the Gc signal peptide is removed at another step of the viral life cycle.

While the M-segment of tick-borne phleboviruses only encodes the glycoproteins Gn and Gc [36,53,105,107,108] the M-segment of insect-borne phleboviruses encodes an additional protein upstream of Gn termed NSm [97,102,104,105] (see Figure 2). Since all M-segment-encoded proteins are expressed from a single mRNA, an NSm-Gn/Gc precursor protein is produced in addition to the Gn/Gc precursor by differential use of an AUG triplet as start codon which is located upstream of the Gn start codons [99,109]. In the case of RVFV, another two AUG triplets—one upstream and one downstream of the NSm start codon—give rise to the expression of a nested set of polyproteins [101,104,110].

The polyprotein precursors are all cleaved by signal peptidase to generate the accessory proteins P78 (Nsm-Gn), P14 (NSm), and P13 (NSm') in addition to the glycoproteins Gn and Gc [67,111]. The role of NSm proteins in the replication of insect-borne phleboviruses is not entirely clear. In vertebrate cells, the P14 protein of RVFV acts as an anti-apoptotic factor [112], however it is not required during viral replication in mammalian or mosquito cell cultures [113,114]. In vivo, P14 appears to be a virulence factor in mammals while P78 seems to be required for the dissemination in the mosquito vector [67], as discussed above. Mutational analysis revealed that the entire NSm region is dispensable for the proper synthesis and processing of the viral glycoproteins although both the NSm-Gn/Gc precursor and the Gn/Gc precursor can contribute to the synthesis of Gn and Gc [102,110,114].

Both Gn and Gc are type I transmembrane proteins, i.e., the N-terminus is orientated towards the ER lumen and the C-terminus is facing the cytoplasm (which corresponds to the interior of the virus after budding) and they span the lipid bilayer only once (although the signal peptide of Gc might serve as second transmembrane domain for Gn as described above) [17,105,115].

3.2. Post-translational Modifications and Subcellular Localization of Gn and Gc

Gn and Gc have a cysteine content of approximately 5% [98,105]. Positions of the cysteine residues are highly conserved among phleboviruses [99], indicating that extensive disulfide-bridge formation may occur and that the positions might be crucial for determining correct polypeptide folding. For Gn and Gc of UUKV it could be demonstrated that both proteins interact with protein disulfide isomerase (PDI) [116], an enzyme ubiquitously found in the ER which breaks up incorrectly formed disulfide bonds and catalyzes the formation of the correct ones leading to the mature, correctly folded three-dimensional protein structure. Other proteins involved in correct folding of UUKV Gn and Gc are the chaperones binding immunoglobulin protein (BiP), calnexin, and calreticulin [116,117] (Figure 1).

Both Gn and Gc contain *N*-glycosylation sites (Asn–X–Ser or Asn–X–Thr) [18,98,109,118], but the exact number of these sites differs between the different phlebovirus species [97,99]. *N*-glycosylation occurs during protein synthesis in the lumen of the ER. Inhibition of *N*-glycosylation decreases the stability of Gn and Gc as demonstrated for the glycoproteins of PTV [119] and prevents the exit of the glycoproteins from the ER [100].

N-glycosylated and correctly folded Gn and Gc form non-covalently linked heterodimers in the ER [13]. The two glycoprotein molecules which associate as a heterodimer do not necessarily originate from the same precursor protein. In the case of UUKV, Gn matures significantly faster than Gc [116]. Therefore, newly synthesized Gn can only dimerize with Gc, which was synthesized earlier [116]. In contrast, in the case of PTV, heterodimers are formed by Gn and Gc molecules synthesized at the same time [119] suggesting that PTV and RVFV Gn and Gc maturate with similar kinetics [13,16].

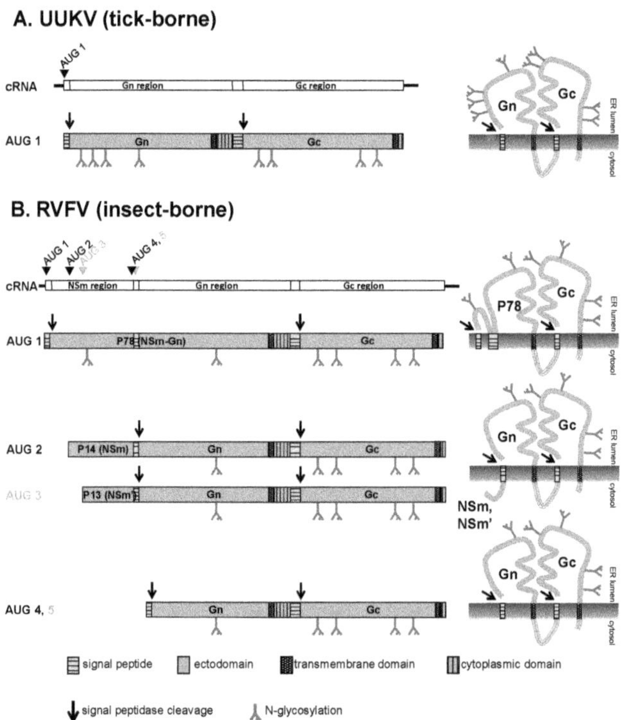

Figure 2. Coding and expression strategy of phlebovirus M-segments. Shown are the M-segments in antigenomic orientation (cRNA), the precursor glycoproteins and the membrane topology of the mature (glyco-) proteins. The antigenomic M-segment RNA serves as a template for viral transcription which results in a single mRNA. (**A**) UUKV as an example for tick-borne phleboviruses. The M-segment of tick-borne phleboviruses encodes only the two glycoproteins Gn and Gc. Translation of the mRNA yields one product, the Gn/Gc precursor. The precursor contains an N-terminal signal sequence preceding Gn and an internal signal sequence preceding Gc. Cleavage by the ER-associated signal peptidase complex yields Gn and Gc. Both Gn and Gc are glycosylated at *N*-glycosylation sites; (**B**) RVFV as an example for insect-borne phleboviruses. The M-segment of insect-borne phleboviruses encodes the non-structural protein NSm followed by the glycoproteins Gn and Gc. In case of RVFV translation initiation at different AUGs results in the expression of a nested set of polyproteins. Translation initiation at AUG 2 yields the NSm-Gn/Gc precursor protein. The precursor contains two internal signal sequences preceding Gn and Gc respectively. Cleavage by signal peptidase yields NSm, Gn and Gc. The Gn signal peptide acts as membrane anchor for NSm. Due to its membrane topology NSm is not glycosylated although it contains a potential *N*-glycosylation site. Translation at AUG 3 results in the expression of an N-terminal truncated NSm protein (NSm') which is functionally equivalent to full-length NSm. Translation at AUG 1 yields the P78-Gc precursor protein. Signal peptidase cleaves the pre-protein after the signal sequences preceding NSm and Gc but not after the signal sequence preceding Gn which might act as membrane anchor instead. P78 is glycosylated at the *N*-glycosylation sites in the NSm and the Gn region. Note the different membrane topology of the NSm region in P78 (translation initiation at AUG 1) compared to NSm or NSm' (translation at AUG 2 or AUG 3). Although P78 and Gc interact with each other, Gc might be unstable in the absence of Gn and therefore might be degraded in the ER. Translation at AUG 4 or 5 yields the Gn/Gc pre-protein. Signal peptidase cleaves the pre-protein after the signal sequences preceding Gn and Gc. Both Gn and Gc are *N*-glycosylated. The in vivo relevance of translation initiation at AUG 3 and 5 is not clear.

An interesting feature of the glycoprotein heterodimers is their intracellular localization. After Gn/Gc dimerization the glycoproteins exit the ER and reach the Golgi apparatus [13,19,22,24,100,102,120]. In contrast to many other viral glycoproteins which are further transported to the plasma membrane the Gn/Gc heterodimers of phleboviruses (and all other bunyaviruses) are retained in the Golgi [13,20,105,120–123]. Consequently, bunyaviruses bud at the Golgi instead of the plasma membrane [24,124–126]. Mutational analysis revealed that only Gn, but not Gc, contains a Golgi retention signal [13,106,115,121,122,127]. The Golgi retention signal seems to be specific for each phlebovirus species since a conserved sequence for this signal could not be identified. In the case of UUKV, the sequence required for Golgi retention is entirely located in the cytoplasmic tail of Gn [115,121] while for RVFV and PTV the Golgi retention signal consists of the Gn transmembrane domain and the adjacent amino acids of the cytoplasmic tail [122,124,127]. As a consequence, all phlebovirus Gn proteins analyzed so far correctly localize to the Golgi in the absence of Gc [122,123,127]. However, Gc does not localize to the Golgi in the absence of Gn [106,122,123]. A lysine-based putative ER retention/retrieval signal is located in the short cytoplasmic tails of phlebovirus Gc proteins. In Gn/Gc heterodimers the ER retention signal of Gc is presumably masked by interaction with the cytoplasmic tail of Gn. Gc is therefore only targeted to the Golgi as long it is associated with Gn. An amino acid alignment of the extreme C-termini of Gc from viruses belonging to the genera *Phlebovirus*, *Hantavirus*, and *Orthobunyavirus* revealed that the lysine at position −3 is conserved across these genera [126]. Furthermore, in some phlebovirus Gc proteins, the conserved lysine is part of a KKXX motif which is the classical ER retention motif for transmembrane proteins [128,129]. Indeed, most phlebovirus Gc proteins are retained in the ER when expressed alone, although PTV Gc has been shown to reach the plasma membrane despite the presence of the conserved lysine at position −3 [106]. A possible explanation might be the fact that ER retrieval signals are not always functional when they are located in short cytoplasmic tails or near amphipathic helices. For simian immunodeficiency virus (SIV) envelope glycoprotein (Env) mutants harboring an additional KKXX-motif in the cytoplasmic tail, it has been demonstrated that only mutants with a cytoplasmic tail longer than 13 amino acids were retained in the ER. In contrast, SIV Env mutants with a cytoplasmic tail length of 13 amino acids or less were transported to the cell surface [130].

In Gn/Gc heterodimers, the conserved lysine in Gc additionally seems to contribute to Golgi retention since heterodimers of UUKV wild-type Gn and Gc with mutations at position −3 were retained in the ER [126]. Furthermore, growth of recombinant RVFV was severely impaired when the conserved lysine in the Gc tail was mutated, because the mutation led to a mislocalization of Gn at the cell surface [124]. Interestingly, the glycosylation pattern of Gn and Gc incorporated into virions reflects their localization signals. Gn carries mostly *N*-linked oligosaccharides of the complex type, indicating extensive oligosaccharide processing in the Golgi, while Gc glycosylation is mainly of the high-mannose or hybrid type [18,118,131,132], in keeping with predominant localization of Gc in the ER. In the case of UUKV, however, the differences in Gn and Gc glycosylation might not result from differential transit of Gn and Gc through the host cell. Instead, steric occlusion seems to prevent processing of *N*-glycans attached to UUKV Gc [133].

3.3. The Role of the Cytoplasmic Tails of Gn and Gc in Virus Assembly and Budding

As mentioned above, the cytoplasmic tail of phlebovirus Gc proteins is very short (e.g., only five amino acids for UUKV) while the cytoplasmic tail of Gn is much longer (e.g., 81 amino acids for UUKV) [105]. The extended length is associated with additional biological functions: the Gn cytoplasmic tail not only contains the Golgi localization signal but is also involved in the initiation of the budding process and the packaging of ribonucleoproteins (RNPs) into virus particles [134]. For UUKV it could be demonstrated that mutation of a di-leucine motif in the cytoplasmic tail of Gn abolished the budding of virus-like particles, although the UUKV glycoproteins were correctly localized to the Golgi [126]. However, the motif required for budding seems to be specific for UUKV since not all phlebovirus Gn proteins contain the di-leucine motif. In the case of RVFV, the di-leucine motif is replaced by phenylalanine and isoleucine [124]. Although mutations of these amino acids affected the release of RVFV virus-like particles, the growth of recombinant virus carrying the mutations was only slightly diminished [124]. Furthermore, for UUKV the Gn and Gc glycoproteins are sufficient for efficient formation and release of virus-like particles [28] whereas for RVFV the formation of virus-like particles in the absence of RNP is inefficient [27].

A distinct feature of phleboviruses (and all other bunyaviruses) is the lack of a matrix protein that typically acts as an anchor between the virus envelope and the genetic core, the RNP [135]. Instead, the cytoplasmic tail of phlebovirus Gn proteins is endowed with matrix protein-like functions. For UUKV, the most C-terminal residues of the cytoplasmic tail of Gn are essential for the incorporation of RNP into virus-like particles [134]. In contrast, for RVFV the N-terminal part of the cytoplasmic tail of Gn is essential [27,124]. In the case of RVFV, the cytoplasmic tail of Gn can bind and package the viral polymerase and the nucleoprotein independently, but the efficient release of virus-like particles requires the nucleoprotein-encapsidated genome-like RNA [27]. For UUKV and PTV it has been observed that interaction of nucleoprotein and glycoproteins only occur in the Golgi and not in the ER although in both compartments the cytoplasmic tail of Gn should be accessible for the nucleoprotein which is synthesized in the soluble fraction of the cytoplasm [22,24,100,120]. Obviously, local accumulation of glycoproteins in the Golgi is a prerequisite for efficient binding of the nucleoprotein or the RNP. It can therefore be assumed that the interaction of RNP and Env proteins is the driving force for the morphogenesis and the budding of phlebovirus particles in the Golgi. When the encapsidation of the ribnucleoproteins and budding of newly formed virus particles in the Golgi are completed, virion containing vesicles are transported via the exocytic pathway to the plasma membrane where the virus particles are released [136].

4. Conclusions

Considerable progress has been made over the last three decades in understanding the role of the glycoproteins in phlebovirus entry. In particular, the finding that phlebovirus Gc proteins are class II viral membrane fusion proteins provided important insights into the membrane fusion reaction and imaging approaches allowed to elucidate the cell biology of phlebovirus entry. Moreover, several attachment factors were identified that might explain viral tropism. However, the expected key determinant of entry and cell tropism, the receptors used by phleboviruses, remain largely elusive. In addition, potential differences between host cell entry of tick-borne and insect-borne phleboviruses and differences in entry into vectors and host cells await further investigation. The emergence of new pathogenic tick-borne phleboviruses, namely Heartland virus (HRTV) and SFTSV, highlights the importance for this type of research. Although the processing of phlebovirus glycoproteins by signal peptidase is a pivotal step of glycoprotein maturation, only limited experimental data concerning this process is currently available. The subsequent steps in phlebovirus glycoprotein maturation, i.e., disulfide bond formation and *N*-glycosylation are even less well characterized. Furthermore, the mechanism of how glycoproteins and RNPs interact during virus assembly is poorly understood. New insights into these topics, together with a better understanding of the phlebovirus entry process, might provide the basis for the rational design of effective countermeasures against highly pathogenic phleboviruses.

Acknowledgments: This work was supported by the Leibniz Graduate School Emerging Infectious Diseases.

Conflicts of Interest: The authors declare no conflict of interest.

Abbreviations

The following abbreviations are used in this manuscript:

ANDV	Andes virus
BiP	Bindung immunoglobulin protein
CavME	Caveolin-1-mediated endocytosis
CIE	Clathrin-independent endocytosis
CME	Clathrin-mediated endocytosis
CNX	Calnexin
CCHFV	Crimean Congo hemorrhagic fever virus
DC	Dendritic cell
DC-SIGN	Dendritic cell-specific intercellular adhesion molecule-3-grabbing non-integrin
Env	Envelope glycoprotein
ER	Endoplasmic reticulum
GAG	Glycosaminoglycan
gRNA	genomic RNA
HDAC 8	Histone deacetylase 8
HRTV	Heartland virus
HIV	Human immunodeficiency virus
HNTV	Hantaan virus
HS	Heparan sulfate
IFITM	Interferon-induced transmembrane protein
L-SIGN	Liver/lymph node-specific intercellular adhesion molecules-3 grabbing non-integrin
LACV	La Crosse virus
Mϕ	Macrophage
NMMHC-IIA	Non-muscle myosin heavy chain IIA
ORF	Open reading frame
PDI	Protein disulfide isomerase
PTV	Punta Toro virus
RNaseK	Ribonuclease kappa
RNP	Ribonucleoprotein
RVFV	Rift Valley fever virus
SFV	Sandfly fever virus
SFTSV	Severe fever with thrombocytopenia virus
SIV	Simian immunodeficiency virus
TOSV	Toscana virus
UUKV	Uukuniemi virus
v-SNARE	Vesicle-soluble NSF attachment protein receptor
VAMP3	Vesicle-associated membrane protein 3
VSV	Vesicular stomatitis virus

References

1. Plyusnin, A.; Beaty, B.J.; Elliott, R.M.; Goldbach, R.; Kormelink, R.; Lundkvist, Å.; Schmaljohn, C.S.; Tesh, R.B. *Bunyaviridae*. In *Virus Taxonomy. Classification and Nomenclature of Viruses. Ninth Report of the International Committee on Taxonomy of Viruses*; King, A.M.Q., Lefkowitz, E., Adams, M.J., Carstens, E.B., Eds.; Academic Press: London/Waltham, UK; San Diego, CA, USA, 2012; pp. 724–741.
2. Horne, K.M.; Vanlandingham, D.L. Bunyavirus-vector interactions. *Viruses* **2014**, *6*, 4373–4397. [CrossRef] [PubMed]
3. Rotenberg, D.; Jacobson, A.L.; Schneweis, D.J.; Whitfield, A.E. Thrips transmission of tospoviruses. *Curr. Opin. Virol.* **2015**, *15*, 80–89. [CrossRef] [PubMed]
4. Klempa, B.; Fichet-Calvet, E.; Lecompte, E.; Auste, B.; Aniskin, V.; Meisel, H.; Barriere, P.; Koivogui, L.; ter Meulen, J.; Kruger, D.H. Novel *Hantavirus* sequences in Shrew, Guinea. *Emerg. Infect. Dis.* **2007**, *13*, 520–522. [CrossRef] [PubMed]

5. Kang, H.J.; Kadjo, B.; Dubey, S.; Jacquet, F.; Yanagihara, R. Molecular evolution of Azagny virus, a newfound *Hantavirus* harbored by the West African pygmy shrew (Crocidura obscurior) in Cote d'Ivoire. *Virol. J.* **2011**, *8*, 373. [CrossRef] [PubMed]
6. Weiss, S.; Witkowski, P.T.; Auste, B.; Nowak, K.; Weber, N.; Fahr, J.; Mombouli, J.V.; Wolfe, N.D.; Drexler, J.F.; Drosten, C.; et al. *Hantavirus* in bat, Sierra Leone. *Emerg. Infect. Dis.* **2012**, *18*, 159–161. [CrossRef] [PubMed]
7. Carey, D.E.; Reuben, R.; Panicker, K.N.; Shope, R.E.; Myers, R.M. Thottapalayam virus: A presumptive arbovirus isolated from a shrew in India. *Indian J. Med. Res.* **1971**, *59*, 1758–1760. [PubMed]
8. Guo, W.P.; Lin, X.D.; Wang, W.; Tian, J.H.; Cong, M.L.; Zhang, H.L.; Wang, M.R.; Zhou, R.H.; Wang, J.B.; Li, M.H.; et al. Phylogeny and origins of hantaviruses harbored by bats, insectivores, and rodents. *PLoS Pathog.* **2013**, *9*, e1003159. [CrossRef] [PubMed]
9. Arai, S.; Ohdachi, S.D.; Asakawa, M.; Kang, H.J.; Mocz, G.; Arikawa, J.; Okabe, N.; Yanagihara, R. Molecular phylogeny of a newfound *Hantavirus* in the Japanese shrew mole (*Urotrichus talpoides*). *Proc. Natl. Acad. Sci. USA* **2008**, *105*, 16296–16301. [CrossRef] [PubMed]
10. Kang, H.J.; Bennett, S.N.; Hope, A.G.; Cook, J.A.; Yanagihara, R. Shared ancestry between a newfound mole-borne *Hantavirus* and hantaviruses harbored by cricetid rodents. *J. Virol.* **2011**, *85*, 7496–7503. [CrossRef] [PubMed]
11. Holmes, E.C.; Zhang, Y.Z. The evolution and emergence of hantaviruses. *Curr. Opin. Virol.* **2015**, *10*, 27–33. [CrossRef] [PubMed]
12. Walter, C.T.; Barr, J.N. Recent advances in the molecular and cellular biology of bunyaviruses. *J. Gen. Virol.* **2011**, *92 (Pt 11)*, 2467–2484. [CrossRef] [PubMed]
13. Chen, S.Y.; Compans, R.W. Oligomerization, transport, and Golgi retention of Punta Toro virus glycoproteins. *J. Virol.* **1991**, *65*, 5902–5909. [PubMed]
14. Antic, D.; Wright, K.E.; Kang, C.Y. Maturation of Hantaan virus glycoproteins G1 and G2. *Virology* **1992**, *189*, 324–328. [CrossRef]
15. Lober, C.; Anheier, B.; Lindow, S.; Klenk, H.D.; Feldmann, H. The Hantaan virus glycoprotein precursor is cleaved at the conserved pentapeptide WAASA. *Virology* **2001**, *289*, 224–229. [CrossRef] [PubMed]
16. Gerrard, S.R.; Nichol, S.T. Synthesis, proteolytic processing and complex formation of N-terminally nested precursor proteins of the Rift Valley fever virus glycoproteins. *Virology* **2007**, *357*, 124–133. [CrossRef] [PubMed]
17. Andersson, A.M.; Melin, L.; Persson, R.; Raschperger, E.; Wikstrom, L.; Pettersson, R.F. Processing and membrane topology of the spike proteins G1 and G2 of Uukuniemi virus. *J. Virol.* **1997**, *71*, 218–225. [PubMed]
18. Kuismanen, E. Posttranslational processing of Uukuniemi virus glycoproteins G1 and G2. *J. Virol.* **1984**, *51*, 806–812. [PubMed]
19. Matsuoka, Y.; Ihara, T.; Bishop, D.H.; Compans, R.W. Intracellular accumulation of Punta Toro virus glycoproteins expressed from cloned cDNA. *Virology* **1988**, *167*, 251–260. [CrossRef]
20. Madoff, D.H.; Lenard, J. A membrane glycoprotein that accumulates intracellularly: Cellular processing of the large glycoprotein of LaCrosse virus. *Cell* **1982**, *28*, 821–829. [CrossRef]
21. Shi, X.; Brauburger, K.; Elliott, R.M. Role of N-linked glycans on bunyamwera virus glycoproteins in intracellular trafficking, protein folding, and virus infectivity. *J. Virol.* **2005**, *79*, 13725–13734. [CrossRef] [PubMed]
22. Kuismanen, E.; Hedman, K.; Saraste, J.; Pettersson, R.F. Uukuniemi virus maturation: Accumulation of virus particles and viral antigens in the Golgi complex. *Mol. Cell. Biol.* **1982**, *2*, 1444–1458. [CrossRef] [PubMed]
23. Fontana, J.; Lopez-Montero, N.; Elliott, R.M.; Fernandez, J.J.; Risco, C. The unique architecture of Bunyamwera virus factories around the Golgi complex. *Cell. Microbiol.* **2008**, *10*, 2012–2028. [CrossRef] [PubMed]
24. Smith, J.F.; Pifat, D.Y. Morphogenesis of sandfly viruses (*Bunyaviridae* family). *Virology* **1982**, *121*, 61–81. [CrossRef]
25. Salanueva, I.J.; Novoa, R.R.; Cabezas, P.; Lopez-Iglesias, C.; Carrascosa, J.L.; Elliott, R.M.; Risco, C. Polymorphism and structural maturation of bunyamwera virus in Golgi and post-Golgi compartments. *J. Virol.* **2003**, *77*, 1368–1381. [CrossRef] [PubMed]

26. Murphy, F.A.; Harrison, A.K.; Whitfield, S.G. *Bunyaviridae*: Morphologic and morphogenetic similarities of Bunyamwera serologic supergroup viruses and several other arthropod-borne viruses. *Intervirology* **1973**, *1*, 297–316. [CrossRef] [PubMed]
27. Piper, M.E.; Sorenson, D.R.; Gerrard, S.R. Efficient cellular release of Rift Valley fever virus requires genomic RNA. *PLoS ONE* **2011**, *6*, e18070. [CrossRef] [PubMed]
28. Overby, A.K.; Popov, V.; Neve, E.P.; Pettersson, R.F. Generation and analysis of infectious virus-like particles of Uukuniemi virus (*Bunyaviridae*): A useful system for studying bunyaviral packaging and budding. *J. Virol.* **2006**, *80*, 10428–10435. [CrossRef] [PubMed]
29. Novoa, R.R.; Calderita, G.; Cabezas, P.; Elliott, R.M.; Risco, C. Key Golgi factors for structural and functional maturation of bunyamwera virus. *J. Virol.* **2005**, *79*, 10852–10863. [CrossRef] [PubMed]
30. Spiropoulou, C.F. *Hantavirus* maturation. In *Hantaviruses*; Schmaljohn, C.S., Nichol, S.T., Eds.; Springer-Verlag: Heidelberg/Berlin, Germany, 2001; pp. 33–46.
31. Cifuentes-Muñoz, N.; Salazar-Quiroz, N.; Tischler, N.D. *Hantavirus* Gn and Gc envelope glycoproteins: Key structural units for virus cell entry and virus assembly. *Viruses* **2014**, *6*, 1801–1822. [CrossRef] [PubMed]
32. Lorenzo, G.; Lopez-Gil, E.; Warimwe, G.M.; Brun, A. Understanding Rift Valley fever: Contributions of animal models to disease characterization and control. *Mol. Immunol.* **2015**, *66*, 78–88. [CrossRef] [PubMed]
33. Liu, S.; Chai, C.; Wang, C.; Amer, S.; Lv, H.; He, H.; Sun, J.; Lin, J. Systematic review of severe fever with thrombocytopenia syndrome: Virology, epidemiology, and clinical characteristics. *Rev. Med. Virol.* **2014**, *24*, 90–102. [CrossRef] [PubMed]
34. Liu, Y.; Li, Q.; Hu, W.; Wu, J.; Wang, Y.; Mei, L.; Walker, D.H.; Ren, J.; Wang, Y.; Yu, X.J. Person-to-person transmission of severe fever with thrombocytopenia syndrome virus. *Vector Borne Zoonotic Dis.* **2012**, *12*, 156–160. [CrossRef] [PubMed]
35. Alkan, C.; Bichaud, L.; de Lamballerie, X.; Alten, B.; Gould, E.A.; Charrel, R.N. Sandfly-borne phleboviruses of Eurasia and Africa: Epidemiology, genetic diversity, geographic range, control measures. *Antivir. Res.* **2013**, *100*, 54–74. [CrossRef] [PubMed]
36. Elliott, R.M.; Brennan, B. Emerging phleboviruses. *Curr. Opin. Virol.* **2014**, *5*, 50–57. [CrossRef] [PubMed]
37. Martin, M.L.; Lindsey-Regnery, H.; Sasso, D.R.; McCormick, J.B.; Palmer, E. Distinction between *Bunyaviridae* genera by surface structure and comparison with Hantaan virus using negative stain electron microscopy. *Arch. Virol.* **1985**, *86*, 17–28. [CrossRef] [PubMed]
38. Overby, A.K.; Pettersson, R.F.; Grunewald, K.; Huiskonen, J.T. Insights into bunyavirus architecture from electron cryotomography of Uukuniemi virus. *Proc. Natl. Acad. Sci. USA* **2008**, *105*, 2375–2379. [CrossRef] [PubMed]
39. Freiberg, A.N.; Sherman, M.B.; Morais, M.C.; Holbrook, M.R.; Watowich, S.J. Three-dimensional organization of Rift Valley fever virus revealed by cryoelectron tomography. *J. Virol.* **2008**, *82*, 10341–10348. [CrossRef] [PubMed]
40. Sherman, M.B.; Freiberg, A.N.; Holbrook, M.R.; Watowich, S.J. Single-particle cryo-electron microscopy of Rift Valley fever virus. *Virology* **2009**, *387*, 11–15. [CrossRef] [PubMed]
41. Huiskonen, J.T.; Overby, A.K.; Weber, F.; Grunewald, K. Electron cryo-microscopy and single-particle averaging of Rift Valley fever virus: Evidence for GN-GC glycoprotein heterodimers. *J. Virol.* **2009**, *83*, 3762–3799. [CrossRef] [PubMed]
42. Rusu, M.; Bonneau, R.; Holbrook, M.R.; Watowich, S.J.; Birmanns, S.; Wriggers, W.; Freiberg, A.N. An assembly model of Rift Valley fever virus. *Front. Microbiol.* **2012**, *3*, 254. [CrossRef] [PubMed]
43. Dessau, M.; Modis, Y. Crystal structure of glycoprotein C from Rift Valley fever virus. *Proc. Natl. Acad. Sci. USA* **2013**, *110*, 1696–1701. [CrossRef] [PubMed]
44. Raftery, M.J.; Lalwani, P.; Krautkrmer, E.; Peters, T.; Scharffetter-Kochanek, K.; Kruger, R.; Hofmann, J.; Seeger, K.; Kruger, D.H.; Schonrich, G. Beta2 integrin mediates hantavirus-induced release of neutrophil extracellular traps. *J. Exp. Med.* **2014**, *211*, 1485–1497. [CrossRef] [PubMed]
45. Gavrilovskaya, I.N.; Shepley, M.; Shaw, R.; Ginsberg, M.H.; Mackow, E.R. Beta3 integrins mediate the cellular entry of hantaviruses that cause respiratory failure. *Proc. Natl. Acad. Sci. USA* **1998**, *95*, 7074–7079. [CrossRef] [PubMed]
46. Gavrilovskaya, I.N.; Brown, E.J.; Ginsberg, M.H.; Mackow, E.R. Cellular entry of hantaviruses which cause hemorrhagic fever with renal syndrome is mediated by beta3 integrins. *J. Virol.* **1999**, *73*, 3951–3959. [PubMed]

47. Garcia-Vallejo, J.J.; van Kooyk, Y. The physiological role of DC-SIGN: A tale of mice and men. *Trends Immunol.* **2013**, *34*, 482–486. [CrossRef] [PubMed]
48. Feinberg, H.; Mitchell, D.A.; Drickamer, K.; Weis, W.I. Structural basis for selective recognition of oligosaccharides by DC-SIGN and DC-SIGNR. *Science* **2001**, *294*, 2163–2166. [CrossRef] [PubMed]
49. Geijtenbeek, T.B.; Kwon, D.S.; Torensma, R.; van Vliet, S.J.; van Duijnhoven, G.C.; Middel, J.; Cornelissen, I.L.; Nottet, H.S.; KewalRamani, V.N.; Littman, D.R.; et al. DC-SIGN, a dendritic cell-specific HIV-1-binding protein that enhances trans-infection of T cells. *Cell* **2000**, *100*, 587–597. [CrossRef]
50. Geijtenbeek, T.B.; van Vliet, S.J.; Koppel, E.A.; Sanchez-Hernandez, M.; Vandenbroucke-Grauls, C.M.; Appelmelk, B.; van Kooyk, Y. Mycobacteria target DC-SIGN to suppress dendritic cell function. *J. Exp. Med.* **2003**, *197*, 7–17. [CrossRef] [PubMed]
51. Tailleux, L.; Schwartz, O.; Herrmann, J.L.; Pivert, E.; Jackson, M.; Amara, A.; Legres, L.; Dreher, D.; Nicod, L.P.; Gluckman, J.C.; et al. DC-SIGN is the major *Mycobacterium tuberculosis* receptor on human dendritic cells. *J. Exp. Med.* **2003**, *197*, 121–127. [CrossRef] [PubMed]
52. Lozach, P.Y.; Kuhbacher, A.; Meier, R.; Mancini, R.; Bitto, D.; Bouloy, M.; Helenius, A. DC-SIGN as a receptor for phleboviruses. *Cell Host Microbe* **2011**, *10*, 75–88. [CrossRef] [PubMed]
53. Hofmann, H.; Li, X.; Zhang, X.; Liu, W.; Kuhl, A.; Kaup, F.; Soldan, S.S.; Gonzalez-Scarano, F.; Weber, F.; He, Y.; et al. Severe fever with thrombocytopenia virus glycoproteins are targeted by neutralizing antibodies and can use DC-SIGN as a receptor for pH-dependent entry into human and animal cell lines. *J. Virol.* **2013**, *87*, 4384–4394. [CrossRef] [PubMed]
54. Navarro-Sanchez, E.; Altmeyer, R.; Amara, A.; Schwartz, O.; Fieschi, F.; Virelizier, J.L.; Arenzana-Seisdedos, F.; Despres, P. Dendritic-cell-specific ICAM3-grabbing non-integrin is essential for the productive infection of human dendritic cells by mosquito-cell-derived dengue viruses. *EMBO Rep.* **2003**, *4*, 723–728. [CrossRef] [PubMed]
55. Tassaneetrithep, B.; Burgess, T.H.; Granelli-Piperno, A.; Trumpfheller, C.; Finke, J.; Sun, W.; Eller, M.A.; Pattanapanyasat, K.; Sarasombath, S.; Birx, D.L.; et al. DC-SIGN (CD209) mediates dengue virus infection of human dendritic cells. *J. Exp. Med.* **2003**, *197*, 823–829. [CrossRef] [PubMed]
56. Cerny, D.; Haniffa, M.; Shin, A.; Bigliardi, P.; Tan, B.K.; Lee, B.; Poidinger, M.; Tan, E.Y.; Ginhoux, F.; Fink, K. Selective susceptibility of human skin antigen presenting cells to productive dengue virus infection. *PLoS Pathog.* **2014**, *10*, e1004548. [CrossRef] [PubMed]
57. Engering, A.; Geijtenbeek, T.B.; van Vliet, S.J.; Wijers, M.; Van, L.E.; Demaurex, N.; Lanzavecchia, A.; Fransen, J.; Figdor, C.G.; Piguet, V.; et al. The dendritic cell-specific adhesion receptor DC-SIGN internalizes antigen for presentation to T cells. *J. Immunol.* **2002**, *168*, 2118–2126. [CrossRef] [PubMed]
58. Bashirova, A.A.; Geijtenbeek, T.B.; van Duijnhoven, G.C.; van Vliet, S.J.; Eilering, J.B.; Martin, M.P.; Wu, L.; Martin, T.D.; Viebig, N.; Knolle, P.A.; et al. A dendritic cell-specific intercellular adhesion molecule 3-grabbing nonintegrin (DC-SIGN)-related protein is highly expressed on human liver sinusoidal endothelial cells and promotes HIV-1 infection. *J. Exp. Med.* **2001**, *193*, 671–678. [CrossRef] [PubMed]
59. Gramberg, T.; Hofmann, H.; Moller, P.; Lalor, P.F.; Marzi, A.; Geier, M.; Krumbiegel, M.; Winkler, T.; Kirchhoff, F.; Adams, D.H.; et al. LSECtin interacts with filovirus glycoproteins and the spike protein of SARS coronavirus. *Virology* **2005**, *340*, 224–236. [CrossRef] [PubMed]
60. Pohlmann, S.; Soilleux, E.J.; Baribaud, F.; Leslie, G.J.; Morris, L.S.; Trowsdale, J.; Lee, B.; Coleman, N.; Doms, R.W. DC-SIGNR, a DC-SIGN homologue expressed in endothelial cells, binds to human and simian immunodeficiency viruses and activates infection in trans. *Proc. Natl. Acad. Sci. USA* **2001**, *98*, 2670–2675. [CrossRef] [PubMed]
61. Leger, P.; Tetard, M.; Youness, B.; Cordes, N.; Rouxel, R.N.; Flamand, M.; Lozach, P.Y. Differential use of the C-type lectins L-SIGN and DC-SIGN for phlebovirus endocytosis. *Traffic* **2016**, *17*, 639–656. [CrossRef] [PubMed]
62. Guo, Y.; Feinberg, H.; Conroy, E.; Mitchell, D.A.; Alvarez, R.; Blixt, O.; Taylor, M.E.; Weis, W.I.; Drickamer, K. Structural basis for distinct ligand-binding and targeting properties of the receptors DC-SIGN and DC-SIGNR. *Nat. Struct. Mol. Biol.* **2004**, *11*, 591–598. [CrossRef] [PubMed]
63. Jin, C.; Liang, M.; Ning, J.; Gu, W.; Jiang, H.; Wu, W.; Zhang, F.; Li, C.; Zhang, Q.; Zhu, H.; et al. Pathogenesis of emerging severe fever with thrombocytopenia syndrome virus in C57/BL6 mouse model. *Proc. Natl. Acad. Sci. USA* **2012**, *109*, 10053–10058. [CrossRef] [PubMed]

64. Chaipan, C.; Soilleux, E.J.; Simpson, P.; Hofmann, H.; Gramberg, T.; Marzi, A.; Geier, M.; Stewart, E.A.; Eisemann, J.; Steinkasserer, A.; et al. DC-SIGN and CLEC-2 mediate human immunodeficiency virus type 1 capture by platelets. *J. Virol.* **2006**, *80*, 8951–8960. [CrossRef] [PubMed]
65. De Boer, S.M.; Kortekaas, J.; de Haan, C.A.; Rottier, P.J.; Moormann, R.J.; Bosch, B.J. Heparan sulfate facilitates Rift Valley fever virus entry into the cell. *J. Virol.* **2012**, *86*, 13767–13771. [CrossRef] [PubMed]
66. Weingartl, H.M.; Zhang, S.; Marszal, P.; McGreevy, A.; Burton, L.; Wilson, W.C. Rift Valley fever virus incorporates the 78 kDa glycoprotein into virions matured in mosquito C6/36 cells. *PLoS ONE* **2014**, *9*, e87385. [CrossRef] [PubMed]
67. Kreher, F.; Tamietti, C.; Gommet, C.; Guillemot, L.; Ermonval, M.; Failloux, A.B.; Panthier, J.J.; Bouloy, M.; Flamand, M. The Rift Valley fever accessory proteins NSm and P78/NSm-GN are distinct determinants of virus propagation in vertebrate and invertebrate hosts. *Emerg. Microbes Infect.* **2014**, *3*, e71. [CrossRef] [PubMed]
68. Riblett, A.M.; Blomen, V.A.; Jae, L.T.; Altamura, L.A.; Doms, R.W.; Brummelkamp, T.R.; Wojcechowskyj, J.A. A haploid genetic screen identifies heparan sulfate proteoglycans supporting Rift Valley fever virus infection. *J. Virol.* **2015**, *90*, 1414–1423. [CrossRef] [PubMed]
69. Pietrantoni, A.; Fortuna, C.; Remoli, M.E.; Ciufolini, M.G.; Superti, F. Bovine lactoferrin inhibits Toscana virus infection by binding to heparan sulphate. *Viruses* **2015**, *7*, 480–495. [CrossRef] [PubMed]
70. Vicente-Manzanares, M.; Ma, X.; Adelstein, R.S.; Horwitz, A.R. Non-muscle myosin II takes centre stage in cell adhesion and migration. *Nat. Rev. Mol. Cell Biol.* **2009**, *10*, 778–790. [CrossRef] [PubMed]
71. Sun, Y.; Qi, Y.; Liu, C.; Gao, W.; Chen, P.; Fu, L.; Peng, B.; Wang, H.; Jing, Z.; Zhong, G.; et al. Nonmuscle myosin heavy chain IIA is a critical factor contributing to the efficiency of early infection of severe fever with thrombocytopenia syndrome virus. *J. Virol.* **2014**, *88*, 237–248. [CrossRef] [PubMed]
72. Althaus, K.; Greinacher, A. MYH9-related platelet disorders. *Semin. Thromb. Hemost.* **2009**, *35*, 189–203. [CrossRef] [PubMed]
73. Arii, J.; Goto, H.; Suenaga, T.; Oyama, M.; Kozuka-Hata, H.; Imai, T.; Minowa, A.; Akashi, H.; Arase, H.; Kawaoka, Y.; et al. Non-muscle myosin IIA is a functional entry receptor for herpes simplex virus-1. *Nature* **2010**, *467*, 859–862. [CrossRef] [PubMed]
74. Lozach, P.Y.; Mancini, R.; Bitto, D.; Meier, R.; Oestereich, L.; Overby, A.K.; Pettersson, R.F.; Helenius, A. Entry of bunyaviruses into mammalian cells. *Cell Host Microbe* **2010**, *7*, 488–499. [CrossRef] [PubMed]
75. Meier, R.; Franceschini, A.; Horvath, P.; Tetard, M.; Mancini, R.; von Mering, C.; Helenius, A.; Lozach, P.Y. Genome-wide small interfering RNA screens reveal VAMP3 as a novel host factor required for Uukuniemi virus late penetration. *J. Virol.* **2014**, *88*, 8565–8578. [CrossRef] [PubMed]
76. Proux-Gillardeaux, V.; Rudge, R.; Galli, T. The tetanus neurotoxin-sensitive and insensitive routes to and from the plasma membrane: Fast and slow pathways? *Traffic* **2005**, *6*, 366–373. [CrossRef] [PubMed]
77. Fader, C.M.; Sanchez, D.G.; Mestre, M.B.; Colombo, M.I. TI-VAMP/VAMP7 and VAMP3/cellubrevin: Two v-SNARE proteins involved in specific steps of the autophagy/multivesicular body pathways. *Biochim. Biophys. Acta* **2009**, *1793*, 1901–1916. [CrossRef] [PubMed]
78. Yamauchi, Y.; Boukari, H.; Banerjee, I.; Sbalzarini, I.F.; Horvath, P.; Helenius, A. Histone deacetylase 8 is required for centrosome cohesion and influenza A virus entry. *PLoS Pathog.* **2011**, *7*, e1002316. [CrossRef] [PubMed]
79. Hackett, B.A.; Yasunaga, A.; Panda, D.; Tartell, M.A.; Hopkins, K.C.; Hensley, S.E.; Cherry, S. RNASEK is required for internalization of diverse acid-dependent viruses. *Proc. Natl. Acad. Sci. USA* **2015**, *112*, 7797–7802. [CrossRef] [PubMed]
80. Harmon, B.; Schudel, B.R.; Maar, D.; Kozina, C.; Ikegami, T.; Tseng, C.T.; Negrete, O.A. Rift Valley fever virus strain MP-12 enters mammalian host cells via caveola-mediated endocytosis. *J. Virol.* **2012**, *86*, 12954–12970. [CrossRef] [PubMed]
81. Hollidge, B.S.; Gonzalez-Scarano, F.; Soldan, S.S. Arboviral encephalitides: Transmission, emergence, and pathogenesis. *J. Neuroimmune Pharmacol.* **2010**, *5*, 428–442. [CrossRef] [PubMed]
82. Santos, R.I.; Rodrigues, A.H.; Silva, M.L.; Mortara, R.A.; Rossi, M.A.; Jamur, M.C.; Oliver, C.; Arruda, E. Oropouche virus entry into HeLa cells involves clathrin and requires endosomal acidification. *Virus Res.* **2008**, *138*, 139–143. [CrossRef] [PubMed]
83. Podbilewicz, B. Virus and cell fusion mechanisms. *Annu. Rev. Cell Dev. Biol.* **2014**, *30*, 111–139. [CrossRef] [PubMed]

84. Vaney, M.C.; Rey, F.A. Class II enveloped viruses. *Cell. Microbiol.* **2011**, *13*, 1451–1459. [CrossRef] [PubMed]
85. Garry, C.E.; Garry, R.F. Proteomics computational analyses suggest that the carboxyl terminal glycoproteins of Bunyaviruses are class II viral fusion protein (beta-penetrenes). *Theor. Biol. Med. Model.* **2004**, *1*, 10. [CrossRef] [PubMed]
86. De Boer, S.M.; Kortekaas, J.; Spel, L.; Rottier, P.J.; Moormann, R.J.; Bosch, B.J. Acid-activated structural reorganization of the Rift Valley fever virus Gc fusion protein. *J. Virol.* **2012**, *86*, 13642–13652. [CrossRef] [PubMed]
87. Filone, C.M.; Heise, M.; Doms, R.W.; Bertolotti-Ciarlet, A. Development and characterization of a Rift Valley fever virus cell-cell fusion assay using alphavirus replicon vectors. *Virology* **2006**, *356*, 155–164. [CrossRef] [PubMed]
88. Bitto, D.; Halldorsson, S.; Caputo, A.; Huiskonen, J.T. Low pH and anionic lipid dependent fusion of Uukuniemi phlebovirus to liposomes. *J. Biol. Chem* **2016**, *291*, 6412–6422. [CrossRef] [PubMed]
89. Smith, S.; Weston, S.; Kellam, P.; Marsh, M. IFITM proteins-cellular inhibitors of viral entry. *Curr. Opin. Virol.* **2014**, *4*, 71–77. [CrossRef] [PubMed]
90. Perreira, J.M.; Chin, C.R.; Feeley, E.M.; Brass, A.L. IFITMs restrict the replication of multiple pathogenic viruses. *J. Mol. Biol.* **2013**, *425*, 4937–4955. [CrossRef] [PubMed]
91. Mudhasani, R.; Tran, J.P.; Retterer, C.; Radoshitzky, S.R.; Kota, K.P.; Altamura, L.A.; Smith, J.M.; Packard, B.Z.; Kuhn, J.H.; Costantino, J.; et al. IFITM-2 and IFITM-3 but not IFITM-1 restrict Rift Valley fever virus. *J. Virol.* **2013**, *87*, 8451–8464. [CrossRef] [PubMed]
92. Li, K.; Markosyan, R.M.; Zheng, Y.M.; Golfetto, O.; Bungart, B.; Li, M.; Ding, S.; He, Y.; Liang, C.; Lee, J.C.; et al. IFITM proteins restrict viral membrane hemifusion. *PLoS Pathog.* **2013**, *9*, e1003124. [CrossRef] [PubMed]
93. Mazzon, M.; Mercer, J. Lipid interactions during virus entry and infection. *Cell. Microbiol.* **2014**, *16*, 1493–1502. [CrossRef] [PubMed]
94. Ikegami, T.; Won, S.; Peters, C.J.; Makino, S. Rift Valley fever virus NSs mRNA is transcribed from an incoming anti-viral-sense S RNA segment. *J. Virol.* **2005**, *79*, 12106–12111. [CrossRef] [PubMed]
95. Barr, J.N. Bunyavirus mRNA synthesis is coupled to translation to prevent premature transcription termination. *RNA* **2007**, *13*, 731–736. [CrossRef] [PubMed]
96. Collett, M.S.; Purchio, A.F.; Keegan, K.; Frazier, S.; Hays, W.; Anderson, D.K.; Parker, M.D.; Schmaljohn, C.; Schmidt, J.; Dalrymple, J.M. Complete nucleotide sequence of the M RNA segment of Rift Valley fever virus. *Virology* **1985**, *144*, 228–245. [CrossRef]
97. Ihara, T.; Smith, J.; Dalrymple, J.M.; Bishop, D.H. Complete sequences of the glycoproteins and M RNA of Punta Toro phlebovirus compared to those of Rift Valley fever virus. *Virology* **1985**, *144*, 246–259. [CrossRef]
98. Ronnholm, R.; Pettersson, R.F. Complete nucleotide sequence of the M RNA segment of Uukuniemi virus encoding the membrane glycoproteins G1 and G2. *Virology* **1987**, *160*, 191–202. [CrossRef]
99. Gro, M.C.; di Bonito, P.; Fortini, D.; Mochi, S.; Giorgi, C. Completion of molecular characterization of Toscana phlebovirus genome: Nucleotide sequence, coding strategy of M genomic segment and its amino acid sequence comparison to other phleboviruses. *Virus Res.* **1997**, *51*, 81–91. [CrossRef]
100. Kuismanen, E.; Bang, B.; Hurme, M.; Pettersson, R.F. Uukuniemi virus maturation: Immunofluorescence microscopy with monoclonal glycoprotein-specific antibodies. *J. Virol.* **1984**, *51*, 137–146. [PubMed]
101. Kakach, L.T.; Wasmoen, T.L.; Collett, M.S. Rift Valley fever virus M segment: Use of recombinant vaccinia viruses to study Phlebovirus gene expression. *J. Virol.* **1988**, *62*, 826–833. [PubMed]
102. Wasmoen, T.L.; Kakach, L.T.; Collett, M.S. Rift Valley fever virus M segment: Cellular localization of M segment-encoded proteins. *Virology* **1988**, *166*, 275–280. [CrossRef]
103. Ulmanen, I.; Seppala, P.; Pettersson, R.F. In vitro translation of Uukuniemi virus-specific RNAs: Identification of a nonstructural protein and a precursor to the membrane glycoproteins. *J. Virol.* **1981**, *37*, 72–79. [PubMed]
104. Suzich, J.A.; Collett, M.S. Rift Valley fever virus M segment: Cell-free transcription and translation of virus-complementary RNA. *Virology* **1988**, *164*, 478–486. [CrossRef]
105. Pettersson, R.F.; Melin, L. Synthesis, assembly, and intracellular transport of *Bunyaviridae* membrane proteins. In *The Bunyaviridae*; Elliott, R., Ed.; Plenum Press: New York, NY, USA, 1996; pp. 159–183.
106. Chen, S.Y.; Matsuoka, Y.; Compans, R.W. Golgi complex localization of the Punta Toro virus G2 protein requires its association with the G1 protein. *Virology* **1991**, *183*, 351–365. [CrossRef]

107. Xu, B.; Liu, L.; Huang, X.; Ma, H.; Zhang, Y.; Du, Y.; Wang, P.; Tang, X.; Wang, H.; Kang, K.; et al. Metagenomic analysis of fever, thrombocytopenia and leukopenia syndrome (FTLS) in Henan province, China: Discovery of a new bunyavirus. *PLoS Pathog.* **2011**, *7*, e1002369. [CrossRef] [PubMed]
108. Yu, X.J.; Liang, M.F.; Zhang, S.Y.; Liu, Y.; Li, J.D.; Sun, Y.L.; Zhang, L.; Zhang, Q.F.; Popov, V.L.; Li, C.; et al. Fever with thrombocytopenia associated with a novel bunyavirus in China. *N. Engl. J. Med.* **2011**, *364*, 1523–1532. [CrossRef] [PubMed]
109. Di Bonito, P.; Mochi, S.; Gro, M.C.; Fortini, D.; Giorgi, C. Organization of the M genomic segment of Toscana phlebovirus. *J. Gen. Virol.* **1997**, *78 (Pt 1)*, 77–81. [CrossRef] [PubMed]
110. Suzich, J.A.; Kakach, L.T.; Collett, M.S. Expression strategy of a phlebovirus: Biogenesis of proteins from the Rift Valley fever virus M segment. *J. Virol.* **1990**, *64*, 1549–1555. [PubMed]
111. Ikegami, T. Molecular biology and genetic diversity of Rift Valley fever virus. *Antivir. Res.* **2012**, *95*, 293–310. [CrossRef] [PubMed]
112. Won, S.; Ikegami, T.; Peters, C.J.; Makino, S. NSm protein of Rift Valley fever virus suppresses virus-induced apoptosis. *J. Virol.* **2007**, *81*, 13335–13345. [CrossRef] [PubMed]
113. Won, S.; Ikegami, T.; Peters, C.J.; Makino, S. NSm and 78-kilodalton proteins of Rift Valley fever virus are nonessential for viral replication in cell culture. *J. Virol.* **2006**, *80*, 8274–8278. [CrossRef] [PubMed]
114. Gerrard, S.R.; Bird, B.H.; Albarino, C.G.; Nichol, S.T. The NSm proteins of Rift Valley fever virus are dispensable for maturation, replication and infection. *Virology* **2007**, *359*, 459–465. [CrossRef] [PubMed]
115. Andersson, A.M.; Pettersson, R.F. Targeting of a short peptide derived from the cytoplasmic tail of the G1 membrane glycoprotein of Uukuniemi virus (*Bunyaviridae*) to the Golgi complex. *J. Virol.* **1998**, *72*, 9585–9596. [PubMed]
116. Persson, R.; Pettersson, R.F. Formation and intracellular transport of a heterodimeric viral spike protein complex. *J. Cell Biol.* **1991**, *112*, 257–266. [CrossRef] [PubMed]
117. Veijola, J.; Pettersson, R.F. Transient association of calnexin and calreticulin with newly synthesized G1 and G2 glycoproteins of Uukuniemi virus (family *Bunyaviridae*). *J. Virol.* **1999**, *73*, 6123–6127. [PubMed]
118. Pesonen, M.; Kuismanen, E.; Pettersson, R.F. Monosaccharide sequence of protein-bound glycans of Uukuniemi virus. *J. Virol.* **1982**, *41*, 390–400. [PubMed]
119. Chen, S.Y.; Matsuoka, Y.; Compans, R.W. Assembly of G1 and G2 glycoprotein oligomers in Punta Toro virus-infected cells. *Virus Res.* **1992**, *22*, 215–225. [PubMed]
120. Gahmberg, N.; Kuismanen, E.; Keranen, S.; Pettersson, R.F. Uukuniemi virus glycoproteins accumulate in and cause morphological changes of the Golgi complex in the absence of virus maturation. *J. Virol.* **1986**, *57*, 899–906. [PubMed]
121. Andersson, A.M.; Melin, L.; Bean, A.; Pettersson, R.F. A retention signal necessary and sufficient for Golgi localization maps to the cytoplasmic tail of a *Bunyaviridae* (Uukuniemi virus) membrane glycoprotein. *J. Virol.* **1997**, *71*, 4717–4727. [PubMed]
122. Gerrard, S.R.; Nichol, S.T. Characterization of the Golgi retention motif of Rift Valley fever virus G(N) glycoprotein. *J. Virol.* **2002**, *76*, 12200–12210. [CrossRef] [PubMed]
123. Ronnholm, R. Localization to the Golgi complex of Uukuniemi virus glycoproteins G1 and G2 expressed from cloned cDNAs. *J. Virol.* **1992**, *66*, 4525–4531. [PubMed]
124. Carnec, X.; Ermonval, M.; Kreher, F.; Flamand, M.; Bouloy, M. Role of the cytosolic tails of Rift Valley fever virus envelope glycoproteins in viral morphogenesis. *Virology* **2014**, *448*, 1–14. [CrossRef] [PubMed]
125. Jäntti, J.; Hildén, P.; Rönkä, H.; Mäkiranta, V.; Keränen, S.; Kuismanen, E. Immunocytochemical analysis of Uukuniemi virus budding compartments: Role of the intermediate compartment and the Golgi stack in virus maturation. *J. Virol.* **1997**, *71*, 1162–1172. [PubMed]
126. Overby, A.K.; Popov, V.L.; Pettersson, R.F.; Neve, E.P. The cytoplasmic tails of Uukuniemi Virus (*Bunyaviridae*) G(N) and G(C) glycoproteins are important for intracellular targeting and the budding of virus-like particles. *J. Virol.* **2007**, *81*, 11381–11391. [CrossRef] [PubMed]
127. Matsuoka, Y.; Chen, S.Y.; Compans, R.W. A signal for Golgi retention in the bunyavirus G1 glycoprotein. *J. Biol. Chem.* **1994**, *269*, 22565–22573. [PubMed]
128. Nilsson, T.; Jackson, M.; Peterson, P.A. Short cytoplasmic sequences serve as retention signals for transmembrane proteins in the endoplasmic reticulum. *Cell* **1989**, *58*, 707–718. [CrossRef]
129. Jackson, M.R.; Nilsson, T.; Peterson, P.A. Identification of a consensus motif for retention of transmembrane proteins in the endoplasmic reticulum. *EMBO J.* **1990**, *9*, 3153–3162. [PubMed]

130. Vincent, M.J.; Martin, A.S.; Compans, R.W. Function of the KKXX motif in endoplasmic reticulum retrieval of a transmembrane protein depends on the length and structure of the cytoplasmic domain. *J. Biol. Chem.* **1998**, *273*, 950–956. [CrossRef] [PubMed]
131. Kakach, L.T.; Suzich, J.A.; Collett, M.S. Rift Valley fever virus M segment: Phlebovirus expression strategy and protein glycosylation. *Virology* **1989**, *170*, 505–510. [CrossRef]
132. Chen, S.Y.; Matsuoka, Y.; Compans, R.W. Assembly and polarized release of Punta Toro virus and effects of brefeldin A. *J. Virol.* **1991**, *65*, 1427–1439. [PubMed]
133. Crispin, M.; Harvey, D.J.; Bitto, D.; Halldorsson, S.; Bonomelli, C.; Edgeworth, M.; Scrivens, J.H.; Huiskonen, J.T.; Bowden, T.A. Uukuniemi Phlebovirus assembly and secretion leave a functional imprint on the virion glycome. *J. Virol.* **2014**, *88*, 10244–10251. [CrossRef] [PubMed]
134. Overby, A.K.; Pettersson, R.F.; Neve, E.P. The glycoprotein cytoplasmic tail of Uukuniemi virus (*Bunyaviridae*) interacts with ribonucleoproteins and is critical for genome packaging. *J. Virol.* **2007**, *81*, 3198–3205. [CrossRef] [PubMed]
135. Strandin, T.; Hepojoki, J.; Vaheri, A. Cytoplasmic tails of bunyavirus Gn glycoproteins-Could they act as matrix protein surrogates? *Virology* **2013**, *437*, 73–80. [CrossRef] [PubMed]
136. Elliott, R.M.; Schmaljohn, C.S. *Bunyaviridae*. In *Fields Virology*, 5th ed.; Knipe, D.M., Howley, P.M., Eds.; Lipincott Williams & Wilkins: Philadelphia, PA, USA, 2013; pp. 1244–1282.

© 2016 by the authors. Licensee MDPI, Basel, Switzerland. This article is an open access article distributed under the terms and conditions of the Creative Commons Attribution (CC BY) license (http://creativecommons.org/licenses/by/4.0/).

Article

Conserved Endonuclease Function of Hantavirus L Polymerase

Sylvia Rothenberger [1], Giulia Torriani [1], Maria U. Johansson [2], Stefan Kunz [1,*] and Olivier Engler [3,*]

1 Institute of Microbiology, University Hospital Center and University of Lausanne, Lausanne CH-1011, Switzerland; Sylvia.Rothenberger-Aubert@chuv.ch (S.R.); Giulia.Torriani@chuv.ch (G.T.)
2 SIB Swiss Institute of Bioinformatics, Lausanne CH-1015, Switzerland; maria.johansson@isb-sib.ch
3 SPIEZ Laboratory, Austrasse, Spiez CH-3700, Switzerland
* Correspondence: Stefan.Kunz@chuv.ch (S.K.); oliver.engler@babs.admin.ch (O.E.); Tel.: +41-21-314-7743 (S.K.); +41-58-468-15-39 (O.E.); Fax: +41-21-314-4060 (S.K.); +41-58-468-14-02 (O.E.)

Academic Editors: Jane Tao and Pierre-Yves Lozach
Received: 27 February 2016; Accepted: 19 April 2016; Published: 2 May 2016

Abstract: Hantaviruses are important emerging pathogens belonging to the Bunyaviridae family. Like other segmented negative strand RNA viruses, the RNA-dependent RNA polymerase (RdRp) also known as L protein of hantaviruses lacks an intrinsic "capping activity". Hantaviruses therefore employ a "cap snatching" strategy acquiring short 5′ RNA sequences bearing 5′cap structures by endonucleolytic cleavage from host cell transcripts. The viral endonuclease activity implicated in cap snatching of hantaviruses has been mapped to the N-terminal domain of the L protein. Using a combination of molecular modeling and structure–function analysis we confirm and extend these findings providing evidence for high conservation of the L endonuclease between Old and New World hantaviruses. Recombinant hantavirus L endonuclease showed catalytic activity and a defined cation preference shared by other viral endonucleases. Based on the previously reported remarkably high activity of hantavirus L endonuclease, we established a cell-based assay for the hantavirus endonuclase function. The robustness of the assay and its high-throughput compatible format makes it suitable for small molecule drug screens to identify novel inhibitors of hantavirus endonuclease. Based on the high degree of similarity to RdRp endonucleases, some candidate inhibitors may be broadly active against hantaviruses and other emerging human pathogenic Bunyaviruses.

Keywords: Bunyaviridae; hantavirus; emerging diseases; endonuclease

1. Introduction

Hantaviruses belong to the *Bunyaviridea* family, a large group of segmented negative strand RNA viruses that include causative agents of severe human diseases [1–3]. Hantaviruses merit significant attention as emerging pathogens with expanding global distribution and incidence on the rise [4–7]. In Asia, the prototypic Hantaan virus (HTNV) and Seoul virus (SEOV) can cause hemorrhagic fever with renal syndrome (HFRS) with fatality rates of up to 3%. In the Americas, the hantaviruses Sin Nombre (SNV) and Andes (ANDV) are associated with hantavirus cardiopulmonary syndrome with up to 40% mortality [7–12]. Puumala virus (PUUV) is endemic in Northern Europe where it causes *nephropathia endemica*, a milder form of HFRS, while Dobrava-Belgrade virus (DOBV) is frequently associated with more severe disease [13]. There is currently no licensed vaccine against hantaviruses and therapeutic options are limited. The development of novel strategies for antiviral therapeutic intervention is therefore an urgent need.

Hantaviruses are enveloped viruses with three negative single-stranded RNA segments, small (S), medium (M) and large (L) [12,14]. The S segment encodes the viral nucleoprotein (N), M the precursor of the envelope glycoprotein (GPC) that gives rise to the mature glycoproteins Gc and Gn, and the L segment codes for the viral RNA-dependent RNA polymerase (RdRp), also known as L protein. The polymerase domains of the hantavirus L protein contain the five conserved motifs seen in other viral RNA polymerases [15]. Sequence analysis has revealed a high degree of conservation of the L polymerase domains among hantaviruses and bunyaviruses at large. In analogy to RdRp of other segmented negative-strand RNA viruses, hantavirus L protein functions as an RNA transcriptase and replicase, but lacks "capping activity", *i.e.*, the capacity to synthesize the 5′cap sequences found in viral transcripts. Similar to RdRp of *Orthomyxoviridae, Arenaviridae*, and *Orthobunyaviridae* hantavirus L protein acquires 5′cap sequence from cellular mRNA transcripts by a mechanism called "cap snatching". Cap snatching, originally described for influenza virus [16,17], involves binding of the 5′cap structure of a cellular mRNA by the viral RdRp followed by cleavage of the mRNA a few nucleotides downstream of the 5′cap structure by a viral endonuclease activity. The resulting short oligonucleotide bearing a 5′cap is then used by the RdRp as a primer for the synthesis of viral transcripts. In influenza virus, a cap-binding domain was found in the PB2 subunit of the polymerase [18], and an endonuclease domain mapped to the N-terminus of the PA subunit [19,20]. At the structural level, influenza PA endonuclease shares characteristics of the two metal-dependent PD (D/E)X K nuclease superfamily [21] with preference for Mn^{2+} ions [22]. Evidence for cap snatching in bunyaviruses was initially reported more than 30 years ago [23]. Newer studies defined an influenza PA-like endonuclease domain in the N-terminal region of the orthobunyavirus La Crosse (LACV) L protein with structural similarities to influenza virus PA endonuclease [24]. A similar endonuclease activity has recently been identified in the N-terminal domain of hantavirus L protein [25]. Expression of recombinant ANDV L protein resulted in a remarkably high endonuclease activity, which resulted in degradation of viral and cellular mRNAs, including L mRNA itself [25]. Accordingly, expression of ANDV L protein could be rescued upon mutations in the catalytic site of the endonuclease.

Due to their essential role in virus multiplication, the conserved endonucleases of RdRp of segmented negative strand RNA virus polymerases are of great interest for basic virus research. Their nature as enzymes makes them further attractive drug targets for therapeutic intervention. Here, we confirm and extend previous studies, providing further evidence for high structural and functional conservation of endonucleases of geographically distant hantaviruses and Bunyaviruses at large. Based on their known remarkable robust activity, we developed a functional cell-based assay for hantavirus endonucleases that is suitable for high-throughput small molecule screens.

2. Materials and Methods

2.1. Modeling

The N-terminal sequences of HTNV L and ANDV L polymerase (accession number X55901 and Q9E005_9VIRU, respectively) were compared to the previously characterized N-terminal endonuclease domains of LACV L protein (accession number A5HC98_BUNLC, residues 1–183) and PA influenza virus (PAN) (Influenza A virus A/VietNam/1203/2004 (H5N1), accession number Q5EP34_9INFA, residues one to 209) [24,26]. The active sites of HTNV and ANDV were modeled using the recently determined structure of LACV (PDB entry 2XI5). The suitability of LACV as a template was established through pair wise comparison of profile hidden Markov models (HMM-HMM alignment) using HHpred [27]. The target-template alignments produced by HHpred were manually inspected and modified when necessary. Model structures were calculated using MODELLER [28]. To search for related endonuclease structures in other life forms, we defined structural motifs corresponding to the conserved spatial arrangement of residues of the active site of LACV endonuclease, specifically H34/D79/D92/K94, H34/D52/D79/D92, and H34/P78/D79/D92 [24], with residue numbering according to PDB entry 2XI5. Using DeepView/Swiss-PdbViewer, as described elsewhere [29], we searched for geometric similarities to these motifs in the subset of all protein chains in the RCSB Protein Data Bank (as of 21 October 2014) that originate from non Cα-only X-ray structures with a resolution of at least 3.0 Å and sequence lengths of 40–10,000, for which the maximum pairwise sequence identity between any two chains in the subset is at most 99%. This subset contains ~45,000 chains, from viral, bacterial, fungal, plant, and animal proteins. Our searches did not yield any close geometric matches between the structural motifs and any existing structure in mammalian cells.

2.2. Construction of Plasmids

The coding region for HTNV L protein was amplified by PCR using the primers 5′-GGGACTAGTGGCACCATGGATAAATATAGAGAAATTCAC-3′ and 5′-GGGGGATCCATAGAAA GAGGAAATAGAATCCTGC-3′, KAPA HiFi™ DNA polymerase (KAPA BIOSYSTEMS, Wilmington, MA, USA) and pWRG/HTNV-L [30], kindly provided by R. Flick, Ames, Iowa, USA, was used as a template. PCR fragments were subcloned into pCR™-Blunt II-TOPO® vector using Zero Blunt® TOPO® PCR Cloning Kit (Invitrogen™, Carlsbad, CA, USA) according to the manufacturer's instructions. The fragment comprising the coding region was subcloned into the corresponding sites of the vector pTM, kindly provided by P.Y. Lozach, Heidelberg, Germany. L protein mutants were generated via a classical two-step mutagenesis approach using pWRG/HTNV L as template, the primers containing the desired mutations (all primer sequences are available upon request) and KAPA HiFi™ DNA polymerase (KAPA BIOSYSTEMS). For immunoblotting, hemagglutinin (HA) tag or enhanced green fluorescent protein (EGFP) were fused to the C-termini of the genes. A linker (amino acids sequence G S, nucleotide sequence *Bam*HI, GGATCC) was placed between the coding sequence of the polymerase and the epitope tag. Two sets of L protein constructs were generated, comprising either the full length HTNV L polymerase (pTM-L) or HTNV L endonuclease domain (amino acids 1–223) (pTM-E), generated using the primers 5′-GGGACTAGTGGCACCATGGATAAATATAGAGAAATTCAC-3′ and 5′-GGGGGATCCCTGGCTTTTGTGTGTACTTAT-3′. All plasmids were verified by sequencing. The plasmid pTM-Nluc was generated by PCR using the appropriate primers, KAPA HiFi™ DNA polymerase (KAPA BIOSYSTEMS) and Nanoluc® Luciferase 1.1 (Promega, Madison, WI, USA) plasmid as template. The constructs were verified by sequencing.

To construct the wild-type and mutant L protein fused to nanoluciferase (NLuc), the coding region comprising amino acid 2 to 171 of Nanoluc® Luciferase 1.1 (Promega) gene reporter, flanked by the restriction sites *Bam*HI and *Pac*I was generated by PCR using the appropriate primers, KAPA HiFi™ DNA polymerase (KAPA BIOSYSTEMS) and Nanoluc® Luciferase 1.1 (Promega) plasmid as template. The *Bam*HI-*Pac*I fragment of pTM L-HA and pTM E-HA was subsequently replaced by the *Bam*HI-*Pac*I fragment comprising Nanoluc® Luciferase 1.1 PUUV L protein was isolated from Vero cells inoculated with PUUV. PUUV lot number 612 from National Collection of Pathogenic Viruses (NCPV), Catalogue number 0504101v was cultivated in the biosafety level (BSL) 3 containment facility at Spiez Laboratory, Spiez, Switzerland. After eight days the supernatant was cleared by low-speed centrifugation, virus RNA was extracted using a EZ1 Virus Mini Kit v2.0 (QIAGEN, Hilden, Germany) and reverse transcribed by using random hexamers and the cDNA amplified using virus-specific primers using PrimeScript™ reverse transcription (RT)-PCR Kit (Takara/Clontech, Mountain View, CA, USA). The fragment encoding amino acid 1 to 223 of PUUV L protein was subcloned into pTM as described above using the primers 5′-GGGACTAGTGGCACCATGGAGAAATACAGAGAGATC-3′ and 5′-GGGGGATCCTCGTACTTTTGGGCCTGTGAC-3′. The mutation D97A was generated via a classical two-step mutagenesis approach. The constructs were verified by sequencing.

2.3. Cells

BSR T7/5 cells stably expressing T7 polymerase [31] were grown in Dulbecco Modified Eagle Medium (DMEM) GlutaMAX™ (Gibco/Thermo Fisher Scientific, Waltham, MA, USA) supplemented with 10% fetal calf serum (Amimed/BioConcept, Allschwil, Switzerland), 1% Tryptose phosphate Broth 1× (Gibco), 1% MEM non-essential amino acids solution (Gibco) and 1% penicillin streptomycin (10,000 U/mL) (Gibco). Cells were grown in an atmosphere of 5% CO_2 and 37 °C. Every passage, 1 mg of Geneticin (Promega) per mL medium was added to the culture.

2.4. Reporter Assays

Cells were seeded at a density of 2 × 10^5 cells per well in 24-wells dishes. Transfections were performed using jetPRIME® reagent (Polyplus-transfection Inc., Illkirch, France) or Lipofectamine® 3000 (Life Technologies, Carlsbad, CA, USA). Twenty-four to 48 hours after transfection, cells were lyzed in 1× Cell Culture Lysis Reagent (Promega). The activity of the Nanoluc® Luciferase (Promega) gene reporter was measured using the Nano Glow® Luciferase Assay System (Promega) according to manufacturer's instructions and using a Lumat LB 9507 (Berthold Technologies, Pforzheim, Germany) luminometer.

2.5. Western Blotting

Cell lysates were resuspended in one volume sodium dodecyl sulfate (SDS)-sample buffer and heated at 95 °C for 5 min before being processed by SDS-polyacrylamide gel electrophoresis (SDS-PAGE). Proteins were transferred to nitrocellulose membranes and then detected using specific primary and secondary antibodies. Rat monoclonal antibody 3F10 against HA was from Roche (Basel, Switzerland), mouse monoclonal antibody JL-8 against EGFP from Living Color/Clontech (Mountain View, CA, USA) and mouse monoclonal antibody B5-1-2 against α-tubulin from Sigma (St. Louis, MO, USA). Polyclonal rabbit anti-mouse or anti-rat antibodies conjugated to horseradish peroxidase (HRP) were from Dako (Glostrup, Denmark). Proteins bands were visualized using a chemiluminescence detection kit (WesternBright™ Sirius chemioluminescent HRP substrate (Advansta, Menlo Park, CA, USA) according to the manufacturer's instructions.

2.6. Protein Expression and Purification

The coding region of the N-terminal 220 residues of HTNV L protein (Accession number X55901) was optimized for expression in *E. coli* and synthetized (DNA2.0). Proteins were expressed in *E. coli* strain BL21 (DE3) C41 (Lucigen Coorporation, Middleton, WI, USA) in lysogeny broth (LB) media at 18 °C overnight after induction with 0.2 mM of isopropyl β-D-1-thiogalactopyranoside (IPTG). Cells were lyzed in Tractor buffer (Clontech, Mountain View, CA, USA) supplemented with lysozyme, DNAse, and EDTA-free protease inhibitor cocktail (Roche, Basel, Switzerland).

Proteins from the soluble fraction were purified using a TALON® Metal Affinity resin (Clontech, Mountain View, CA, USA) as recommended by the manufacturer.

2.7. In Vitro Endonuclease Assay

For nuclease experiments, 5 to 10 µM of the purified protein were incubated with 25 ng/µL single stranded M13 mp18 DNA (Bayou Biolabs, Metairie, LA, USA) in 20 mM Tris-HCl pH 8.0, 150 mM NaCl, 2.5 mM β-mercaptoethanol and at 37 °C for 60 min. Divalent cations were added to a 2 mM concentration. As a control, the reaction was performed in EDTA 20 mM.

2.8. Cell Viability

Cytotoxicity of candidate compounds was assessed using CellTiter-Glo® Luminescent Cell Viability Assay (Promega), which is used to determine the number of viable cells in a culture based on quantification of ATP. Briefly, 2 × 10^5 cells were plated per well of a 24-well tissue culture plate and transfected with the indicated constructs as described in 2.4. After 24 h, CellTiter-Glo® reagent was added and the assay performed according to the manufacturer's instructions.

2.9. Statistical Analysis

Nanoluciferase assay data were analyzed using one-way analysis of variance (ANOVA) as indicated in the figures and figure legends.

3. Results

3.1. Conservation of the N-Terminal Endonuclease of Hantavirus L Polymerase

Previous studies demonstrated the existence of an endonuclease function in the N-terminal domain of the New World hantavirus ANDV and other hantaviruses [25]. Here we sought to compare the endonuclease function of Old World and New World hantaviruses that differ in geographic distribution, genetics, and disease potential. In a first step, we compared the N-terminal sequences of ANDV L protein to the prototypic Old World hantavirus HTNV and the previously characterized endonuclease domains of LACV L protein (residues 1–183) and influenza virus PD (residues 1–209, H5N1 A/VietNam/1203/2004) [24,26]. As expected, based on the high mutation rate of the viruses, our alignment revealed only low sequence identity of the HTNV L protein N-terminal region with LACV (13.9% at the amino acid level) and with influenza virus PA (9.5%). Sequence homology between the endonuclease domains from LACV and influenza virus was likewise low (8.8%). Secondary structure predictions resulted in a good match between residues 1–183 of LACV L protein and the N-terminal region of HTNV L protein encoded by amino acids 1–220 (data not shown). Next, we modeled the active site of the putative N-terminal endonuclease domains of HTNV and ANDV (residues 1–163) (Figure 1A) using the high-resolution structure of LACV [24] as a starting point, as detailed in Materials and Methods. According to the resulting model shown in Figure 1B, the LACV L endonuclease domain contains a cation-binding fold similar to the one found in influenza virus PA and other members of the PD (*D*/*E*)X K nuclease superfamily. Based on the model, we propose that residues H36, E54, D97, E110 and T112 of HTNV L protein correspond to previously identified key residues of the active site of the LACV endonuclease, namely H34, D52, D79, D92 and K94 (Figure 1B,C). Our structure-based alignment differed only slightly from previous alignments with E54 likely representing a residue within the active site, rather than the originally proposed E75 [24,32]. In sum, our modeling suggested a high degree of conservation of the endonuclease domain between HTNV, ANDV, and LACV.

To validate our model of the putative active site of the hantavirus endonuclease, we undertook structure function analysis. Expression of recombinant hantavirus endonuclease results in degradation of its proper mRNA, reducing expression to nearly undetectable levels [25]. Mutations of the putative active site restored expression of the recombinant protein [25], allowing the identification of key residues implicated in catalytic activity. Based on the HTNV L protein model (Figure 1B), we generated a recombinant HTNV L protein fragment comprised of the N-terminal 223 amino acids corresponding to the putative endonuclease domain, containing a C-terminal HA-tag or EGFP HTNV Ewt-HA and HTNV Ewt-GFP (Figure 2A). Constructs allowed expression under the control of a T7 promoter allowing efficient expression in the cytosol, where hantavirus replication takes place. In a first step, we mutated residue D97 of HTNV L protein that corresponds to the previously identified residue D79 residing in the PD sequence of LACV L protein [24]. To monitor endonuclease activity of our constructs, we co-transfected a NLuc reporter plasmid, allowing us to measure degradation of reporter transcript via luciferase assay, as previously described [25]. Mutant and wild-type HTNV E-HA and HTNV E-GFP were transiently transfected into BSR T7/5 cells that stably express T7 polymerase. After 48 h of expression, wild-type HTNV Ewt-HA and HTNV Ewt-GFP were barely detectable by Western blot, whereas strong signals were observed for the mutants HTNV ED97A-HA, HTNV ED97A-GFP (Figure 2B). Co-expression of the NLuc reporter construct with HTNV Ewt-HA and HTNV Ewt-GFP, but not the mutants HTNV ED97A-HA and HTNV ED97A-GFP markedly reduced the luminescence signal (Figure 2C).

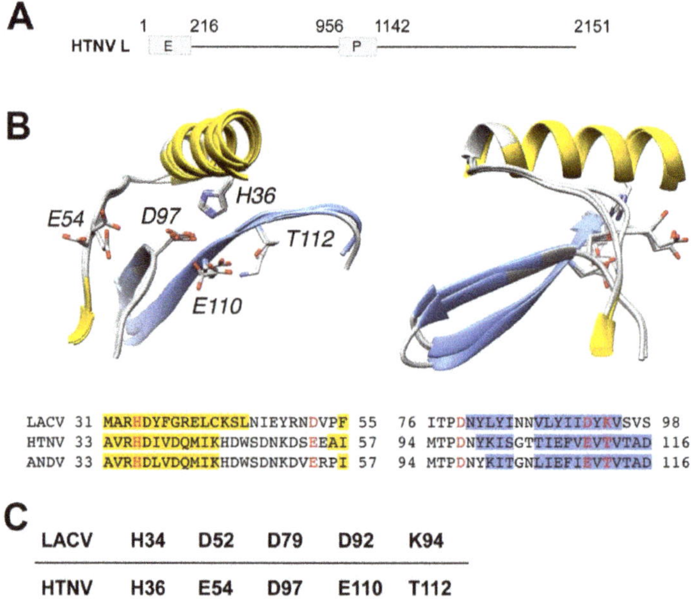

Figure 1. Model of the putative active sites of prototypic Hantaan virus (HTNV) and Andes (ANDV) superposed on the active site of orthobunyavirus La Crosse (LACV). (**A**) Schematic representation of HTNV L protein. The L segment of the prototypic HTNV strain 76/118 has 6533 nucleotides and encodes the L protein of 2151 amino acids (aa). HTNV L contains a polymerase (P) domain (aa 956–1142) and a putative endonuclease (E) domain (aa 1–216). (**B**) Models of the putative active sites of HTNV and ANDV superposed on the active site of LACV. Ribbon diagrams of LACV (PDB entry 2XI5), HTNV and ANDV after structural superposition of key active site residues. Secondary structure classifications was calculated using definition of secondary structure proteins (DSSP) [33] for LACV or predicted using the method/program PSI-PRED [34]. Helix residues have been colored yellow and β-strand residues have been colored blue in the ribbon diagrams as well as in the alignment, according to DSSP or PSI-PRED results, respectively. Annotations in ribbon diagrams follow the HTNV residue numbering in the alignment. Key residues are colored red in the alignment and shown with side chains in the ribbon diagrams. The ribbon diagrams to the right have been rotated 90 degrees around the vertical coordinate axis. (**C**) Amino acid residues of the active site of HTNV L endonuclease derived from the model in (**A**).

Next, we subjected HTNV E-HA to site directed mutagenesis performing alanine replacement of additional residues predicted to be located within or close to the putative active site. Specifically, we mutated residues H36, V34, R35, D37, E54, and the previously identified residues K44 [25] and E75 [24,32]. Examination by Western blot revealed that mutations V34A and E75A hardly increased protein expression, suggesting intact endonuclease activity (Figure 2D). In contrast, mutations R35A, H36A, D37A, K44A, E54A, and D97A resulted in at least partial rescue of protein expression, indicating reduced endonuclease activity (Figure 2D). Co-expression of mutants and wild-type HTNV E-HA with the NLuc expression plasmid resulted in reporter activities that correlated with the expression levels of the HTNV E-HA variants (Figure 2D,E).

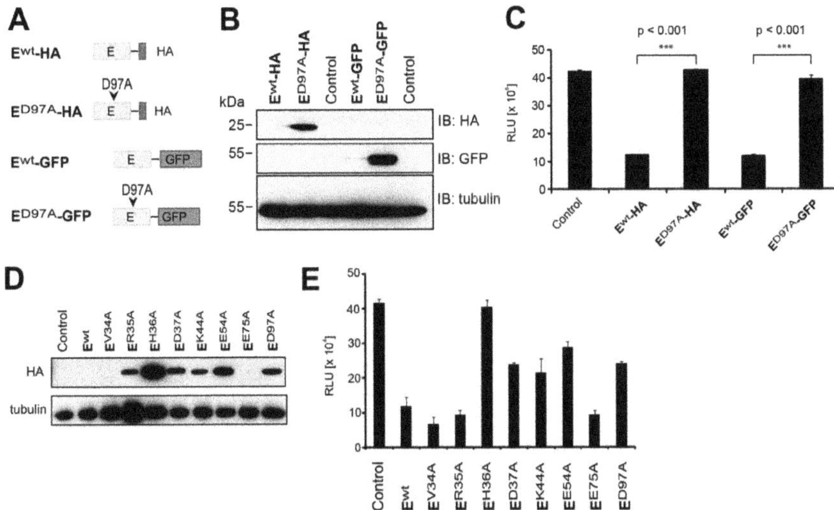

Figure 2. Mutational analysis of HTNV endonuclease domain. (**A**) Schematic representation of the constructs. (**B**) BSR T7/5 cells were transfected with wild-type (wt) or D97A HTNV L protein N-terminal constructs along with an nanoluciferase (NLuc) reporter control plasmid. Cells were lyzed 48 h post-transfection. Proteins were separated by sodium dodecyl sulfate-polyacrylamide gel electrophoresis (SDS-PAGE) and transferred on membranes for immunoblotting. (**C**) NLuc activity was measured using the Nano Glow® Luciferase Assay System (Promega). Data represent mean ± SD (n = 3). One representative experiment out of three is shown. Data were analyzed using one-way ANOVA with p-values indicated. (**D**) BSR T7/5 cells were co-transfected with HA-tagged WT L protein N-terminal constructs and the indicated mutants together with an NLuc reporter plasmid. Cells were lyzed 48 h post-transfection and HA-tagged endonuclease constructs detected in Western blot. (**E**) Nanoluciferase activity was measured using the Nano Glow® Luciferase Assay System (Promega). Data represent mean ± SD (n = 3). One representative experiment out of three is shown. Data were analyzed using one-way ANOVA with p-values *** $p < 0.001$, The significant difference with mutant E75A ($p < 0.05$) was not consistently observed between independent experiments.

The remarkably high endonuclease activity of L protein had so far only been demonstrated for hantaviruses associated with severe human diseases. We next compared the endonuclease activity of the pathogenic HTNV with PUUV, which is associated with only mild human disease. At the amino acid sequence level, the putative endonucleases of HTNV and PUUV are similar (Figure 3A), allowing expression cloning of the putative PUUV endonuclease. A side-by-side comparison of the wild-type and D97 mutants of HTNV E-HA and PUUV E-HA revealed similar low expression levels of the wild-type forms with significant rescue upon mutation D97A (Figure 3B). When compared to HTNV ED97A-HA, PUUV ED97A-HA was consistently expressed at lower levels and migrated slightly different. The reasons for this are currently unclear. To quantitatively assess the relative catalytic activity of the endonucleases of PUUV and HTNV, the constructs were co-expressed with our NLuc reporter and lucifease activity detected as described above. The direct comparison showed similar degradation of the NLuc transcript by HTNV Ewt-HA, PUUV Ewt-HA with rescue by the mutation D97A (Figure 3C) indicating that the remarkably high endonuclease activity is conserved between the two hantavirus species.

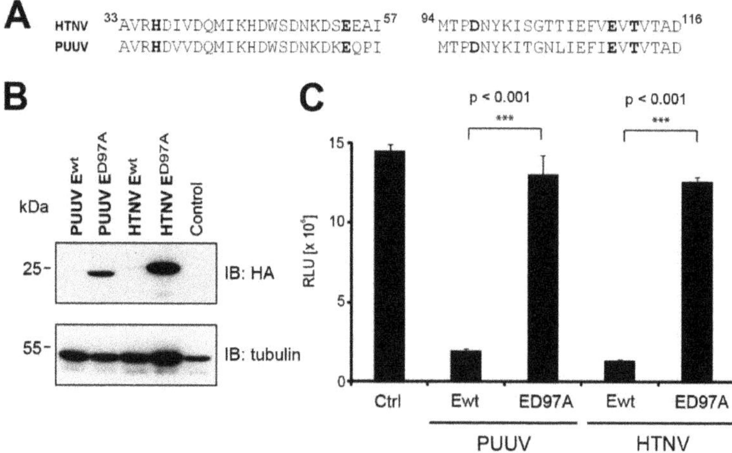

Figure 3. Puumala virus (PUUV) L protein possesses a functional endonuclease domain. (**A**) Sequence alignment of the N-terminal domains of HTNV and PUUV L protein. (**B**) BSR T7/5 cells were transfected with wt or D97A PUUV and HTNV L protein N-terminal constructs along with an NLuc reporter control plasmid. Cells were lyzed 48 hours post-transfection. Proteins were separated by SDS-PAGE and transferred on membranes for immunoblotting. (**C**) NLuc activity was measured using the Nano Glow® Luciferase Assay System (Promega). Data are expressed as mean $\pm$ SD (n = 3) and were analyzed by one-way ANOVA with p-values indicated.

3.2. *In Vitro Activity and Cation-Dependence of HTNV L Endonuclease*

Biochemical studies on the viral endonucleases of influenza virus and LACV *in vitro* revealed enzymatic activity against single-stranded-RNA (ssRNA) and ssDNA with strong dependence on Mn^{2+} [19,22,24,26]. In order to demonstrate catalytic activity of HTNV endonuclease *in vitro* under defined conditions, we expressed the 220 N-terminal residues of HTNV L protein containing the putative endonuclease domain in a bacterial system as detailed in Section 2.6. As a control, we engineered a "catalytic dead" version. Upon induction of bacterial expression both the wild-type protein and the catalytically inactive mutant were initially produced. However, during subsequent purification, the catalytically inactive variant was consistently lost, resulting in markedly reduced yields. A possible reason may be an overall reduced stability of the protein due to the presence of four mutations H36A, E75A, D97A and E110A, but this remains speculative at this point. The wild-type protein was purified via immobilized metal affinity chromatography (IMAC), resulting in a >90% pure protein as detected by Coomassie brilliant blue (Figure 4A). Using ssDNA, previously identified as a suitable substrate for both influenza and LACV endonucleases [19,24,26], we were able to detect the catalytic activity of our recombinant HTNV endonuclease *in vitro* (Figure 4B). Addressing divalent cation-dependence, we found high activity of HTNV endonuclease against ssDNA in presence of Mn^{2+}, partial activity in presence of Mg^{2+}, but none upon addition of Zn^{2+}, or Ca^{2+} (Figure 4B). Double-stranded DNA (dsDNA) was found to be a poor substrate, excluding non-specific nuclease contamination (data not shown). Together, our *in vitro* studies revealed for the first time enzymatic activity of HTNV endonuclease with a strong preference for Mn^{2+} shared with the endonucleases of LACV and influenza. The low residual activity of HTNV endonuclease in presence of Mg^{2+} resembles influenza PA endonuclease and differs from LACV [22,24]. The reasons for these differences are unknown but may be related to the specific nature of the metal-binding residues in hantaviruses *vs.* orthobunyaviruses.

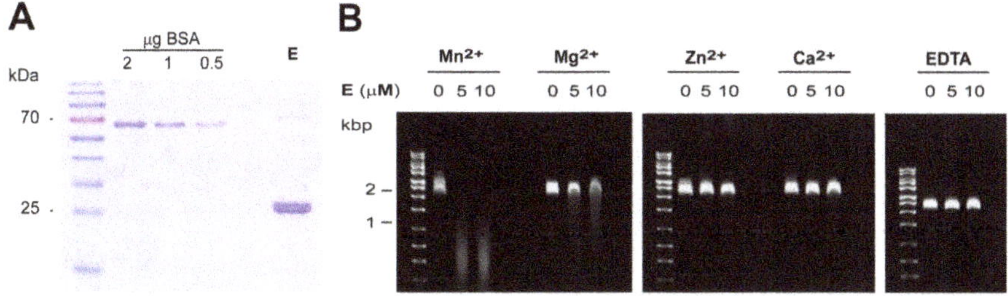

Figure 4. *In vitro* analysis of HTNV endonuclease. (**A**) The N-terminal 220 residues of HTNV L protein were expressed in *E. coli* and purified using TALON® metal affinity resin as described in the methods section. A fraction of the purified material was separated by SDS-PAGE and stained with Coomassie blue. (**B**) Divalent cation-dependent nuclease activity. Single-stranded M13mp18 DNA (25 ng/μL) was incubated at 37 °C during 60 min in presence of 0, 5 or 10 μM of purified HTNV N-terminal domain and 2 mM of the indicated divalent cations, or 10 mM ethylenediaminetetraacetic acid (EDTA).

3.3. Development of a Cell-Based Assay for Hantavirus Endonuclease

The remarkably robust endonuclease activity of hantavirus L protein resulting in degradation of transcripts in *cis* and in *trans* opened the possibility to establish a cell-based functional assay for the endonuclease activity based on an NLuc reporter. Previous studies revealed that the efficiency of degradation of mRNAs by hantavirus endonucleases seems proportional to the length of the transcript [25]. We hypothesized that longer transcripts may therefore enhance the sensitivity of our assay. To test this possibility, we fused wild-type and endonuclease dead (D97A) full-length HTNV L protein and HTNV Ewt at the C-terminus to NLuc, resulting in the constructs HTNV Lwt-NLuc, LD97A-NLuc, Ewt-NLuc, and ED97A-NLuc (Figure 5A). Constructs were expressed via a T7-driven cytoplasmic expression plasmid in BSR T7/5 cells. Due to the lack of a specific antibody to NLuc, we were unable to detect our NLuc fusion proteins in Western blot. We therefore relied on the more sensitive NLuc reporter to detect the presence of our fusion proteins. As expected, we observed marked reduction of the NLuc reporter signal in presence of active endonucleases (Figure 5B). The presence of the mutation D97A increased the NLuc signal in the context of the full-length L protein construct HTNV Lwt-NLuc *vs.* LD97A-NLuc consistently by >40-fold, whereas *circa* 8-fold enhancement was observed with HTNV Ewt-NLuc compared to ED97A-NLuc (Figure 5B). The *circa* 5-fold difference in degradation between full-length L protein and the isolated endonuclease domain correlated well with the relative length of their transcripts (Figure 5A). Taken together, our results show that both endonuclease and NLuc activities are maintained in our fusion constructs. The mutation D97A located in the active site of the endonuclease domain has more impact in the context of the full-length L protein polymerase than in the context of the N-terminal domain, likely due to the different length of the mRNA, in line with previous studies [25]. Although the dynamic range of the assay based on the full-length polymerase seemed higher, the absolute signal intensity was considerably lower, resulting in more variability. We therefore opted for the shorter E-NLuc construct for further development of the assay.

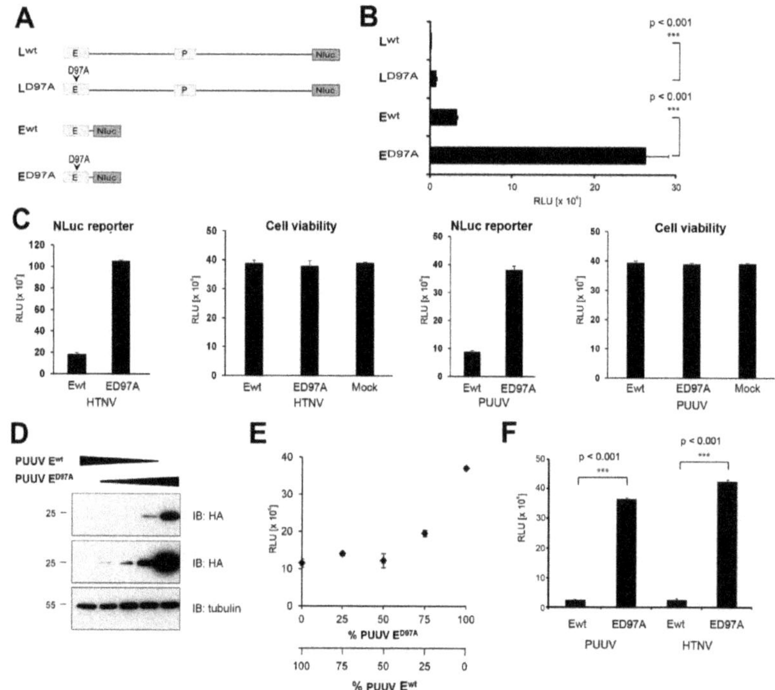

Figure 5. Development of a cell-based assay for endonuclease activity. (**A**) Schematic representation of the constructs. Full-length L protein wt and D97A mutant, as well as the endonuclease domain (E) wt and D97A mutant were fused to NLuc at the C-terminus. (**B**) The constructs were expressed in BSR T7/5 cells. Cells were lyzed 48 h post-transfection and NLuc activity measured using the Nano Glow® Luciferase Assay System (Promega). Data represent mean ± SD (n = 3) and were analyzed by one-way ANOVA. One representative experiment out of three is shown. (**C**) Cell viability assay. The indicated NLuc fusion constructs were transfected into BSR T7/5 cells. After 24 h, NLuc reporter activity was measured as in (**B**). Cell viability was assessed by Cell TiterGlo® assay. Data represent mean ± SD (n = 3). (**D**) Different relative amounts of PUUV Ewt-HA and PUUV ED97A-HA were co-transfected into BSR T7/5 cells. At 24 h post transfection, protein expression was detected in Western blot as in Figure 2B. (**E**) Different relative amounts of PUUV Ewt-NLuc and PUUV ED97A-NLuc were co-transfected into BSR T7/5 cells and NLuc reporter activity detected after 24 h as in (**B**). Data represent mean ± SD (n = 3). (**F**) PUUV Ewt and HTNV Ewt and the corresponding D97A mutants were fused to NLuc at the C-terminus. All constructs were expressed as in (**B**). Cells were lyzed 24 h post-transfection and NLuc activity measured. Data represent mean ± SD (n = 3) and were analyzed by one-way ANOVA with p-values indicated. One representative experiment out of three is shown.

A possible concern was unspecific toxicity of our hantavirus E-NLuc construct due to its capacity to degrade cellular transcripts. To address this issue, we expressed wild-type and D97A mutant E-NLuc constructs of HTNV and PUUV under assay conditions for 24 h and assessed cell viability by Cell TiterGlo® assay that detects cellular ATP levels. As shown in Figure 5C, none of the constructs caused significant reduction in cell viability under these conditions. The reduced expression of reporter plasmids co-transfected with wild-type endonuclease (Figure 5C) suggests that the endonuclease has no absolute specificity for viral RNA and may degrade cellular transcripts as well. Of note, in our assay format, expression of wild-type endonuclease was only of short duration (24 h). At this point, the impact on cell viability was only mild (Figure 5C). However, prolonged expression of high levels of recombinant wild-type endonuclease may be detrimental for cells and the duration of the assay has to be kept at a minimum to prevent unspecific off-target effects. A crucial aspect of a functional cell-based assay is its dose–response characteristic. To get a first estimate of the dynamic range of our assay, we transfected cells with different ratios of wild-type and D97A mutant PUUV E-HA and E-NLuc constructs and determined protein expression and reporter activity, respectively. As expected, increased proportion of the D97A mutant construct progressively rescued protein expression of the PUUV E-HA constructs (Figure 5D). Detection of reporter activity in cells transfected with different ratios of wild-type and D97A mutant PUUV E-NLuc resulted in a dose–response characterized by a negligible increase of NLuc activity up to a wild-type to D97A mutant ratio of *circa* 1. With wild-type to D97A mutant ratios of >1, the reporter activity increased inversely proportional to the amount of residual wild-type, as expected (Figure 5E). This particular dose–response characteristic was observed independently of the absolute amount of constructs expressed, suggesting an effect of the ratio, rather than expression level.

In a last step, we assessed the robustness of our assay. To this end, we determined its Z' value ($Z' = 1 - (3\sigma_{c+} + 3\sigma_{c-})/(\mu_{c+} - \mu_{c-})$), which depends on the sum of the standard deviations of positive and negative controls (σ_{c+} and σ_{c-} respectively) as well as the difference between the mean activity of these controls (μ_{c+} and μ_{c-}) (35). For the determination of the Z' value of our assay, wild-type and D97A mutant E-NLuc constructs of HTNV and PUUV were used as positive and negative controls, respectively (Figure 5E). Data from three independent experiments performed in triplicates yielded Z' values of >0.9 (Table 1), indicating sufficient robustness for implementation in high-throughput formats [35].

Table 1. Evaluation of the quality of the cell-based endonuclease assay.

Experiment	PUUV	HTNV
Number 1	0.94	0.93
Number 2	0.93	0.90
Number 3	0.94	0.90

Calculated values of the Z'-factor from three independent experiments performed in triplicates. The Z'-factor was calculated as described in Zhang *et al.* [35].

4. Discussion

In the present study, we confirmed and extended previous studies providing evidence for a high conservation of the endonuclease activity found in the N-terminal domain of the L polymerase of hantaviruses. We expressed active recombinant hantavirus endonuclease and were able to show catalytic activity *in vitro* and specific preference for divalent cations. Based on its remarkably robust activity, we developed a cell-based functional assay for hantavirus endonuclease suitable for high throughput formats.

Based on the existing high-resolution structure of the LACV endonuclease domain [24], we modeled the active sites of the putative endonuclease domain of HTNV and ANDV L polymerases. The overall fold of hantavirus endonuclease resembled the structure of the LACV homologue, which was expected based on the predicted similar secondary structures and our modeling approach. According to the model, residues H36, E54, D97, E110 and T112 of HTNV L protein correspond to the key residues of the LACV endonucleases active site, H34, D52, D79, D92 and K94. Performing structure–function analysis, using site-directed mutagenesis, we validated our model of the active site demonstrating a crucial role of residues H36, E54 or D97 for the activity of HTNV L endonuclease, in line with previous studies performed on the endonuclease of ANDV [25]. Residues H36, E54, D97, E110 and T112 are fully conserved in all known hantavirus isolates, suggesting that this activity is highly conserved.

The past years have seen the advent of efficient rescue systems for several bunyaviruses [36–41]. However, for hantaviruses, progress has been limited to minireplicon systems [30,42] and to the best of our knowledge, no hantavirus rescue system has so far been reported. The robust endonuclease activity of recombinant hantavirus L protein results in degradation of viral and cellular transcripts in *cis* and *trans* [25]. As pointed out earlier [25], the resulting inability to express recombinant hantavirus L protein in sufficient quantities in mammalian cells may represent an obstacle for the development of hantavirus reverse genetics systems. In our present study, we identified mutants of L protein's endonuclease with partially reduced activity. Such variants may retain sufficient endonuclease activity to fulfill essential functions required for transcription, e.g., cap-snatching, without causing excessive degradation of viral and cellular transcripts. We are currently testing this approach in the context of a minireplicon system.

The robust endonuclease activity of hantavirus L protein resulting in degradation of its own transcript was observed with the N-terminal domain alone and in the context of the full-length L protein. However, it is unclear if L protein reaches sufficiently high expression levels in infected cells to cause a similar effect. Interestingly, recent insights into the structure of the bunyavirus polymerase revealed that the complementary 3′ and 5′ ends of the viral RNA do not form a "panhandle" structure in the pre-initiation complex, as previously anticipated, but are bound to separate sites within L protein distant from the N-terminal endonuclease domain [43]. In the proposed model of viral RNA replication, the 3′ and 5′ termini remain bound to L protein at all times, whereas nascent viral RNA is protected by association with the N protein [43]. If this model of L protein is applicable to hantaviruses as well, L protein's endonuclease would be unable to degrade viral RNA as long as it remains associated with the polymerase or N. The endonuclease activity of L protein's N-terminal domain may specifically degrade "naked" viral RNA that may leak out of the replication-transcription complex and may otherwise interfere with efficient replication.

Many potent antiviral drugs currently used for treatment of human infections with negative stranded RNA viruses target viral replication and transcription. The most important drug target within the viral replication-transcription complex is the viral polymerase, which frequently shows a high degree of conservation within a given virus family and catalyzes both transcription and replication. Considering its crucial role in viral transcription and its nature as an enzyme, the endonuclease activity found in RdRp of segmented negative strand RNA viruses appears to be a possible drug target for therapeutic intervention. The data at hand suggest a high degree of conservation of the endonucleases found in RdRp of hantaviruses, bunyaviruses, and orthomyxoviruses. The identification of broadly-active inhibitors of viral endonucleases by small molecule drug screens appears therefore conceivable. Considering the existence of endogenous endonucleases in human cells, whose function is essential for normal physiology, possible cross-reactivity of candidate inhibitors is a concern. To address this potential pitfall, we performed extensive searches based on structural motifs defined by the conserved spatial arrangement of residues of the active site of LACV as detailed in Materials and Methods. Screening of >100,000 structures of proteins from all kingdoms of life, we found a large distance between the conserved viral endonucleases and any mammalian structure, suggesting that the identification of specific inhibitors that block the activity of viral endonuclease, but not human counterparts may be feasible. The main purpose of our current screening assay is to identify specific inhibitors that block the viral endonuclease with minimal cross-reactivity with endogenous human endonucleases. However, the differential inhibition of viral *vs.* human endonuclease would of course represent an essential step of counter-screening following the identification of initial candidate inhibitors that block the viral endonuclease.

Based on the conserved robust endonuclease activity of hantavirus L protein, we developed a reliable cell-based reporter assay suitable for high-throughput screening of collections of synthetic compounds in a rapid and cost-effective manner. The negligible toxicity of our reporter constructs minimizes unwanted off-target effects and therefore the probability of false positive hits. We examined robustness of our assay and found a Z′-value of 0.9, whereas Z′-values > 0.5 are generally considered as "excellent assays" with a low probability of false positive or negative hits [35]. The high throughput format of our assay will allow us to cover large "chemical space", opening the possibility to identify candidate inhibitors with broad activity targeting conserved structures present in viral endonucleases with possible therapeutic application against a range of known and newly emerging viruses. In addition, candidate small molecule inhibitors identified in our studies may serve as "molecular probes" to dissect the largely unknown mechanisms underlying hantavirus multiplication in human cells and advance our knowledge about this important emerging virus family.

Acknowledgments: The authors thank Stephan Günther from the Bernhard-Nocht-Institute for Tropical Medicine, Hamburg, Germany, for invaluable inputs and stimulating discussions. We further acknowledge Ramon Flick for providing the cDNA of HTNV L protein, Pierre-Yves Lozach for the expression vector pTM, and Klaus Conzelmann for the BSR T7/5 cells. This research was supported by Swiss Federal Office for Civil Protection (Grant Nr. 353004328/STM to O.E. and S.K.), and funds from the University of Lausanne (S.K.).

Author Contributions: Olivier Engler, Stefan Kunz and Sylvia Rothenberger conceived and designed the study. Maria U. Johansson conducted the modeling. Sylvia Rothenberger and Giulia Torriani conducted the experiments. Stefan Kunz and Sylvia Rothenberger wrote the manuscript. All authors have read and approved the manuscript.

Conflicts of Interest: The authors declare no conflict of interest.

References

1. Walter, C.T.; Barr, J.N. Recent advances in the molecular and cellular biology of bunyaviruses. *J. Gen. Virol.* **2011**, *92*, 2467–2484. [CrossRef] [PubMed]
2. Elliott, R.M.; Brennan, B. Emerging phleboviruses. *Curr. Opin. Virol.* **2014**, *5*, 50–57. [CrossRef] [PubMed]
3. Elliott, R.M. Orthobunyaviruses: Recent genetic and structural insights. *Nat. Rev. Microbiol.* **2014**, *12*, 673–685. [CrossRef] [PubMed]
4. Manigold, T.; Vial, P. Human hantavirus infections: Epidemiology, clinical features, pathogenesis and immunology. *Schweiz. Med. Wochenschr.* **2014**, *144*, w13937. [CrossRef] [PubMed]
5. Watson, D.C.; Sargianou, M.; Papa, A.; Chra, P.; Starakis, I.; Panos, G. Epidemiology of Hantavirus infections in humans: A comprehensive, global overview. *Crit. Rev. Microbiol.* **2014**, *40*, 261–272. [CrossRef] [PubMed]
6. Heyman, P.; Ceianu, C.S.; Christova, I.; Tordo, N.; Beersma, M.; Joao Alves, M.; Lundkvist, A.; Hukic, M.; Papa, A.; Tenorio, A.; *et al.* A five-year perspective on the situation of haemorrhagic fever with renal syndrome and status of the hantavirus reservoirs in Europe, 2005–2010. *Euro Surveill.* **2011**, *16*, 15–22.
7. Vaheri, A.; Henttonen, H.; Voutilainen, L.; Mustonen, J.; Sironen, T.; Vapalahti, O. Hantavirus infections in Europe and their impact on public health. *Rev. Med. Virol.* **2013**, *23*, 35–49. [CrossRef] [PubMed]
8. Jonsson, C.B.; Figueiredo, L.T.; Vapalahti, O. A global perspective on hantavirus ecology, epidemiology, and disease. *Clin. Microbiol. Rev.* **2010**, *23*, 412–441. [CrossRef] [PubMed]
9. Kruger, D.H.; Schonrich, G.; Klempa, B. Human pathogenic hantaviruses and prevention of infection. *Hum. Vaccines* **2011**, *7*, 685–693. [CrossRef]
10. Macneil, A.; Nichol, S.T.; Spiropoulou, C.F. Hantavirus pulmonary syndrome. *Virus Res.* **2011**, *162*, 138–147. [CrossRef] [PubMed]
11. Muranyi, W.; Bahr, U.; Zeier, M.; van der Woude, F.J. Hantavirus infection. *J. Am. Soc. Nephrol.* **2005**, *16*, 3669–3679. [CrossRef] [PubMed]
12. Schmaljohn, C.S.; Nichol, S.T. Bunyaviridae. In *Fields Virology*, 5th ed.; Knipe, D.M., Howley, P.M., Eds.; Lippincott Williams & Wilkins: Philadelphia, PA, USA, 2007; pp. 1741–1790.
13. Papa, A. Dobrava-Belgrade virus: Phylogeny, epidemiology, disease. *Antivir. Res.* **2012**, *95*, 104–117. [CrossRef] [PubMed]
14. Vaheri, A.; Strandin, T.; Hepojoki, J.; Sironen, T.; Henttonen, H.; Makela, S.; Mustonen, J. Uncovering the mysteries of hantavirus infections. *Nat. Rev. Microbiol.* **2013**, *11*, 539–550. [CrossRef] [PubMed]
15. Kukkonen, S.K.; Vaheri, A.; Plyusnin, A. L protein, the RNA-dependent RNA polymerase of hantaviruses. *Arch. Virol.* **2005**, *150*, 533–556. [CrossRef] [PubMed]
16. Plotch, S.J.; Bouloy, M.; Krug, R.M. Transfer of 5′-terminal cap of globin mRNA to influenza viral complementary RNA during transcription *in vitro*. *Proc. Natl. Acad. Sci. USA* **1979**, *76*, 1618–1622. [CrossRef] [PubMed]
17. Plotch, S.J.; Bouloy, M.; Ulmanen, I.; Krug, R.M. A unique cap(m7GpppXm)-dependent influenza virion endonuclease cleaves capped RNAs to generate the primers that initiate viral RNA transcription. *Cell* **1981**, *23*, 847–858. [CrossRef]

18. Guilligay, D.; Tarendeau, F.; Resa-Infante, P.; Coloma, R.; Crepin, T.; Sehr, P.; Lewis, J.; Ruigrok, R.W.; Ortin, J.; Hart, D.J.; *et al.* The structural basis for cap binding by influenza virus polymerase subunit PB2. *Nat. Struct. Mol. Biol.* **2008**, *15*, 500–506. [CrossRef] [PubMed]
19. Dias, A.; Bouvier, D.; Crepin, T.; McCarthy, A.A.; Hart, D.J.; Baudin, F.; Cusack, S.; Ruigrok, R.W. The cap-snatching endonuclease of influenza virus polymerase resides in the PA subunit. *Nature* **2009**, *458*, 914–918. [CrossRef] [PubMed]
20. Yuan, P.; Bartlam, M.; Lou, Z.; Chen, S.; Zhou, J.; He, X.; Lv, Z.; Ge, R.; Li, X.; Deng, T.; *et al.* Crystal structure of an avian influenza polymerase PA(N) reveals an endonuclease active site. *Nature* **2009**, *458*, 909–913. [CrossRef] [PubMed]
21. Steczkiewicz, K.; Muszewska, A.; Knizewski, L.; Rychlewski, L.; Ginalski, K. Sequence, structure and functional diversity of PD-(D/E)XK phosphodiesterase superfamily. *Nucleic Acids Res.* **2012**, *40*, 7016–7045. [CrossRef] [PubMed]
22. Crepin, T.; Dias, A.; Palencia, A.; Swale, C.; Cusack, S.; Ruigrok, R.W. Mutational and metal binding analysis of the endonuclease domain of the influenza virus polymerase PA subunit. *J. Virol.* **2010**, *84*, 9096–9104. [CrossRef] [PubMed]
23. Bishop, D.H.; Gay, M.E.; Matsuoko, Y. Nonviral heterogeneous sequences are present at the 5′ ends of one species of snowshoe hare bunyavirus S complementary RNA. *Nucleic Acids Res.* **1983**, *11*, 6409–6418. [CrossRef] [PubMed]
24. Reguera, J.; Weber, F.; Cusack, S. Bunyaviridae RNA polymerases (L-protein) have an N-terminal, influenza-like endonuclease domain, essential for viral cap-dependent transcription. *PLoS Pathog.* **2010**, *6*, e1001101. [CrossRef] [PubMed]
25. Heinemann, P.; Schmidt-Chanasit, J.; Gunther, S. The N terminus of Andes virus L protein suppresses mRNA and protein expression in mammalian cells. *J. Virol.* **2013**, *87*, 6975–6985. [CrossRef] [PubMed]
26. DuBois, R.M.; Slavish, P.J.; Baughman, B.M.; Yun, M.K.; Bao, J.; Webby, R.J.; Webb, T.R.; White, S.W. Structural and biochemical basis for development of influenza virus inhibitors targeting the PA endonuclease. *PLoS Pathog.* **2012**, *8*, e1002830. [CrossRef] [PubMed]
27. Remmert, M.; Biegert, A.; Hauser, A.; Soding, J. HHblits: Lightning-fast iterative protein sequence searching by HMM-HMM alignment. *Nat. Methods* **2012**, *9*, 173–175. [CrossRef] [PubMed]
28. Eswar, N.; Webb, B.; Marti-Renom, M.A.; Madhusudhan, M.S.; Eramian, D.; Shen, M.Y.; Pieper, U.; Sali, A. Comparative protein structure modeling using MODELLER. *Curr. Protoc. Protein Sci.* **2007**, Chapter 2, Unit 2.9. [CrossRef]
29. Johansson, M.U.; Zoete, V.; Michielin, O.; Guex, N. Defining and searching for structural motifs using DeepView/Swiss-PdbViewer. *BMC Bioinform.* **2012**, *13*. [CrossRef] [PubMed]
30. Flick, K.; Hooper, J.W.; Schmaljohn, C.S.; Pettersson, R.F.; Feldmann, H.; Flick, R. Rescue of Hantaan virus minigenomes. *Virology* **2003**, *306*, 219–224. [CrossRef]
31. Buchholz, U.J.; Finke, S.; Conzelmann, K.K. Generation of bovine respiratory syncytial virus (BRSV) from cDNA: BRSV NS2 is not essential for virus replication in tissue culture, and the human RSV leader region acts as a functional BRSV genome promoter. *J. Virol.* **1999**, *73*, 251–259. [PubMed]
32. Klemm, C.; Reguera, J.; Cusack, S.; Zielecki, F.; Kochs, G.; Weber, F. Systems to establish bunyavirus genome replication in the absence of transcription. *J. Virol.* **2013**, *87*, 8205–8212. [CrossRef] [PubMed]
33. Kabsch, W.S. Dictionary of protein secondary structure: Pattern recognition of hydrogen-bonded and geometrical features. *Biopolymers* **1983**, *22*, 2577–2637.
34. Jones, D.T. Protein secondary structure prediction based on position-specific scoring matrices. *J. Mol. Biol.* **1999**, *292*, 195–202. [CrossRef] [PubMed]
35. Zhang, J.H.; Chung, T.D.; Oldenburg, K.R. A Simple Statistical Parameter for Use in Evaluation and Validation of High Throughput Screening Assays. *J. Biomol. Screen.* **1999**, *4*, 67–73. [CrossRef] [PubMed]
36. Bridgen, A.; Elliott, R.M. Rescue of a segmented negative-strand RNA virus entirely from cloned complementary DNAs. *Proc. Natl. Acad. Sci. USA* **1996**, *93*, 15400–15404. [CrossRef] [PubMed]
37. Rezelj, V.V.; Overby, A.K.; Elliott, R.M. Generation of mutant Uukuniemi viruses lacking the nonstructural protein NSs by reverse genetics indicates that NSs is a weak interferon antagonist. *J. Virol.* **2015**, *89*, 4849–4856. [CrossRef] [PubMed]
38. Brennan, B.; Li, P.; Zhang, S.; Li, A.; Liang, M.; Li, D.; Elliott, R.M. Reverse genetics system for severe fever with thrombocytopenia syndrome virus. *J. Virol.* **2015**, *89*, 3026–3037. [CrossRef] [PubMed]

39. Ikegami, T.; Won, S.; Peters, C.J.; Makino, S. Rescue of infectious rift valley fever virus entirely from cDNA, analysis of virus lacking the NSs gene, and expression of a foreign gene. *J. Virol.* **2006**, *80*, 2933–2940. [CrossRef] [PubMed]
40. Bergeron, E.; Albarino, C.G.; Khristova, M.L.; Nichol, S.T. Crimean-Congo hemorrhagic fever virus-encoded ovarian tumor protease activity is dispensable for virus RNA polymerase function. *J. Virol.* **2010**, *84*, 216–226. [CrossRef] [PubMed]
41. Bergeron, E.; Zivcec, M.; Chakrabarti, A.K.; Nichol, S.T.; Albarino, C.G.; Spiropoulou, C.F. Recovery of Recombinant Crimean Congo Hemorrhagic Fever Virus Reveals a Function for Non-structural Glycoproteins Cleavage by Furin. *PLoS Pathog.* **2015**, *11*, e1004879. [CrossRef] [PubMed]
42. Brown, K.S.; Ebihara, H.; Feldmann, H. Development of a minigenome system for Andes virus, a New World hantavirus. *Arch. Virol.* **2012**, *157*, 2227–2233. [CrossRef] [PubMed]
43. Gerlach, P.; Malet, H.; Cusack, S.; Reguera, J. Structural Insights into Bunyavirus Replication and Its Regulation by the vRNA Promoter. *Cell* **2015**, *161*, 1267–1279. [CrossRef] [PubMed]

© 2016 by the authors. Licensee MDPI, Basel, Switzerland. This article is an open access article distributed under the terms and conditions of the Creative Commons Attribution (CC BY) license (http://creativecommons.org/licenses/by/4.0/).

Review

Early Bunyavirus-Host Cell Interactions

Amelina Albornoz [1,†], Anja B. Hoffmann [2,†], Pierre-Yves Lozach [2,*] and Nicole D. Tischler [1,*]

1 Molecular Virology Laboratory, Fundación Ciencia & Vida, Av. Zañartu 1482, 7780272 Santiago, Chile; aalbornoz@cienciavida.org

2 CellNetworks—Cluster of Excellence and Department of Infectious Diseases, Virology, University Hospital Heidelberg, Im Neuenheimer Feld 324, 69120 Heidelberg, Germany; anja.hoffmann@uni-heidelberg.de

* Correspondence: pierre-yves.lozach@med.uni-heidelberg.de (P.Y.L.); ntischler@cienciavida.org (N.D.T.); Tel: +49-(0)6221-561-328 (P.Y.L.); +56-22-367-2015 (N.D.T.)

† These authors contributed equally to this work.

Academic Editor: Eric O. Freed
Received: 30 March 2016; Accepted: 15 May 2016; Published: 24 May 2016

Abstract: The *Bunyaviridae* is the largest family of RNA viruses, with over 350 members worldwide. Several of these viruses cause severe diseases in livestock and humans. With an increasing number and frequency of outbreaks, bunyaviruses represent a growing threat to public health and agricultural productivity globally. Yet, the receptors, cellular factors and endocytic pathways used by these emerging pathogens to infect cells remain largely uncharacterized. The focus of this review is on the early steps of bunyavirus infection, from virus binding to penetration from endosomes. We address current knowledge and advances for members from each genus in the *Bunyaviridae* family regarding virus receptors, uptake, intracellular trafficking and fusion.

Keywords: bunyavirus; cell entry; endocytosis; hantavirus; RNA virus; tospovirus; virus membrane fusion; virus receptor

---

1. Introduction

The *Bunyaviridae* is a large family of RNA viruses, which comprises five genera (*Hantavirus*, *Nairovirus*, *Orthobunyavirus*, *Phlebovirus* and *Tospovirus*) [1]. With over 350 identified isolates distributed worldwide, these viruses represent a global threat to livestock, agricultural productivity and human public health. Many cause serious diseases with high mortality rates in domestic animals and humans, such as fatal hepatitis, encephalitis and hemorrhagic fever. Bunyaviruses are unique in the way they infect a large range of hosts, including vertebrates, invertebrates and plants. Recently, novel genera in the family have been proposed based on the identification of new bunyavirus members. However their host range has not yet been determined [2]. The increasing frequency of bunyavirus outbreaks over the last decade makes these viruses potential emerging agents of disease. No vaccines or treatments are currently approved for human use. Some are classified as potential biological weapons and listed as high-priority pathogens by the World Health Organization.

Most of the available information on bunyaviruses comes from studies of a limited number of isolates. However, it is apparent that there is a wide variety of viruses, vectors, hosts, diseases and geographical distributions. Hantaviruses are harbored in small mammals like rodents, shrews, moles, and bats, and are mainly transmitted to humans through inhalation of contaminated aerosols from the feces of infected rodents [3]. The other known bunyaviruses are all arthropod-borne viruses, which for convenience will be referred to as arbo-bunyaviruses; orthobunyaviruses, nairoviruses and phleboviruses spread to vertebrates by blood-feeding arthropods [4–6], while tospoviruses are plant-specific and are transmitted via non-hematophagous vectors, namely thrips [7]. For a more complete picture of bunyaviruses, we recommend recent books and reviews [1–10].

The diversity among the *Bunyaviridae* family is also manifested at the cellular and molecular levels in the genomic organization, virion structure, tropism, cellular receptors and cell entry. In this review, we address current knowledge and advances regarding early bunyavirus-host cell interactions, from virus binding to penetration into the cytosol.

2. Bunyavirus Genome Organization and Virion Structure

Bunyaviruses are enveloped with a tri-segmented single-stranded RNA genome, which replicates in the cytosol [1]. The three viral RNA segments code for a minimum of four structural proteins in a negative-sense orientation (Figure 1) [1]. The largest genomic RNA segment (L) encodes the RNA-dependent RNA polymerase L, which is required for the initiation of viral replication after the virus genome is released into the cytosol. The medium virus RNA segment (M) codes for a precursor polypeptide that is further processed into two envelope glycoproteins, G_N and G_C, in the endoplasmic reticulum (ER) or Golgi apparatus (Figure 2), from where virions acquire their lipid bilayer membrane and assemble [1,11]. The precise location and mechanisms for the glycoprotein maturation and virus budding in the ER-Golgi machinery can differ among bunyavirus isolates and cell types and very often remain to be defined. The smallest segment (S) encodes the nucleoprotein N, which associates with the viral RNA genome and together with the viral polymerase L constitutes the pseudo-helical ribonucleoproteins (RNPs) [1]. Bunyaviruses do not possess any classical matrix protein or rigid inner structure. The N protein thus has an important role in protecting the viral genetic information. In the past five years, the crystal structure of N has been solved for several bunyavirus members, providing new insights into the mechanism of RNP assembly and showing some distinctions in the N proteins among the different genera [12–27]. Bunyaviruses also encode some non-structural proteins [28–31], but thus far, none have been found to be involved in virus entry and, therefore, will not be discussed here.

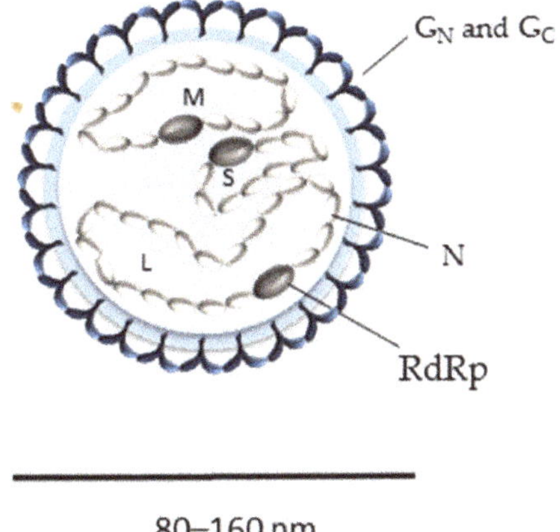

Figure 1. Schematic representation of a bunyavirus particle. The three viral genomic segments are termed according to their size: S (small), M (medium) and L (large). Abbreviations: G_N: glycoprotein G_N; G_C: glycoprotein G_C; N: nucleoprotein; RdRp: RNA-dependent RNA polymerase.

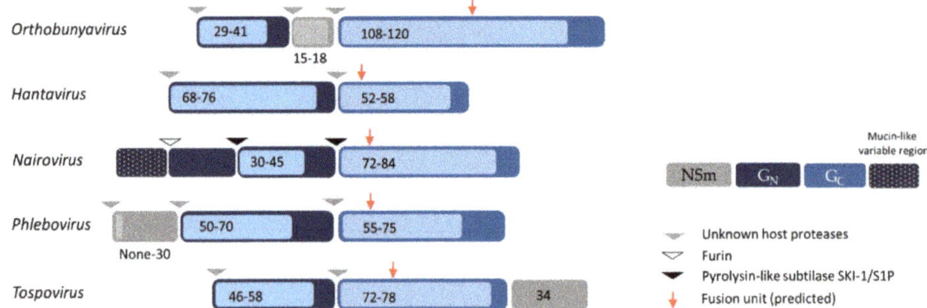

Figure 2. Schematic representation of bunyavirus G_N and G_C precursor glycoprotein sequences of each genus. Light and dark colored boxes indicate the smallest and highest molecular weight (kDa) of each protein in each genus, respectively. Arrow heads indicate the proteolytic cleavage sites within the glycoprotein precursor by host proteases [32,33]. Red arrows show the localization of the fusion peptide for each genus based on the crystal structure obtained from Rift Valley fever virus Gc [34] and on bioinformatics predictions and biochemical analysis of the glycoproteins from the orthobunyavirus La Crosse, the hantavirus Andes, the nairovirus Crimean-Congo hemorrhagic fever and the tospovirus tomato spotted wilt [35–37].

On particles, the two envelope glycoproteins G_N and G_C are responsible for virus attachment to target cells and acid-activated penetration [8,38]. Electron micrographs of bunyaviruses show particles that are roughly spherical, heterogeneous in size with an average diameter of 80–160 nm and with spike-like projections of 5–10 nm composed of G_N and G_C heteromultimers [1]. Recent cryo-electron tomography studies confirmed the high degree of pleomorphism previously observed for bunyaviruses [39–45]. Ultrastructural analyses of the phleboviruses Rift Valley fever (RVFV) and Uukuniemi (UUKV) revealed that the most regular particles exhibited surface glycoprotein protrusions arranged on an icosahedral lattice, with an atypical T = 12 triangulation [39–41,43]. In contrast, tomography data obtained for the orthobunyavirus Bunyamwera displayed non-icosahedral viral particles with glycoprotein spikes exhibiting a unique tripod-like arrangement, while spikes from hantavirus glycoproteins arrange with local symmetry into tetramers (Figure 3) [42,44,45].

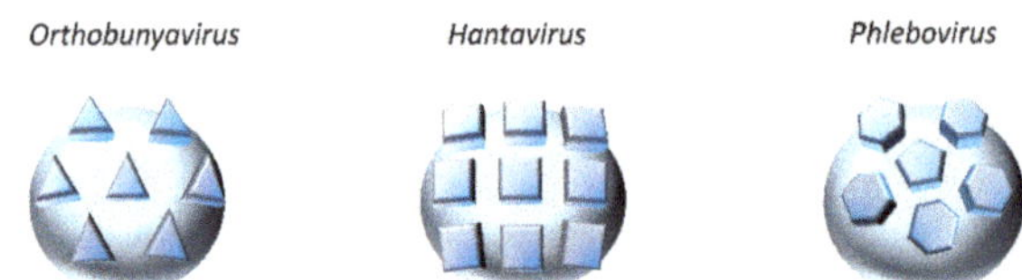

Figure 3. Schematic representation of the bunyavirus G_N and G_C glycoprotein arrangement on the surface of viral particles. The symmetries shown here were obtained by cryo-electron tomography and image reconstruction from Bunyamwera virus (*Orthobunyavirus*, left panel), Tula virus (TULV) and Hantaan virus (HTNV) (*Hantavirus*, middle panel), as well as Rift Valley fever (RVFV) and Uukuniemi (UUKV) viruses (*Phlebovirus*, right panel) [39–45]. Images were adapted from [44].

3. Receptors for Arbo-Bunyaviruses in Mammalian Hosts

During natural transmission to mammalian hosts, arbo-bunyaviruses are introduced into the skin dermis by infected arthropods. Due to their presence in the anatomical site of initial infection, dermal macrophages and dendritic cells (DCs) are among the first cells to encounter the incoming viruses [8]. To establish infection and replicate, viruses need to gain access to the intracellular environment. This very first step is strictly dependent on surface-exposed cellular receptors that include proteins, lipids and glycans and to which virus particles bind [46,47]. Some surface receptors can mediate virus entry into cells, without the requirement of additional co-receptors and molecules. Alternatively, some primary receptors limit the free diffusion of viral particles and/or promote interactions with secondary receptor complexes, which are responsible for virus entry into the cytoplasm [46,48]. When viruses rely on many cellular surface factors for entry and infection, the primary receptor is often referred to as the attachment factor. Only a few surface attachment factors and receptors have been reported for bunyaviruses, and very often, their role in cell entry remains to be uncovered.

Virus-receptor interactions are often specific and multivalent. Binding to multiple receptor molecules clustered within microdomains can enhance the avidity of low-affinity interactions [46]. Polysaccharides on glycoproteins, as well as glycolipids in the extracellular matrix found at the surface of most mammalian cells are highly polar structures. They can serve as a first docking site for many viruses, including bunyaviruses, through electrostatic interactions, which are in general of low affinity [48,49]. Glycosaminoglycans (GAGs), such as heparan sulfate, have been shown to promote infection by RVFV and another phlebovirus, Toscana virus (TOSV) [50–52]. When heparin was used to compete with GAGs on the cell surface, infection by both viruses was significantly reduced. Similar results were obtained when heparan sulfate molecules were removed from cells by enzymatic digestion prior to being exposed to these viruses. Furthermore, cells deficient in the synthesis of heparan sulfate were shown to be less sensitive to RVFV infection [50,52]. Interestingly, glycoproteins from the cell-cultured RVFV used in one of these studies did not show any distinction in basic amino acids with those from the virus found in the serum or organs of infected animals [50]. This suggests that the heparan sulfate-dependence of RVFV does not seem to result from virus culture adaptation. However, the fact that cells lacking GAGs on their surface remain sensitive to infection, even at lower levels, indicates that these viruses can use alternative receptors to attach to and enter cells.

Recent work has shown that RVFV and UUKV target and infect DCs by subverting the C-type lectin Dendritic Cell-Specific Intercellular adhesion molecule-3-Grabbing Non-integrin (DC-SIGN; also known as CD209) [53]. In the presence of neutralizing antibodies, dermal DCs remained resistant to RVFV and UUKV infection [53]. When DC-SIGN was expressed at the surface of cells, which are usually poorly infected by bunyaviruses, a significant fraction of cells was infected by the Germiston orthobunyavirus and by many phleboviruses, including RVFV, UUKV, Punta Toro and TOSV [53]. The list of proposed bunyaviruses that are able to use DC-SIGN has since been extended (Table 1). Recent studies have shown that the lectin enhances infection by rhabdoviral particles pseudotyped with the glycoproteins of the orthobunyavirus La Crosse (LACV) and with those of severe fever with thrombocytopenia syndrome virus (SFTSV), an important emerging human tick-borne phlebovirus pathogen [54]. A similar approach has recently been used to investigate the role of DC-SIGN in infection by the Crimean-Congo hemorrhagic fever virus (CCHFV) [55], a tick-borne nairovirus that infects endothelial cells and macrophages, but also dermal-like DCs [56–58]. DC-SIGN provides an interesting bridge between arbo-bunyaviruses amplified in arthropod vectors and initial infection in humans. This immune receptor is: (1) expressed on immature dermal DCs, which are present in the anatomical site of virus transmission; and (2) specialized in capturing pathogens with an *N*-glycan coat of high-mannose residues, such as those on the glycoproteins of virions derived from insects [8,59]. For these reasons, interactions between DC-SIGN and insect-borne pathogens are thought to be the most relevant, although several studies have suggested a role for this lectin in infection by various microbes that are not transmitted by arthropods [59]. With a lower extension, DC-SIGN is also expressed on alveolar macrophages and DCs in the lungs [60] and may represent an interesting receptor candidate for aerosol-transmitted bunyaviruses. So far, hantavirus-DC-SIGN interactions remain to be investigated.

Table 1. Potential receptors documented for bunyaviruses.

Transmission	Receptor/Cofactor	Bunyavirus	Genus	Ref.
Arthropod bite	DC-SIGN	SFTSV, UUKV, RVFV	*Phlebovirus*	[50–55,61–65]
		LACV	*Orthobunyavirus*	
		CCHFV	*Nairovirus*	
	L-SIGN	RVFV, SFTSV, TOSV, UUKV	*Phlebovirus*	
	NMMHC-IIA	SFTSV		
	Heparan Sulfate	RVFV, TOSV		
	Nucleolin	CCHFV	*Nairovirus*	
	50-kDa protein	TSWV	*Tospovirus*	
Aerosol inhalation	β_3 integrin	ANDV, SNV, HTNV, PUUV, SEOV, NY-1V	*Hantavirus*	[66–72]
	β_1 integrin	PHV, TULV		
	β_2 integrin	HTNV		
	gC1qR	HTNV		
	DAF	HTNV, PUUV		
	70-kDa protein	HTNV		

DC-SIGN: Dendritic Cell-Specific Intercellular adhesion molecule-3-Grabbing Non-integrin; L-SIGN: Liver-Specific Intercellular adhesion molecule-3-Grabbing Non-integrin; NMMHC-IIA: non-muscle myosin heavy chain IIA; DAF: decay-accelerated factor; SFTSV: severe fever with thrombocytopenia syndrome virus; UUKV: Uukuniemi virus; RVFV: Rift Valley fever virus; LACV: La Crosse virus; CCHFV: Crimean-Congo hemorrhagic fever virus; TOSV: Toscana virus; TSWV: tomato spotted wilt virus; ANDV: Andes virus; SNV: Sin Nombre virus; PUUV: Puumala virus; SEOV: Seoul virus; NY-1V: New York-1 virus; PHV: Prospect Hill virus; TULV: Tula virus; HTNV: Hantaan virus.

In addition to DC-SIGN, the phleboviruses RVFV, TOSV and UUKV have been shown to subvert Liver-Specific Intercellular adhesion molecule-3-Grabbing Non-integrin (L-SIGN) [62], a closely-related C-type lectin expressed in liver sinusoidal endothelial cells [59]. Others have established that rhabdoviral particles pseudotyped with the envelope glycoproteins of SFTSV, but not with those of RVFV and LACV, can subvert L-SIGN [54]. The engagement of multiple viral glycoproteins by homo-tetrameric lectins is critical for high-avidity interactions between the closely related C-type lectin DC-SIGN and the arbovirus dengue from the *Flaviviridae* family [73]. Rhabdoviruses and bunyaviruses differ significantly with regard to the assembly and maturation of viral progeny. The structural organization of *N*-glycans on their surface may not exactly reflect the organization on RVFV and LACV viral particles, which in turn could favor less efficient virus binding to C-type lectins and infection. It is tempting to postulate that by acting as an attachment receptor on liver sinusoidal endothelial cells, L-SIGN plays a role in the liver tropism of some bunyaviruses.

Nucleolin has been identified as a potential binding factor for the nairovirus CCHFV [63]. This factor has been identified following a strategy combining co-immunoprecipitation with a G_C ectodomain fragment as bait and mass spectrometry analysis. However, there is currently no evidence of surface nucleolin as an entry factor for CCHFV, although the protein was shown to be present on the surface of cells susceptible to the virus. Following a similar strategy, but instead using the G_N ectodomain as bait, it has been proposed that the non-muscle myosin heavy chain IIA (NMMHC-IIA) is involved in the infectious entry of SFTSV [61]. NMMHC-IIA is an intracellular protein that has been shown to reach the cell surface of human umbilical vein endothelial cells (HUVECs) and Vero cells [61]. Silencing of the NMMHC-IIA gene in HUVECs resulted in a decrease in SFTSV infection, while ectopic expression of NMMHC-IIA in HeLa cells, which lack the protein expression, but are sensitive to infection, resulted in a ~20-fold increased sensitivity to the virus [61]. It remains unknown whether this protein serves as an entry receptor or merely as an attachment factor. Recently, SFTSV was seen in secreted vesicles positive for CD63, a cellular marker associated with extracellular vesicles [74]. The virions within these vesicles were efficiently delivered to uninfected cells. This is the first evidence for bunyaviruses that an isolate hijacks the exocytic machinery for receptor-independent transmission and entry into host cells.

4. Receptors for Plant-Specific Bunyaviruses

The fact that tospoviruses are plant-specific sets these bunyaviruses apart from others. Little is known about the initial steps of tospovirus infection, both in arthropod vectors and plants. The tomato spotted wilt virus (TSWV), which is vectored by the western flower thrips *Frankliniella occidentalis*, is by far the most investigated tospovirus pathogen-vector system. TSWV represents a threat to agricultural productivity in the Southwestern United States, and now in South America, Europe and Australia, where the thrips *F. occidentalis* has recently spread [75,76]. The tospovirus glycoproteins are thought to have been conserved during evolution only to disseminate the virus in arthropod vector populations, but not in plants. Mutations in the TSWV glycoproteins that make the thrips *F. occidentalis* resistant to infection do not affect the spread of the virus in plant cells [77,78].

In insects, TSWV glycoproteins were found to interact with a protein of 50 kDa expressed in the larval midguts of *F. occidentalis*, the function of which remains unknown [65]. The 50-kDa protein was recognized by anti-idiotypic antibodies against G_N and G_C [65], though only the latter blocked infection [64]. Interestingly, adult thrips seem to lose the expression of this 50-kDa protein, and it is tempting to correlate the expression of the protein with the capacity of TSWV to infect larval, but not adult thrips through the midgut barrier [79–81].

In plants, tospoviruses are believed to propagate by active transport under the form of non-enveloped viral RNP structures through cell wall-embedded pores, the plasmodesmata [82]. The non-structural protein NSm of TSWV appears to be an important player in RNP cell-to-cell transfer in plants [83,84]. NSm has been shown to interact with a protein located on both orifices of the plasmodesmata pores in *Arabidopsis thaliana*, the At-4/1 protein [85]. Such plasmodesmial proteins are believed to have receptor-like properties for TSWV [86]. However, no direct evidence has been reported so far of At-4/1's role in cell-to-cell TSWV spread. Not much is known about tospovirus receptors and early interactions with host cells.

5. Receptors for Aerosol-Transmitted Bunyaviruses: How Hantaviruses Target Cells

The aerosol transmission of hantaviruses to humans implies that upon inhalation, these bunyaviruses encounter the epithelium of the lung. In the terminal lung alveoli, epithelial cells are tightly attached to endothelial cells derived from a capillary network, with the particularity that their basement membranes are fused into a single layer, thereby forming a thin air-blood barrier [87]. Hantavirus infection of human endothelial cells causes dramatic changes in the barrier function of the endothelium [3,88–90]. As a consequence, increased capillary permeability and vascular leakage associated with pulmonary edema or hemorrhagic fevers are often observed in infected patients [91–93]. A substantial amount of work has been done to identify potential cellular factors involved in hantavirus infection of endothelial cells. Natural ligands of endothelial cell surface receptors, including vitronectin and fibronectin, but not heparin and laminin, were found to antagonize hantavirus entry [67]. Screening antibodies against different integrin subunits serving as receptors for vitronectin and fibronectin [94] revealed that anti-β1 integrin antibodies specifically impair the infection of Vero cells by the Prospect Hill (PHV) and Tula (TULV) viruses, while anti-β3 integrin antibodies reduced by 60%–70% the sensitivity of these cells to Hantaan virus (HTNV), Andes virus (ANDV) and other pathogenic hantaviruses (Table 1) [67,68]. When the human β3 integrin was overexpressed as heterodimers with the α_{IIb} or α_V subunits in CHO cells, a 20–30-fold increase in infection was observed for the Sin Nombre (SNV) and New York-1 virus (NY-1V), though these cells are otherwise not very sensitive to hantaviruses [67].

Integrins and endothelial cells represent an interesting model to explain the molecular and cellular basis of the pathogenesis associated with hantavirus infections in humans [88–90]. Pathogenic NY-1V and HTNV were shown to bind β3 integrins through a specific N-terminal PSI (plexin-semaphorin-integrin) domain [95]. This domain is only exposed and accessible when the receptor adapts an inactive conformation in which it exhibits low affinity to its ligand [96]. In this study, the aspartate in position 39 of human β3 integrin was found to be critical for promoting the infection of CHO cells. Interestingly, the murine β3 integrin carries an asparagine at position 39 instead and is unable to mediate infection unless it is substituted by aspartate [95]. This suggests that β3 integrin promotes infection by these viruses in humans, but may not do so in mice. In this regard, it remains to be determined whether integrins from natural rodent reservoirs may promote infection.

Epithelial and endothelial cells form polarized monolayers, an aspect that is often neglected in *in vitro* investigations of viral cell entry into different cell lines. The endothelial cell polarity can impact at large the distribution of the surface receptors, as well as the global organization of the exocytic-endocytic machinery and, therefore, the virus entry program [97–99]. For instance, β3 integrin seems to be mainly located on the basolateral side of polarized epithelial cells, allowing attachment to the extracellular matrix [100,101]. However, it has also been found on apical surfaces, where it detects soluble ligands and circulating cells in luminal compartments [102–105]. Efficient infection of these cells by the Puumula virus (PUUV) and HTNV occurs from the apical side only, while ANDV and the Black Creek Canal virus (BCCV) can infect epithelial cells from both surfaces [106–108]. Additional cellular factors may bring hantaviruses to the appropriate side of the cell for entry. The decay-accelerated factor (DAF or CD55), an inhibitor of the complement cascade, seems an interesting candidate. DAF is involved in the apical entry of PUUV and HTNV into polarized cells, and SNV binds to DAF with high affinity [66,109]. More recently, DAF has been shown to participate in the transport of unrelated viruses in polarized cells, from the apical side to the lateral tight junctions, an environment believed to promote virus-receptor interactions that are otherwise inaccessible for viral particles docking on apical surfaces [46,110,111]. Although DAF does not seem to be required for hantavirus infection of non-polarized cells *in vitro* [67,70], the factor might be critical *in vivo*.

It is apparent that hantaviruses, as other unrelated bunyaviruses, can exploit additional entry factors or even use alternative cellular receptors for infection. A 70-kDa cellular protein on β3 integrin-expressing CHO cells has been found to interact with HTNV [71]. However, in the absence of further work, the identity of this factor remains unknown. Ectopic expression of heterodimers between the human β2 and different α chains of integrins was shown to make CHO cells 4–8-times more susceptible to lentiviral particles pseudotyped with the glycoproteins of HTNV [72]. To ascertain the physiological relevance of this finding, it remains to be assessed whether the virus uses β2 integrin to enter cells that naturally express this protein. HTNV was also found to bind to the receptor of the globular head domain of the complement protein C1q (gC1qR) in Vero cells [70]. The role of gC1qR in HTNV infection was confirmed in approaches based either on silencing in Vero cells or on overexpression of the protein in CHO cells [70]. Nevertheless, it is not clear whether gC1qR acts as a virus receptor or merely as a cellular factor that is important for virus entry beyond attachment to cells. Finally, two independent haploid genetic screens and one small interfering RNA (siRNA)-based screen highlighted the importance of host genes related to cholesterol sensing, regulation and biosynthesis for ANDV infection [112,113]. None of the cellular factors whose role in hantavirus infection is described above were mentioned in these screens, and no new candidate receptor was reported [112,113].

6. Bunyavirus Uptake

To enter host cells, it is apparent that bunyaviruses rely on the physical uptake of particles into the endocytic cellular machinery. While the number of reports on receptors and subsequent endocytic pathways used by bunyaviruses has increased over the past decade, the transition processes between the extracellular and intracellular stages remain largely uncovered. However, using fluorescently-labeled UUKV and enhanced green fluorescent protein-tagged DC-SIGN, it was possible for the first time to visualize virus-receptor interactions in live cells and to analyze their dynamics [53]. Using this powerful model system, it was possible to observe receptor recruitment to cell-associated virus particles. This confirmed the hypothesis that viruses collect receptors at the site of contact and thus generate a receptor-rich microdomain in the plasma membrane. Such a series of events is arguably a prerequisite for local plasma membrane curvature and receptor-mediated signal transduction, which in turn results in the sorting of viral particles into the endosomal vesicles [46]. Cholesterol and other lipids also play an important role in these mechanisms by promoting the formation of docking sites for specific proteins. Transduction of cells with lentiviral or rhabdoviral particles pseudotyped with ANDV glycoproteins appears to be sensitive to cholesterol depletion [113,114]. Infection by CCHFV and two orthobunyaviruses, Oropouche (OROV) and Akabane, is abolished in cells depleted of cholesterol by methyl-β-cyclodextrin [115–118].

Sequence motifs in receptors' cytosolic tails generally define the identity of the endocytic route in which the cargo is taken up. These motifs serve as docking sites for specific adaptor proteins with functions in signaling, endocytic internalization and intracellular trafficking [46]. The cytosolic tail of DC-SIGN carries several motifs, including two leucines (LL), which are critical for the endocytosis of cargo by the lectin [119,120]. When a mutant of DC-SIGN lacking the LL motif in the cytosolic tail was expressed, UUKV still attached to the cells [53,121]. However, viruses were not internalized, and there was no infection, indicating that DC-SIGN serves as an endocytic receptor, not only as an attachment factor. In contrast to DC-SIGN, the endocytic function of L-SIGN was not required for UUKV infection [62]. Similar levels of infection were obtained in cells expressing either the wild-type lectin or its endocytic-defective mutant. This indicates a fundamental distinction in the use of DC-SIGN and L-SIGN by these viruses for entry; *i.e.* DC-SIGN as an endocytic receptor *versus* L-SIGN as an attachment factor. It is not known whether other signal motifs in receptors identified for bunyaviruses have a function in virus internalization and infection.

Several lines of data suggest that orthobunyaviruses and nairoviruses mainly subvert clathrin-mediated endocytosis (CME) to penetrate and infect cells (Figure 4). Studies based on the use of siRNAs, dominant-negative mutants and chemical inhibitors against adaptor protein 2 (AP2) and clathrin suggest that the orthobunyaviruses Akabane, LACV, OROV and Tahyna, as well as the nairovirus CCHFV depend on CME for infection [115–118,122,123]. Additional work is still needed to better understand the uptake of phleboviruses. In cells expressing DC-SIGN, electron microscopy pictures did not exclusively show particles of UUKV in clathrin-coated endosomes, and in cells lacking the lectin expression, viral particles could be seen in rare cases associating with clathrin-coated pits and vesicles [53,124]. In addition, clathrin silencing had no significant impact on UUKV infection [124]. Whether RVFV uses clathrin remains unclear. It has been shown that a genetically modified, non-spreading strain of the virus relies on clathrin for successful entry while two independent studies suggest that the RVFV vaccine strain MP12 enters cells both through caveolin-dependent mechanisms and macropinocytosis [125–127]. In the case of hantaviruses, the use of CME for virus entry seems to be isolate-specific (Figure 4). A first study based on chemical inhibitors indicates that the HTNV and BCCV, as well as Seoul virus (SEOV) depend on clathrin-mediated uptake, while other studies propose that the SNV and ANDV hijack an alternative endocytic pathway that does not involve clathrin [128–130].

Altogether, these reports most likely underline the ability of bunyaviruses to use alternative endocytic routes in a single cell or distinct tissues, as suggested by a growing amount of data obtained for unrelated viruses, such as influenza virus. The divergent endocytic processes by which virus receptors are internalized into the cells, as well as the expression pattern of virus receptors on the cell surface certainly influence the capacity of these viruses to enter one or more endocytic pathways to infect cells and tissues.

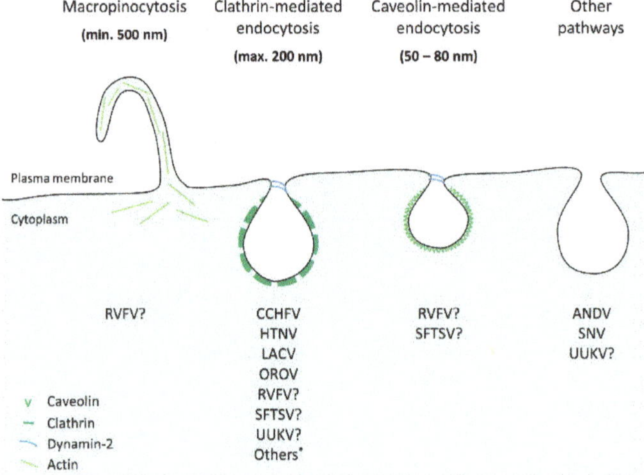

Figure 4. Bunyavirus endocytosis. Endocytic internalization of bunyaviruses into animal cells occurs via various pinocytic pathways, which involve several cellular factors, such as adaptor and coat proteins. A growing body of evidence however indicates that several bunyaviruses use clathrin-mediated endocytosis. * Akabane, Black Creek Canal, California encephalitis, Inkoo, Jamestown, Keystone, Melao, Serra do Navio, Snowshoe Hare, Seoul, Tahyna and Trivittatus viruses. SFTSV: severe fever with thrombocytopenia syndrome virus; UUKV: Uukuniemi virus; RVFV: Rift Valley fever virus; LACV: La Crosse virus; CCHFV: Crimean-Congo hemorrhagic fever virus; ANDV: Andes virus; SNV: Sin Nombre virus; HTNV: Hantaan virus; OROV: Oropouche virus.

7. Bunyavirus Intracellular Trafficking

Upon uptake, bunyaviral particles are sorted into vesicles and traffic through the endocytic machinery until reaching the appropriate endosomal compartments for fusion and penetration into the cytosol. Transport from early (EEs) to late (LEs) endosomes is a complex, sensitive cell biological process that is not yet thoroughly understood and involves hundreds of cellular factors with a wide range of functions [131,132]. It is accompanied by major protein and lipid remodeling and concomitant changes in the endosomal luminal milieu [133]. The endosomes provide an environment in which the decreasing pH, from ~6.5 in EEs down to ~5.5–5.0 in LEs and lysosomes, provides a convenient cue for virus activation [134]. Many studies have clearly established the dependence of bunyaviruses on endosomal acidification for infection [54,115–118,124–126,135,136]. Several members from the different genera of the *Bunyaviridae* family are sensitive to extremely low concentrations of lysomotropic weak bases, such as ammonium chloride (in the range of mM), or inhibitors of vacuolar H+ ATPases, such as bafilomycin A1 and concanamycin B (in the range of nM), which all neutralize the endosomal pH.

Various data support the view that bunyaviruses transit through EEs during their journey in the endocytic machinery (Figure 5). The expression of dominant negative (DN) and constitutively active mutants against endogenous Rab5, a small GTPase required for the trafficking and maturation of EEs, blocks infection by many bunyaviruses, including UUKV, CCHFV and LACV [117,122–124]. Confocal microscopy pictures show the transit of UUKV through Rab5-positive EEs [124], while OROV, CCHFV and HTNV enter vesicles positive for EEA1 [116,117,128], a Rab5 effector protein that exclusively localizes to EEs. An important body of data indicates that bunyaviruses are late-penetrating viruses, a large group of viruses that share dependence on late endosomal maturation for infection [134]. Acid-activated penetration of UUKV and RVFV occurs 20–40 min after internalization, which is compatible with the timing of LE maturation [124,125]. Microtubules are known to drive the trafficking of LEs from the cellular periphery towards the nucleus, a process concomitant with late endosomal maturation. Drug-based inhibitory studies demonstrated that UUKV, CCHFV and some hantaviruses, such as ANDV, BCCV, HTNV and SEOV, require an intact microtubule network for productive infection [124,130,137], which suggests the involvement of LE mobility in virus entry. Viral membrane fusion of UUKV, RVFV, CCHFV and ANDV takes place at pH levels below 6.0, typical for late endosomal vesicles [123–125,135,138]. Recently, the phospholipid bis(monoacylglycerol)phosphate (BMP), located in LEs, has been shown to facilitate UUKV membrane fusion [139]. Finally, UUKV and, possibly, HTNV were found in late endosomal compartments [124,128].

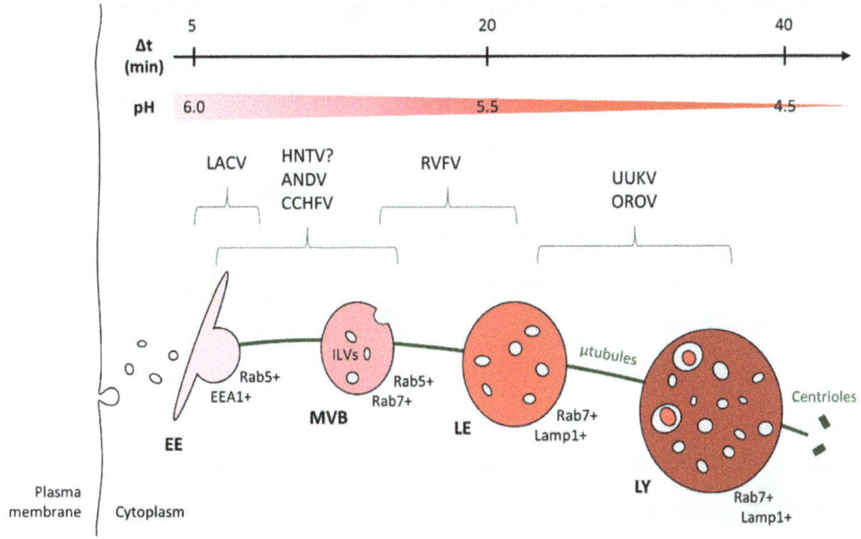

Figure 5. Bunyavirus intracellular trafficking. This figure shows an overview of the different potential locations of bunyavirus penetration. On the top, the scales indicate the time required by a cargo to traffic from the plasma membrane to an organelle (Δt) and the pH inside the endosomes (pH). Abbreviations: EE: early endosome; ILV: intraluminal vesicle; LE: late endosome; LY: lysosome; µtubule: microtubule; MVB: multivesicular body; LAMP: lysosome-associated membrane glycoproteins.

While much evidence supports the idea that bunyaviruses penetrate cells from late endosomal compartments and can be confidently considered late penetrating viruses, many of these viruses appear to infect cells independently of active Rab7, the most critical small GTPase for LE trafficking and maturation [140]. Expression of the Rab7 DN mutant has no impact on infection by CCHFV [117,123], and Rab7 does not seem to be essential either for infection by LACV [122]. However, little is known on intracellular trafficking of this latter virus. Though OROV and UUKV were seen in endosomes positive for Rab7, the expression of Rab7 DN mutant T22N has no significant effect on UUKV infection [116,124]. In contrast, the constitutive active mutant of Rab7 seems to promote UUKV infection [124]. Many reasons could explain why Rab7 is dispensable for infection, such as the mislocalization of the Rab mutants and the presence of multiple Rab7 isoforms, among others. Alternatively, viruses might simply escape the endocytic machinery earlier, for example during sorting from EEs to the nascent multivesicular bodies (MVBs), an initial stage in the LE maturation process driven in part by a switch of GTPases from Rab5 to Rab7 [131,133]. This seems to be the case for CCHFV, which has recently been shown to penetrate cells from MVBs [117].

8. Bunyavirus-Cell Membrane Fusion

As the ultimate step of the entry program, endocytosed viruses need to cross the membrane of endosomal vesicles to release their genome and accessory proteins into the cytosol. Bunyaviruses accomplish this by membrane fusion, a process intricately coordinated in time and space that is mediated by the viral fusion envelope protein [38,134]. Acidification has been shown to be sufficient to trigger the fusion of mature RVFV, UUKV and ANDV particles in cell-free *in vitro* approaches [113,125,139,141]. While the endosomal pH is a critical signal for triggering the fusion of many enveloped viruses, it is sometimes not sufficient. Among others, interactions with receptors, proteolytic cleavage in viral envelope glycoproteins or specific lipids in the target endosomal membranes may also be required [132]. In addition to the endosomal acidification, SFTSV and LACV penetration seems to involve the cleavage of viral envelope glycoproteins G_N and G_C by endosomal cathepsin proteases [54]. Using virus-liposome-based assays, the fusion of hantaviruses was shown to also depend on the strict presence of cholesterol in the target membrane [113], while the late endosomal phospholipid BMP was demonstrated to facilitate UUKV fusion [139].

Typically, upon fusion activation, viral fusion proteins undergo multiple conformational changes, target and harpoon the endosome lipid bilayer via the fusion subunit and progressively pull the target and viral membranes through stages of close apposition, hemifusion and fusion pore formation [132,142,143]. The viral material is released through the pore, resulting in the infection of the cell. The Gc glycoproteins show major modifications in their biochemical properties when RVFV, TSWV and ANDV particles are exposed to low-pH buffers [125,141,144]. In the case of ANDV and RVFV, the acid-triggered rearrangement of the G_C glycoprotein did not require interactions with target membranes [125,141].

At least three different classes of viral fusion proteins (classes I-III) have been proposed, each with specific features from structural and mechanistic perspectives [142,143]. The only X-ray structure available for the ectodomain of a bunyavirus glycoprotein is the pre-fusion state of the RVFV G_C protein, recently solved at 1.9 Å [34]. The overall fold shows a strong resemblance to class II membrane fusion proteins. Although the degree of amino acid identity is rarely above 30% among the bunyavirus glycoproteins, early bioinformatics predictions and analysis involving peptides and site-directed mutagenesis suggest that the peptide responsible for bunyavirus membrane fusion is carried within the Gc glycoprotein (Figure 2) [35–37,135,145–148].

The ectodomain of class II fusion proteins is composed of three sub-domains (I–III) and is connected to a transmembrane cytosolic tail by a stem region [143,149–151]. Exogenous peptides analogous to domain III and the stem region of class II enveloped alpha- and flavi-viruses were successfully employed to block virus fusion and infection [152,153]. It has been demonstrated that such peptides interfere in intramolecular interactions that occur upon acid-activated rearrangement of the viral fusion protein. The presence of these peptides arguably maintains the fusion protein in an intermediate conformation that precedes the post-fusion stage, thereby preventing membrane fusion and virus infection. Similar strategies were used for some bunyaviruses with identical results. Stem peptides that are derived from G_C block RVFV infection [154]. Fusion and infection by ANDV and PUUV were also inhibited in the presence of exogenous stem peptides and domain III from G_C [155], the sequences of which were predicted from a molecular model for ANDV Gc [36]. In these latter experiments, the fusion was detained at a late stage in the process, before membrane hemifusion [155]. Furthermore, following treatment at low pH, the ANDV G_C protein resembled the stable homotrimer post-fusion form of class II fusion proteins [141]. Together, these data suggest that in addition to phleboviruses, hantaviruses also seem to share structural and mechanistic properties with class II enveloped viruses.

Specific histidine residues in fusion proteins, including those from class II proteins, serve as sensors for acidification in endosomal lumen and often define the optimal pH value for virus fusion [156–159]. The variability of optimal pH can be attributed to the local environment of histidines, which influences their pKa in a range as wide as pH 4.5–7.3 [160]. Such essential histidines have been identified in the RVFV G_C by mutational analysis [125], and the optimal pH for penetration has been determined for several bunyavirus isolates using diverse fusion models, from virus glycoprotein-mediated cell-cell fusion assays to virus-liposome and virus-cell-based fusion approaches. Acid-activated viral membrane fusion occurs in a range of pH 5.4–5.6 for RVFV and UUKV [124,125,139], 5.5–6.0 for CCHFV [123], 5.8–6.0 for LACV and California encephalitis virus, another orthobunyavirus [161], 5.8–6.1 for TSWV [144] and at pH ~5.8–5.9 for ANDV [135]. HTNV seems to be the only exception, with a slightly higher activation pH, ~6.3 [136]. Yet, its entry has been observed to also depend on microtubules and LEs, and thus, the cellular factors and location of endosomal escape need more thorough analysis. Ultimately, the major cue for the fusion activation of most bunyaviruses appears to be the endosomal pH.

9. Concluding Remarks and Future Perspectives

In this review, we have summarized current knowledge of the early interactions between bunyaviruses and host cells, from virus attachment on the cell surface and uptake to intracellular trafficking and fusion. It is apparent that hundreds of cellular factors with a wide range of functions are involved in bunyavirus entry and penetration into cells. Though each isolate in the family most likely presents specificities, requirements and distinct mechanisms for the very first steps of infection, it seems that many bunyaviruses rely on late, even if partial, endosomal maturation for infection. It is, however, clear that many aspects of the cell biology of bunyavirus endocytosis and penetration require further investigation. High throughput screens involving the use of siRNAs or haploid cells, such as those recently reported for RVFV, UUKV and ANDV [52,112,113,162,163], or ultimately, gene knock-out by the clustered regularly interspaced short palindromic repeats/CRISPR associated protein 9 (CRISPR/Cas9) system may help to identify new cellular factors and processes that are important for bunyavirus endocytosis.

While it is clear that bunyaviruses use many receptors to target and infect a large number of different species and tissues, only a few receptors have been documented in humans and other vertebrates and not a single one in arthropod vectors. Progress will require a detailed cell biological analysis of receptors and the infection process in different types of tissues, not only originating from mammals, but also from other hosts and vectors and all of these under more relevant physiological conditions. Therefore, the combination of new *in vitro* models with *ex vivo* and *in vivo* approaches will certainly allow for novel findings on bunyavirus transmission, entry and spread [164–167]. Much further work is also needed in the characterization of viral particles originating from different hosts and vectors [167]. The lipid composition of the viral envelope, adaptive mutations in the virus genome and the nature of oligosaccharides in virus glycoproteins arguably influence the identity of first-target cells and the initial stages of infection, including interactions with receptors, sorting into the endocytic machinery and acid-activated fusion.

Ideally, preventing bunyavirus dissemination requires approaches that target the early steps of infection. While identification of the host range is paramount, at the molecular level single inhibitors cannot accurately define a cellular pathway. Perturbants have many side effects or simply impair different processes in cells. Only a combination of well-defined inhibitor profiles, in quantitative and qualitative assays that allow for the monitoring and analysis of the very first minutes of infection, will make the identification of specific cellular factors and mechanisms involved in bunyavirus entry possible.

Virus fusion proteins and uncoating also remain insufficiently characterized given the central role played by acid-activated membrane fusion in bunyavirus penetration. Structural information on the post-fusion form of the G_C protein is still missing. It remains to be examined whether other bunyaviruses also rely on class II fusion proteins for penetration or rather additional conformations may be found, as in the *Flaviviridae* family [143,168]. There is evidence that the N-terminal half of the orthobunyavirus G_C protein is dispensable for fusion and seems to correspond to an additional functional subunit [145]. A number of outstanding questions also remain regarding the global, highly ordered arrangement and interactions of the G_N and G_C glycoproteins on the virions [169]. To this extent, the X-ray structures of G_N ectodomains are still to be discovered.

These are the keys to broadening our knowledge of bunyaviral dissemination and tissue tropism and, ultimately, helping to develop new anti-bunyavirus strategies. In this regard, all of the factors and mechanisms that have been shown to be involved in virus entry, from virus or cell perspectives, can potentially be used as targets to block the initial steps of transmission and the subsequent spread throughout hosts.

Acknowledgments: This work was supported by the FONDECYT 1140050 and Basal PFB-16 grants from CONICYT (Chile) to Nicole Tischler, the FONDECYT 3150695 postdoctoral grant to Amelina Albornoz and the CellNetworks Research Group funds to Pierre-Yves Lozach.

Author Contributions: All authors discussed the manuscript content and illustrations and also wrote this review.

Conflicts of Interest: The authors declare no conflict of interest.

Abbreviations

The following abbreviations are used in this manuscript:

ANDV	Andes virus
BCCV	Black Creek Canal virus
BMP	phospholipid bis(monoacylglycerol)
CCHFV	Crimean-Congo hemorrhagic fever virus
CME	clathrin-mediated endocytosis
DAF	decay-accelerated factor
DC	dendritic cell
DC-SIGN	dendritic cell-specific intercellular adhesion molecule 3-grabbing non-integrin
DN	dominant negative
EE	early endosome
ER	endoplasmic reticulum
GAG	glycosaminoglycan
HTNV	Hantaan virus
HUVEC	human umbilical vein endothelial cell
LACV	La Crosse virus
LE	late endosome
LL	di-leucine
L-SIGN	liver/lymph node-specific intercellular adhesion molecule 3-grabbing non-integrin
MVB	multivesicular body
NMMHC-IIA	non-muscle myosin heavy chain IIA
NY-1V	New York-1 virus
OROV	Oropouche virus
PHV	Prospect Hill virus
PSI	plexin-semaphorin-integrin
PUUV	Puumula virus
RNP	ribonucleoprotein
RVFV	Rift Valley fever virus
SEOV	Seoul virus
SFTSV	severe fever with thrombocytopenia syndrome virus
siRNA	small interfering RNA
SNV	Sin Nombre virus
TOSV	Toscana virus
TSWV	tomato spotted wilt virus
UUKV	Uukuniemi virus

References

1. Schmaljohn, C.; Elliott, R.M. Bunyaviridae. In *Fields Virology*, 6th ed.; Knipe, D.M., Howley, P.M., Eds.; Lippincott Williams & Wilkins: Philadelphia, PA, USA, 2014; Volume 1, pp. 1244–1282.
2. Junglen, S.; Drosten, C. Virus discovery and recent insights into virus diversity in arthropods. *Curr. Opin. Microbiol.* **2013**, *16*, 507–513. [CrossRef] [PubMed]
3. Vaheri, A.; Strandin, T.; Hepojoki, J.; Sironen, T.; Henttonen, H.; Makela, S.; Mustonen, J. Uncovering the mysteries of hantavirus infections. *Nat. Rev. Microbiol.* **2013**, *11*, 539–550. [CrossRef] [PubMed]
4. Elliott, R.M.; Brennan, B. Emerging phleboviruses. *Curr. Opin. Virol.* **2014**, *5*, 50–57. [CrossRef] [PubMed]
5. Elliott, R.M. Orthobunyaviruses: recent genetic and structural insights. *Nat. Rev. Microbiol.* **2014**, *12*, 673–685. [CrossRef] [PubMed]
6. Lasecka, L.; Baron, M.D. The molecular biology of nairoviruses, an emerging group of tick-borne arboviruses. *Arch. Virol.* **2014**, *159*, 1249–1265. [CrossRef] [PubMed]
7. Rotenberg, D.; Jacobson, A.L.; Schneweis, D.J.; Whitfield, A.E. Thrips transmission of tospoviruses. *Curr. Opin. Virol.* **2015**, *15*, 80–89. [CrossRef] [PubMed]
8. Leger, P.; Lozach, P.Y. Bunyaviruses: from transmission by arthropods to virus entry into the mammalian host first-target cells. *Future Virol.* **2015**, *10*, 859–881. [CrossRef]
9. Horne, K.M.; Vanlandingham, D.L. Bunyavirus-vector interactions. *Viruses* **2014**, *6*, 4373–4397. [CrossRef] [PubMed]
10. Holmes, E.C.; Zhang, Y.Z. The evolution and emergence of hantaviruses. *Curr. Opin. Virol.* **2015**, *10*, 27–33. [CrossRef] [PubMed]
11. Pettersson, R.; Melin, L. Synthesis, assembly and intracellular transport of *Bunyaviridae.* membrane proteins. In *The Bunyaviridae*; Elliott, R.M., Ed.; Plenum Press: New York, NY, USA, 1996; pp. 159–188.
12. Ferron, F.; Li, Z.; Danek, E.I.; Luo, D.; Wong, Y.; Coutard, B.; Lantez, V.; Charrel, R.; Canard, B.; Walz, T.; Lescar, J. The hexamer structure of Rift Valley fever virus nucleoprotein suggests a mechanism for its assembly into ribonucleoprotein complexes. *PLoS Pathog.* **2011**, *7*, e1002030. [CrossRef] [PubMed]
13. Raymond, D.D.; Piper, M.E.; Gerrard, S.R.; Skiniotis, G.; Smith, J.L. Phleboviruses encapsidate their genomes by sequestering RNA bases. *Proc. Natl. Acad. Sci. USA* **2012**, *109*, 19208–19213. [CrossRef] [PubMed]
14. Reguera, J.; Cusack, S.; Kolakofsky, D. Segmented negative strand RNA virus nucleoprotein structure. *Curr. Opin. Virol.* **2014**, *5*, 7–15. [CrossRef] [PubMed]
15. Wang, Y.; Dutta, S.; Karlberg, H.; Devignot, S.; Weber, F.; Hao, Q.; Tan, Y.J.; Mirazimi, A.; Kotaka, M. Structure of Crimean-Congo hemorrhagic fever virus nucleoprotein: superhelical homo-oligomers and the role of caspase-3 cleavage. *J. Virol.* **2012**, *86*, 12294–12303. [CrossRef] [PubMed]
16. Carter, S.D.; Surtees, R.; Walter, C.T.; Ariza, A.; Bergeron, E.; Nichol, S.T.; Hiscox, J.A.; Edwards, T.A.; Barr, J.N. Structure, function, and evolution of the Crimean-Congo hemorrhagic fever virus nucleocapsid protein. *J. Virol.* **2012**, *86*, 10914–10923. [CrossRef] [PubMed]
17. Guo, Y.; Wang, W.; Ji, W.; Deng, M.; Sun, Y.; Zhou, H.; Yang, C.; Deng, F.; Wang, H.; Hu, Z.; Lou, Z.; Rao, Z. Crimean-Congo hemorrhagic fever virus nucleoprotein reveals endonuclease activity in bunyaviruses. *Proc. Natl. Acad. Sci. USA* **2012**, *109*, 5046–5051. [CrossRef] [PubMed]
18. Ariza, A.; Tanner, S.J.; Walter, C.T.; Dent, K.C.; Shepherd, D.A.; Wu, W.; Matthews, S.V.; Hiscox, J.A.; Green, T.J.; Luo, M.; *et al.* Nucleocapsid protein structures from orthobunyaviruses reveal insight into ribonucleoprotein architecture and RNA polymerization. *Nucleic Acids Res.* **2013**, *41*, 5912–5926. [CrossRef] [PubMed]
19. Dong, H.; Li, P.; Elliott, R.M.; Dong, C. Structure of Schmallenberg orthobunyavirus nucleoprotein suggests a novel mechanism of genome encapsidation. *J. Virol.* **2013**, *87*, 5593–5601. [CrossRef] [PubMed]
20. Jiao, L.; Ouyang, S.; Liang, M.; Niu, F.; Shaw, N.; Wu, W.; Ding, W.; Jin, C.; Peng, Y.; Zhu, Y.; *et al.* Structure of severe fever with thrombocytopenia syndrome virus nucleocapsid protein in complex with suramin reveals therapeutic potential. *J. Virol.* **2013**, *87*, 6829–6839. [CrossRef] [PubMed]
21. Li, B.; Wang, Q.; Pan, X.; Fernandez de Castro, I.; Sun, Y.; Guo, Y.; Tao, X.; Risco, C.; Sui, S.F.; Lou, Z. Bunyamwera virus possesses a distinct nucleocapsid protein to facilitate genome encapsidation. *Proc. Natl. Acad. Sci. USA* **2013**, *110*, 9048–9053. [CrossRef] [PubMed]

22. Niu, F.; Shaw, N.; Wang, Y.E.; Jiao, L.; Ding, W.; Li, X.; Zhu, P.; Upur, H.; Ouyang, S.; Cheng, G.; *et al.* Structure of the Leanyer orthobunyavirus nucleoprotein-RNA complex reveals unique architecture for RNA encapsidation. *Proc. Natl. Acad. Sci. USA* **2013**, *110*, 9054–9059. [CrossRef] [PubMed]
23. Olal, D.; Dick, A.; Woods, V.L., Jr.; Liu, T.; Li, S.; Devignot, S.; Weber, F.; Saphire, E.O.; Daumke, O. Structural insights into RNA encapsidation and helical assembly of the Toscana virus nucleoprotein. *Nucleic Acids Res.* **2014**, *42*, 6025–6037. [CrossRef] [PubMed]
24. Raymond, D.D.; Piper, M.E.; Gerrard, S.R.; Smith, J.L. Structure of the Rift Valley fever virus nucleocapsid protein reveals another architecture for RNA encapsidation. *Proc. Natl. Acad. Sci. USA* **2010**, *107*, 11769–11774. [CrossRef] [PubMed]
25. Reguera, J.; Malet, H.; Weber, F.; Cusack, S. Structural basis for encapsidation of genomic RNA by La Crosse orthobunyavirus nucleoprotein. *Proc. Natl. Acad. Sci. USA* **2013**, *110*, 7246–7251. [CrossRef] [PubMed]
26. Olal, D.; Daumke, O. Structure of the Hantavirus nucleoprotein provides insights into the mechanism of RNA encapsidation. *Cell Rep.* **2016**, *14*, 2092–2099. [CrossRef] [PubMed]
27. Surtees, R.; Ariza, A.; Punch, E.K.; Trinh, C.H.; Dowall, S.D.; Hewson, R.; Hiscox, J.A.; Barr, J.N.; Edwards, T.A. The crystal structure of the Hazara virus nucleocapsid protein. *BMC Struct. Biol.* **2015**, *15*, 24. [CrossRef] [PubMed]
28. Eifan, S.; Schnettler, E.; Dietrich, I.; Kohl, A.; Blomstrom, A.L. Non-structural proteins of arthropod-borne bunyaviruses: roles and functions. *Viruses* **2013**, *5*, 2447–2468. [CrossRef] [PubMed]
29. Vera-Otarola, J.; Solis, L.; Soto-Rifo, R.; Ricci, E.P.; Pino, K.; Tischler, N.D.; Ohlmann, T.; Darlix, J.L.; Lopez-Lastra, M. The Andes hantavirus NSs protein is expressed from the viral small mRNA by a leaky scanning mechanism. *J. Virol.* **2012**, *86*, 2176–2187. [CrossRef] [PubMed]
30. Plyusnin, A. Genetics of hantaviruses: implications to taxonomy. *Arch. Virol.* **2002**, *147*, 665–682. [CrossRef] [PubMed]
31. Jaaskelainen, K.M.; Plyusnina, A.; Lundkvist, A.; Vaheri, A.; Plyusnin, A. Tula hantavirus isolate with the full-length ORF for nonstructural protein NSs survives for more consequent passages in interferon-competent cells than the isolate having truncated NSs ORF. *Virol J.* **2008**, *5*, 3. [CrossRef] [PubMed]
32. Plyusnin, A.; Beaty, B.J.; Elliott, R.M.; Goldbach, R.; Kormelink, R.; Lundkvist, Å.; Schmaljohn, C.S.; Tesh, R.B. The *Bunyaviridae*. In virus taxonomy: ninth report of the international committee on taxonomy of viruses. 2012 International committee on taxonomy of viruses. *Elsevier Inc.* **2012**.
33. Sanchez, A.J.; Vincent, M.J.; Nichol, S.T. Characterization of the glycoproteins of Crimean-Congo hemorrhagic fever virus. *J. Virol.* **2002**, *76*, 7263–7275. [CrossRef] [PubMed]
34. Dessau, M.; Modis, Y. Crystal structure of glycoprotein C from Rift Valley fever virus. *Proc. Natl. Acad. Sci. USA* **2013**, *110*, 1696–1701. [CrossRef] [PubMed]
35. Garry, C.E.; Garry, R.F. Proteomics computational analyses suggest that the carboxyl terminal glycoproteins of Bunyaviruses are class II viral fusion protein (beta-penetrenes). *Theor. Biol. Med. Model.* **2004**, *1*, 10. [CrossRef] [PubMed]
36. Tischler, N.D.; Gonzalez, A.; Perez-Acle, T.; Rosemblatt, M.; Valenzuela, P.D. Hantavirus Gc glycoprotein: evidence for a class II fusion protein. *J. Gen. Virol.* **2005**, *86*, 2937–2947. [CrossRef] [PubMed]
37. Plassmeyer, M.L.; Soldan, S.S.; Stachelek, K.M.; Martin-Garcia, J.; Gonzalez-Scarano, F. California serogroup Gc (G1) glycoprotein is the principal determinant of pH-dependent cell fusion and entry. *Virology* **2005**, *338*, 121–132. [CrossRef] [PubMed]
38. Cifuentes-Munoz, N.; Salazar-Quiroz, N.; Tischler, N.D. Hantavirus Gn and Gc envelope glycoproteins: key structural units for virus cell entry and virus assembly. *Viruses* **2014**, *6*, 1801–1822. [CrossRef] [PubMed]
39. Overby, A.K.; Pettersson, R.F.; Grunewald, K.; Huiskonen, J.T. Insights into bunyavirus architecture from electron cryotomography of Uukuniemi virus. *Proc. Natl Acad. Sci. U S A* **2008**, *105*, 2375–2379. [CrossRef] [PubMed]
40. Freiberg, A.N.; Sherman, M.B.; Morais, M.C.; Holbrook, M.R.; Watowich, S.J. Three-dimensional organization of Rift Valley fever virus revealed by cryoelectron tomography. *J. Virol.* **2008**, *82*, 10341–10348. [CrossRef] [PubMed]
41. Huiskonen, J.T.; Overby, A.K.; Weber, F.; Grunewald, K. Electron cryo-microscopy and single-particle averaging of Rift Valley fever virus: evidence for Gn-Gc glycoprotein heterodimers. *J. Virol.* **2009**, *83*, 3762–3769. [CrossRef] [PubMed]

42. Huiskonen, J.T.; Hepojoki, J.; Laurinmaki, P.; Vaheri, A.; Lankinen, H.; Butcher, S.J.; Grunewald, K. Electron cryotomography of Tula hantavirus suggests a unique assembly paradigm for enveloped viruses. *J. Virol.* **2010**, *84*, 4889–4897. [CrossRef] [PubMed]
43. Sherman, M.B.; Freiberg, A.N.; Holbrook, M.R.; Watowich, S.J. Single-particle cryo-electron microscopy of Rift Valley fever virus. *Virology* **2009**, *387*, 11–15. [CrossRef] [PubMed]
44. Bowden, T.A.; Bitto, D.; McLees, A.; Yeromonahos, C.; Elliott, R.M.; Huiskonen, J.T. Orthobunyavirus ultrastructure and the curious tripodal glycoprotein spike. *PLoS Pathog.* **2013**, *9*, e1003374. [CrossRef] [PubMed]
45. Battisti, A.J.; Chu, Y.K.; Chipman, P.R.; Kaufmann, B.; Jonsson, C.B.; Rossmann, M.G. Structural studies of Hantaan virus. *J. Virol.* **2011**, *85*, 835–841. [CrossRef] [PubMed]
46. Boulant, S.; Stanifer, M.; Lozach, P.Y. Dynamics of virus-receptor interactions in virus binding, signaling, and endocytosis. *Viruses* **2015**, *7*, 2794–2815. [CrossRef] [PubMed]
47. Helenius, A. Virus Entry and Uncoating. In *Fields Virology*, 6th ed.; Knipe, D.M., Howley, P.M., Eds.; Lippincott Williams & Wilkins: Philadelphia, PA, USA, 2014.
48. Jolly, C.L.; Sattentau, Q.J. Attachment factors. *Adv. Exp. Med. Biol.* **2013**, *790*, 1–23. [PubMed]
49. Grove, J.; Marsh, M. The cell biology of receptor-mediated virus entry. *J. Cell. Biol.* **2011**, *195*, 1071–1082. [CrossRef] [PubMed]
50. De Boer, S.M.; Kortekaas, J.; de Haan, C.A.; Rottier, P.J.; Moormann, R.J.; Bosch, B.J. Heparan sulfate facilitates Rift Valley fever virus entry into the cell. *J. Virol.* **2012**, *86*, 13767–13771. [CrossRef] [PubMed]
51. Pietrantoni, A.; Fortuna, C.; Remoli, M.E.; Ciufolini, M.G.; Superti, F. Bovine lactoferrin inhibits Toscana virus infection by binding to heparan sulphate. *Viruses* **2015**, *7*, 480–495. [CrossRef] [PubMed]
52. Riblett, A.M.; Blomen, V.A.; Jae, L.T.; Altamura, L.A.; Doms, R.W.; Brummelkamp, T.R.; Wojcechowskyj, J.A. A Haploid genetic screen identifies heparan sulfate proteoglycans supporting Rift Valley fever virus infection. *J. Virol.* **2015**, *90*, 1414–1423. [CrossRef] [PubMed]
53. Lozach, P.Y.; Kuhbacher, A.; Meier, R.; Mancini, R.; Bitto, D.; Bouloy, M.; Helenius, A. DC-SIGN as a receptor for phleboviruses. *Cell Host Microbe* **2011**, *10*, 75–88. [CrossRef] [PubMed]
54. Hofmann, H.; Li, X.; Zhang, X.; Liu, W.; Kuhl, A.; Kaup, F.; Soldan, S.S.; Gonzalez-Scarano, F.; Weber, F.; He, Y.; *et al.* Severe fever with thrombocytopenia virus glycoproteins are targeted by neutralizing antibodies and can use DC-SIGN as a receptor for pH-dependent entry into human and animal cell lines. *J. Virol.* **2013**, *87*, 4384–4394. [CrossRef] [PubMed]
55. Suda, Y.; Fukushi, S.; Tani, H.; Murakami, S.; Saijo, M.; Horimoto, T.; Shimojima, M. Analysis of the entry mechanism of Crimean-Congo hemorrhagic fever virus, using a vesicular stomatitis virus pseudotyping system. *Arch. Virol.* **2016**. in press. [CrossRef] [PubMed]
56. Connolly-Andersen, A.M.; Douagi, I.; Kraus, A.A.; Mirazimi, A. Crimean-Congo hemorrhagic fever virus infects human monocyte-derived dendritic cells. *Virology* **2009**, *390*, 157–162. [CrossRef] [PubMed]
57. Connolly-Andersen, A.M.; Moll, G.; Andersson, C.; Akerstrom, S.; Karlberg, H.; Douagi, I.; Mirazimi, A. Crimean-Congo hemorrhagic fever virus activates endothelial cells. *J. Virol.* **2011**, *85*, 7766–7774. [CrossRef] [PubMed]
58. Peyrefitte, C.N.; Perret, M.; Garcia, S.; Rodrigues, R.; Bagnaud, A.; Lacote, S.; Crance, J.M.; Vernet, G.; Garin, D.; Bouloy, M.; *et al.* Differential activation profiles of Crimean-Congo hemorrhagic fever virus- and Dugbe virus-infected antigen-presenting cells. *J. Gen. Virol.* **2010**, *91*, 189–198. [CrossRef] [PubMed]
59. Svajger, U.; Anderluh, M.; Jeras, M.; Obermajer, N. C-type lectin DC-SIGN: an adhesion, signalling and antigen-uptake molecule that guides dendritic cells in immunity. *Cell. Signal.* **2010**, *22*, 1397–1405. [CrossRef] [PubMed]
60. Gillespie, L.; Roosendahl, P.; Ng, W.C.; Brooks, A.G.; Reading, P.C.; Londrigan, S.L. Endocytic function is critical for influenza A virus infection via DC-SIGN and L-SIGN. *Sci. Rep.* **2016**, *6*, 19428. [CrossRef] [PubMed]
61. Sun, Y.; Qi, Y.; Liu, C.; Gao, W.; Chen, P.; Fu, L.; Peng, B.; Wang, H.; Jing, Z.; Zhong, G.; *et al.* Nonmuscle myosin heavy chain IIA is a critical factor contributing to the efficiency of early infection of severe fever with thrombocytopenia syndrome virus. *J. Virol.* **2014**, *88*, 237–248. [CrossRef] [PubMed]
62. Leger, P.; Tetard, M.; Youness, B.; Cordes, N.; Rouxel, R.N.; Flamand, M.; Lozach, P.Y. Differential use of the C-type lectins L-SIGN and DC-SIGN for phlebovirus endocytosis. *Traffic* **2016**, *17*, 639–656. [CrossRef] [PubMed]

63. Xiao, X.; Feng, Y.; Zhu, Z.; Dimitrov, D.S. Identification of a putative Crimean-Congo hemorrhagic fever virus entry factor. *Biochem. Biophys. Res. Commun.* **2011**, *411*, 253–258. [CrossRef] [PubMed]
64. Medeiros, R.B.; Ullman, D.E.; Sherwood, J.L.; German, T.L. Immunoprecipitation of a 50-kDa protein: a candidate receptor component for tomato spotted wilt tospovirus (*Bunyaviridae*) in its main vector, *Frankliniella occidentalis*. *Virus Res.* **2000**, *67*, 109–118. [CrossRef]
65. Bandla, M.D.; Campbell, L.R.; Ullman, D.E.; Sherwood, J.L. Interaction of tomato spotted wilt tospovirus (TSWV) glycoproteins with a thrips midgut protein, a potential cellular receptor for TSWV. *Phytopathology* **1998**, *88*, 98–104. [CrossRef] [PubMed]
66. Krautkramer, E.; Zeier, M. Hantavirus causing hemorrhagic fever with renal syndrome enters from the apical surface and requires decay-accelerating factor (DAF/CD55). *J. Virol.* **2008**, *82*, 4257–4264. [CrossRef] [PubMed]
67. Gavrilovskaya, I.N.; Shepley, M.; Shaw, R.; Ginsberg, M.H.; Mackow, E.R. Beta3 Integrins mediate the cellular entry of hantaviruses that cause respiratory failure. *Proc. Natl Acad. Sci. U S A* **1998**, *95*, 7074–7079. [CrossRef] [PubMed]
68. Gavrilovskaya, I.N.; Brown, E.J.; Ginsberg, M.H.; Mackow, E.R. Cellular entry of hantaviruses which cause hemorrhagic fever with renal syndrome is mediated by beta3 integrins. *J. Virol.* **1999**, *73*, 3951–3959. [PubMed]
69. Matthys, V.S.; Gorbunova, E.E.; Gavrilovskaya, I.N.; Mackow, E.R. Andes virus recognition of human and Syrian hamster beta3 integrins is determined by an L33P substitution in the PSI domain. *J. Virol.* **2010**, *84*, 352–360. [CrossRef] [PubMed]
70. Choi, Y.; Kwon, Y.C.; Kim, S.I.; Park, J.M.; Lee, K.H.; Ahn, B.Y. A hantavirus causing hemorrhagic fever with renal syndrome requires gC1qR/p32 for efficient cell binding and infection. *Virology* **2008**, *381*, 178–183. [CrossRef] [PubMed]
71. Mou, D.L.; Wang, Y.P.; Huang, C.X.; Li, G.Y.; Pan, L.; Yang, W.S.; Bai, X.F. Cellular entry of Hantaan virus A9 strain: specific interactions with beta3 integrins and a novel 70kDa protein. *Biochem. Biophys. Res. Commun.* **2006**, *339*, 611–617. [CrossRef] [PubMed]
72. Raftery, M.J.; Lalwani, P.; Krautkrmer, E.; Peters, T.; Scharffetter-Kochanek, K.; Kruger, R.; Hofmann, J.; Seeger, K.; Kruger, D.H.; Schonrich, G. Beta2 integrin mediates hantavirus-induced release of neutrophil extracellular traps. *J. Exp. Med.* **2014**, *211*, 1485–1497. [CrossRef] [PubMed]
73. Pokidysheva, E.; Zhang, Y.; Battisti, A.J.; Bator-Kelly, C.M.; Chipman, P.R.; Xiao, C.; Gregorio, G.G.; Hendrickson, W.A.; Kuhn, R.J.; Rossmann, M.G. Cryo-EM reconstruction of dengue virus in complex with the carbohydrate recognition domain of DC-SIGN. *Cell* **2006**, *124*, 485–493. [CrossRef] [PubMed]
74. Silvas, J.A.; Popov, V.L.; Paulucci-Holthauzen, A.; Aguilar, P.V. Extracellular vesicles mediate receptor-independent transmission of novel tick-borne bunyavirus. *J. Virol.* **2015**, *90*, 873–886. [CrossRef] [PubMed]
75. Gilbertson, R.L.; Batuman, O.; Webster, C.G.; Adkins, S. Role of the insect supervectors *Bemisia tabaci* and *Frankliniella occidentalis* in the emergence and global spread of plant viruses. *Annu. Rev. Virol.* **2015**, *2*, 67–93. [CrossRef] [PubMed]
76. Tentchev, D.; Verdin, E.; Marchal, C.; Jacquet, M.; Aguilar, J.M.; Moury, B. Evolution and structure of tomato spotted wilt virus populations: evidence of extensive reassortment and insights into emergence processes. *J. Gen. Virol.* **2011**, *92*, 961–973. [CrossRef] [PubMed]
77. Nagata, T.; Inoue-Nagata, A.K.; Prins, M.; Goldbach, R.; Peters, D. Impeded thrips transmission of defective tomato spotted wilt virus isolates. *Phytopathology* **2000**, *90*, 454–459. [CrossRef] [PubMed]
78. Resende Rde, O.; de Haan, P.; de Avila, A.C.; Kitajima, E.W.; Kormelink, R.; Goldbach, R.; Peters, D. Generation of envelope and defective interfering RNA mutants of tomato spotted wilt virus by mechanical passage. *J. Gen. Virol.* **1991**, *72*, 2375–2383. [CrossRef] [PubMed]
79. Ullman, D.E.; Westcot, D.M.; Chenault, K.D.; Sherwood, J.L.; German, T.L.; Bandla, M.D.; Cantone, F.A.; Duer, H.L. Compartmentalization, intracellular transport, and autophagy of tomato spotted wilt tospovirus proteins in infected thrips cells. *Phytopathology* **1996**, *85*, 644–654. [CrossRef]
80. Sakimura, K. The present status of thrips-borne viruses. In *Biological Tranmission of Disease Agents*; Maramorosch, K., Ed.; Academic Press: New York, NY, USA, 1962; pp. 33–40.
81. Ullman, D.E.; Cho, J.J.; Mau, R.F.L.; Westcot, D.M.; Cantone, D.M. A midgut barrier to TSWV acquisition by adult western flower thrips. *Phytopathology* **1992**, *82*, 1333–1342. [CrossRef]

82. Ritzenthaler, C. Parallels and distinctions in the direct cell-to-cell spread of the plant and animal viruses. *Curr. Opin. Virol.* **2011**, *1*, 403–409. [CrossRef] [PubMed]
83. Storms, M.M.; Kormelink, R.; Peters, D.; Van Lent, J.W.; Goldbach, R.W. The nonstructural NSm protein of tomato spotted wilt virus induces tubular structures in plant and insect cells. *Virology* **1995**, *214*, 485–493. [CrossRef] [PubMed]
84. Kormelink, R.; Storms, M.; Van Lent, J.; Peters, D.; Goldbach, R. Expression and subcellular location of the NSm protein of tomato spotted wilt virus (TSWV), a putative viral movement protein. *Virology* **1994**, *200*, 56–65. [CrossRef] [PubMed]
85. Paape, M.; Solovyev, A.G.; Erokhina, T.N.; Minina, E.A.; Schepetilnikov, M.V.; Lesemann, D.E.; Schiemann, J.; Morozov, S.Y.; Kellmann, J.W. At-4/1, an interactor of the tomato spotted wilt virus movement protein, belongs to a new family of plant proteins capable of directed intra- and intercellular trafficking. *Mol. Plant. Microbe Interact.* **2006**, *19*, 874–883. [CrossRef] [PubMed]
86. Amari, K.; Boutant, E.; Hofmann, C.; Schmitt-Keichinger, C.; Fernandez-Calvino, L.; Didier, P.; Lerich, A.; Mutterer, J.; Thomas, C.L.; Heinlein, M.; *et al.* A family of plasmodesmal proteins with receptor-like properties for plant viral movement proteins. *PLoS Pathog.* **2010**, *6*, e1001119. [CrossRef] [PubMed]
87. Murray, J.F. 2010: the year of the lung. *Int. J. Tuberc. Lung Dis.* **2010**, *14*, 1–4. [PubMed]
88. Dalrymple, N.A.; Mackow, E.R. Virus interactions with endothelial cell receptors: implications for viral pathogenesis. *Curr. Opin. Virol.* **2014**, *7*, 134–140. [CrossRef] [PubMed]
89. Gavrilovskaya, I.; Gorbunova, E.; Matthys, V.; Dalrymple, N.; Mackow, E. The role of the endothelium in HPS pathogenesis and potential therapeutic approaches. *Adv. Virol.* **2012**, *2012*, 467059. [CrossRef] [PubMed]
90. Hepojoki, J.; Vaheri, A.; Strandin, T. The fundamental role of endothelial cells in hantavirus pathogenesis. *Front. Microbiol.* **2014**, *5*, 727. [CrossRef] [PubMed]
91. Macneil, A.; Nichol, S.T.; Spiropoulou, C.F. Hantavirus pulmonary syndrome. *Virus Res.* **2011**, *162*, 138–147. [CrossRef] [PubMed]
92. Rasmuson, J.; Andersson, C.; Norrman, E.; Haney, M.; Evander, M.; Ahlm, C. Time to revise the paradigm of hantavirus syndromes? Hantavirus pulmonary syndrome caused by European hantavirus. *Eur. J. Clin. Microbiol. Infect. Dis.* **2011**, *30*, 685–690. [CrossRef] [PubMed]
93. Schmaljohn, C.; Hjelle, B. Hantaviruses: a global disease problem. *Emerg. Infect. Dis.* **1997**, *3*, 95–104. [CrossRef] [PubMed]
94. Johansson, S.S.G.; Wennerberg, K.; Armulik, A.; Lohikangas, L. Fibronectin-integrin interactions. *Front. Biosci.* **1995**, *2*, 126–146. [CrossRef]
95. Raymond, T.; Gorbunova, E.; Gavrilovskaya, I.N.; Mackow, E.R. Pathogenic hantaviruses bind plexin-semaphorin-integrin domains present at the apex of inactive, bent alphavbeta3 integrin conformers. *Proc. Natl. Acad. Sci. U S A* **2005**, *102*, 1163–1168. [CrossRef] [PubMed]
96. Askari, J.A.; Buckley, P.A.; Mould, A.P.; Humphries, M.J. Linking integrin conformation to function. *J. Cell Sci.* **2009**, *122*, 165–170. [CrossRef] [PubMed]
97. Tucker, S.P.; Compans, R.W. Virus infection of polarized epithelial cells. *Adv. Virus Res.* **1993**, *42*, 187–247. [PubMed]
98. Bomsel, M.; Alfsen, A. Entry of viruses through the epithelial barrier: pathogenic trickery. *Nat. Rev. Mol. Cell. Biol.* **2003**, *4*, 57–68. [CrossRef] [PubMed]
99. Fletcher, N.F.; Howard, C.; McKeating, J.A. Over the fence or through the gate: how viruses infect polarized cells. *Immunotherapy* **2012**, *4*, 249–251. [CrossRef] [PubMed]
100. Schoenenberger, C.A.; Zuk, A.; Zinkl, G.M.; Kendall, D.; Matlin, K.S. Integrin expression and localization in normal MDCK cells and transformed MDCK cells lacking apical polarity. *J. Cell Sci.* **1994**, *107*, 527–541. [PubMed]
101. Manninen, A. Epithelial polarity–generating and integrating signals from the ECM with integrins. *Exp. Cell. Res.* **2015**, *334*, 337–349. [CrossRef] [PubMed]
102. Conforti, G.; Dominguez-Jimenez, C.; Zanetti, A.; Gimbrone, M.A., Jr.; Cremona, O.; Marchisio, P.C.; Dejana, E. Human endothelial cells express integrin receptors on the luminal aspect of their membrane. *Blood* **1992**, *80*, 437–446. [PubMed]
103. Aplin, J.D.; Spanswick, C.; Behzad, F.; Kimber, S.J.; Vicovac, L. Integrins beta 5, beta 3 and alpha v are apically distributed in endometrial epithelium. *Mol. Hum. Reprod.* **1996**, *2*, 527–534. [CrossRef] [PubMed]

104. Gawaz, M.; Neumann, F.J.; Dickfeld, T.; Reininger, A.; Adelsberger, H.; Gebhardt, A.; Schomig, A. Vitronectin receptor (alpha(v)beta3) mediates platelet adhesion to the luminal aspect of endothelial cells: implications for reperfusion in acute myocardial infarction. *Circulation* **1997**, *96*, 1809–1818. [CrossRef] [PubMed]
105. Byzova, T.V.; Plow, E.F. Activation of alphaVbeta3 on vascular cells controls recognition of prothrombin. *J. Cell. Biol.* **1998**, *143*, 2081–2092. [CrossRef] [PubMed]
106. Raftery, M.J.; Kraus, A.A.; Ulrich, R.; Kruger, D.H.; Schonrich, G. Hantavirus infection of dendritic cells. *J. Virol.* **2002**, *76*, 10724–10733. [CrossRef] [PubMed]
107. Ravkov, E.V.; Nichol, S.T.; Compans, R.W. Polarized entry and release in epithelial cells of Black Creek Canal virus, a New World hantavirus. *J. Virol.* **1997**, *71*, 1147–1154. [PubMed]
108. Rowe, R.K.; Pekosz, A. Bidirectional virus secretion and nonciliated cell tropism following Andes virus infection of primary airway epithelial cell cultures. *J. Virol.* **2006**, *80*, 1087–1097. [CrossRef] [PubMed]
109. Buranda, T.; Swanson, S.; Bondu, V.; Schaefer, L.; Maclean, J.; Mo, Z.; Wycoff, K.; Belle, A.; Hjelle, B. Equilibrium and kinetics of Sin Nombre hantavirus binding at DAF/CD55 functionalized bead surfaces. *Viruses* **2014**, *6*, 1091–1111. [CrossRef] [PubMed]
110. Coyne, C.B.; Bergelson, J.M. Virus-induced ABL and FYN kinase signals permit coxsackievirus entry through epithelial tight junctions. *Cell* **2006**, *124*, 119–131. [CrossRef] [PubMed]
111. Sobo, K.; Rubbia-Brandt, L.; Brown, T.D.; Stuart, A.D.; McKee, T.A. Decay-accelerating factor binding determines the entry route of echovirus 11 in polarized epithelial cells. *J. Virol.* **2011**, *85*, 12376–12386. [CrossRef] [PubMed]
112. Petersen, J.; Drake, M.J.; Bruce, E.A.; Riblett, A.M.; Didigu, C.A.; Wilen, C.B.; Malani, N.; Male, F.; Lee, F.H.; Bushman, F.D.; *et al.* The major cellular sterol regulatory pathway is required for Andes virus infection. *PLoS Pathog.* **2014**, *10*, e1003911. [CrossRef] [PubMed]
113. Kleinfelter, L.M.; Jangra, R.K.; Jae, L.T.; Herbert, A.S.; Mittler, E.; Stiles, K.M.; Wirchnianski, A.S.; Kielian, M.; Brummelkamp, T.R.; Dye, J.M.; *et al.* Haploid genetic screen reveals a profound and direct dependence on cholesterol for hantavirus membrane fusion. *MBio* **2015**, *6*, e00801. [CrossRef] [PubMed]
114. Cifuentes-Munoz, N.; Darlix, J.L.; Tischler, N.D. Development of a lentiviral vector system to study the role of the Andes virus glycoproteins. *Virus Res.* **2010**, *153*, 29–35. [CrossRef] [PubMed]
115. Bangphoomi, N.; Takenaka-Uema, A.; Sugi, T.; Kato, K.; Akashi, H.; Horimoto, T. Akabane virus utilizes alternative endocytic pathways to entry into mammalian cell lines. *J. Vet. Med. Sci.* **2014**, *76*, 1471–1478. [CrossRef] [PubMed]
116. Santos, R.I.; Rodrigues, A.H.; Silva, M.L.; Mortara, R.A.; Rossi, M.A.; Jamur, M.C.; Oliver, C.; Arruda, E. Oropouche virus entry into HeLa cells involves clathrin and requires endosomal acidification. *Virus Res.* **2008**, *138*, 139–143. [CrossRef] [PubMed]
117. Shtanko, O.; Nikitina, R.A.; Altuntas, C.Z.; Chepurnov, A.A.; Davey, R.A. Crimean-Congo hemorrhagic fever virus entry into host cells occurs through the multivesicular body and requires ESCRT regulators. *PLoS Pathog.* **2014**, *10*, e1004390. [CrossRef] [PubMed]
118. Simon, M.; Johansson, C.; Mirazimi, A. Crimean-Congo hemorrhagic fever virus entry and replication is clathrin-, pH- and cholesterol-dependent. *J. Gen. Virol.* **2009**, *90*, 210–215. [CrossRef] [PubMed]
119. Lozach, P.Y.; Burleigh, L.; Staropoli, I.; Navarro-Sanchez, E.; Harriague, J.; Virelizier, J.L.; Rey, F.A.; Despres, P.; Arenzana-Seisdedos, F.; Amara, A. Dendritic cell-specific intercellular adhesion molecule 3-grabbing non-integrin (DC-SIGN)-mediated enhancement of dengue virus infection is independent of DC-SIGN internalization signals. *J. Biol. Chem.* **2005**, *280*, 23698–23708. [CrossRef] [PubMed]
120. Azad, A.K.; Torrelles, J.B.; Schlesinger, L.S. Mutation in the DC-SIGN cytoplasmic triacidic cluster motif markedly attenuates receptor activity for phagocytosis and endocytosis of mannose-containing ligands by human myeloid cells. *J. Leukoc. Biol.* **2008**, *84*, 1594–1603. [CrossRef] [PubMed]
121. Goncalves, A.R.; Moraz, M.L.; Pasquato, A.; Helenius, A.; Lozach, P.Y.; Kunz, S. Role of DC-SIGN in Lassa virus entry into human dendritic cells. *J. Virol.* **2013**, *87*, 11504–11515. [CrossRef] [PubMed]
122. Hollidge, B.S.; Nedelsky, N.B.; Salzano, M.V.; Fraser, J.W.; Gonzalez-Scarano, F.; Soldan, S.S. Orthobunyavirus entry into neurons and other mammalian cells occurs via clathrin-mediated endocytosis and requires trafficking into early endosomes. *J. Virol.* **2012**, *86*, 7988–8001. [CrossRef] [PubMed]
123. Garrison, A.R.; Radoshitzky, S.R.; Kota, K.P.; Pegoraro, G.; Ruthel, G.; Kuhn, J.H.; Altamura, L.A.; Kwilas, S.A.; Bavari, S.; Haucke, V.; *et al.* Crimean-Congo hemorrhagic fever virus utilizes a clathrin- and early endosome-dependent entry pathway. *Virology* **2013**, *444*, 45–54. [CrossRef] [PubMed]

124. Lozach, P.Y.; Mancini, R.; Bitto, D.; Meier, R.; Oestereich, L.; Overby, A.K.; Pettersson, R.F.; Helenius, A. Entry of bunyaviruses into mammalian cells. *Cell. Host Microbe* **2010**, *7*, 488–499. [CrossRef] [PubMed]
125. De Boer, S.M.; Kortekaas, J.; Spel, L.; Rottier, P.J.; Moormann, R.J.; Bosch, B.J. Acid-activated structural reorganization of the Rift Valley fever virus Gc fusion protein. *J. Virol.* **2012**, *86*, 13642–13652. [CrossRef] [PubMed]
126. Harmon, B.; Schudel, B.R.; Maar, D.; Kozina, C.; Ikegami, T.; Tseng, C.T.; Negrete, O.A. Rift Valley fever virus strain MP-12 enters mammalian host cells via caveola-mediated endocytosis. *J. Virol.* **2012**, *86*, 12954–12970. [CrossRef] [PubMed]
127. Filone, C.M.; Hanna, S.L.; Caino, M.C.; Bambina, S.; Doms, R.W.; Cherry, S. Rift valley fever virus infection of human cells and insect hosts is promoted by protein kinase C epsilon. *PLoS ONE* **2010**, *5*, e15483. [CrossRef] [PubMed]
128. Jin, M.; Park, J.; Lee, S.; Park, B.; Shin, J.; Song, K.J.; Ahn, T.I.; Hwang, S.Y.; Ahn, B.Y.; Ahn, K. Hantaan virus enters cells by clathrin-dependent receptor-mediated endocytosis. *Virology* **2002**, *294*, 60–69. [CrossRef] [PubMed]
129. Buranda, T.; BasuRay, S.; Swanson, S.; Agola, J.; Bondu, V.; Wandinger-Ness, A. Rapid parallel flow cytometry assays of active GTPases using effector beads. *Anal. Biochem.* **2013**, *442*, 149–157. [CrossRef] [PubMed]
130. Ramanathan, H.N.; Jonsson, C.B. New and Old World hantaviruses differentially utilize host cytoskeletal components during their life cycles. *Virology* **2008**, *374*, 138–150. [CrossRef] [PubMed]
131. Scott, C.C.; Vacca, F.; Gruenberg, J. Endosome maturation, transport and functions. *Semin. Cell. Dev. Biol.* **2014**, *31*, 2–10. [CrossRef] [PubMed]
132. White, J.M.; Whittaker, G.R. Fusion of Enveloped Viruses in Endosomes. *Traffic* **2016**, *17*, 593–614. [CrossRef] [PubMed]
133. Scott, C.C.; Gruenberg, J. Ion flux and the function of endosomes and lysosomes: pH is just the start: the flux of ions across endosomal membranes influences endosome function not only through regulation of the luminal pH. *Bioessays.* **2011**, *33*, 103–110. [CrossRef] [PubMed]
134. Lozach, P.Y.; Huotari, J.; Helenius, A. Late-penetrating viruses. *Curr. Opin. Virol.* **2011**, *1*, 35–43. [CrossRef] [PubMed]
135. Cifuentes-Muñoz, N.; Barriga, G.P.; Valenzuela, P.D.T.; Tischler, N.D. Aromatic and polar residues spanning the candidate fusion peptide of Andes virus are essential for membrane fusion and infection. *J. Gen. Virol.* **2011**, *92*, 552–563. [CrossRef] [PubMed]
136. Arikawa, J.; Takashima, I.; Hashimoto, N. Cell fusion by haemorrhagic fever with renal syndrome (HFRS) viruses and its application for titration of virus infectivity and neutralizing antibody. *Arch. Virol.* **1985**, *86*, 303–313. [CrossRef] [PubMed]
137. Simon, M.; Johansson, C.; Lundkvist, A.; Mirazimi, A. Microtubule-dependent and microtubule-independent steps in Crimean-Congo hemorrhagic fever virus replication cycle. *Virology* **2009**, *385*, 313–322. [CrossRef] [PubMed]
138. Liu, L.; Celma, C.C.; Roy, P. Rift Valley fever virus structural proteins: expression, characterization and assembly of recombinant proteins. *Virol J.* **2008**, *5*, 82. [CrossRef] [PubMed]
139. Bitto, D.; Halldorsson, S.; Caputo, A.; Huiskonen, J.T. Low pH and anionic lipid dependent fusion of Uukuniemi phlebovirus to liposomes. *J. Biol. Chem.* **2016**, *291*, 6412–6422. [CrossRef] [PubMed]
140. Wang, T.; Ming, Z.; Xiaochun, W.; Hong, W. Rab7: role of its protein interaction cascades in endo-lysosomal traffic. *Cell. Signal.* **2011**, *23*, 516–521. [CrossRef] [PubMed]
141. Acuna, R.; Bignon, E.; Mancini, R.; Lozach, P.Y.; Tischler, N.D. Acidification triggers Andes hantavirus membrane fusion and rearrangement of Gc into a stable post-fusion homotrimer. *J. Gen. Virol.* **2015**, *96*, 3192–3197. [CrossRef] [PubMed]
142. Harrison, S.C. Viral membrane fusion. *Virology* **2015**, 498–507. [CrossRef] [PubMed]
143. Kielian, M. Mechanisms of Virus Membrane Fusion Proteins. *Annu. Rev. Virol.* **2014**, *1*, 171–189. [CrossRef] [PubMed]
144. Whitfield, A.E.; Ullman, D.E.; German, T.L. tomato spotted wilt virus glycoprotein G(c) is cleaved at acidic pH. *Virus Res.* **2005**, *110*, 183–186. [CrossRef] [PubMed]
145. Shi, X.; Goli, J.; Clark, G.; Brauburger, K.; Elliott, R.M. Functional analysis of the Bunyamwera orthobunyavirus Gc glycoprotein. *J. Gen. Virol.* **2009**, *90*, 2483–2492. [CrossRef] [PubMed]

146. Plassmeyer, M.L.; Soldan, S.S.; Stachelek, K.M.; Roth, S.M.; Martin-Garcia, J.; Gonzalez-Scarano, F. Mutagenesis of the La Crosse Virus glycoprotein supports a role for Gc (1066–1087) as the fusion peptide. *Virology* **2007**, *358*, 273–282. [CrossRef] [PubMed]
147. Soldan, S.S.; Hollidge, B.S.; Wagner, V.; Weber, F.; Gonzalez-Scarano, F. La Crosse virus (LACV) Gc fusion peptide mutants have impaired growth and fusion phenotypes, but remain neurotoxic. *Virology* **2010**, *404*, 139–147. [CrossRef] [PubMed]
148. Jacoby, D.R.; Cooke, C.; Prabakaran, I.; Boland, J.; Nathanson, N.; Gonzalez-Scarano, F. Expression of the La Crosse M segment proteins in a recombinant vaccinia expression system mediates pH-dependent cellular fusion. *Virology* **1993**, *193*, 993–996. [CrossRef] [PubMed]
149. Kielian, M. Class II virus membrane fusion proteins. *Virology* **2006**, *344*, 38–47. [CrossRef] [PubMed]
150. Vaney, M.C.; Rey, F.A. Class II enveloped viruses. *Cell. Microbiol.* **2011**, *13*, 1451–1459. [CrossRef] [PubMed]
151. Modis, Y. Class II fusion proteins. *Adv. Exp. Med. Biol.* **2013**, *790*, 150–166. [PubMed]
152. Liao, M.; Kielian, M. Domain III from class II fusion proteins functions as a dominant-negative inhibitor of virus membrane fusion. *J. Cell. Biol.* **2005**, *171*, 111–120. [CrossRef] [PubMed]
153. Schmidt, A.G.; Yang, P.L.; Harrison, S.C. Peptide inhibitors of dengue-virus entry target a late-stage fusion intermediate. *PLoS Pathog.* **2010**, *6*, e1000851. [CrossRef] [PubMed]
154. Koehler, J.W.; Smith, J.M.; Ripoll, D.R.; Spik, K.W.; Taylor, S.L.; Badger, C.V.; Grant, R.J.; Ogg, M.M.; Wallqvist, A.; Guttieri, M.C.; *et al.* A fusion-inhibiting peptide against Rift Valley fever virus inhibits multiple, diverse viruses. *PLoS Negl. Trop. Dis.* **2013**, *7*, e2430. [CrossRef] [PubMed]
155. Barriga, G.P.; Villalón-Letelier, F.; Márquez, C.L.; Bignon, E.A.; Acuña, R.; Ross, B.H.; Monasterio, O.; Mardones, G.A.; Vidal, S.E.; Tischler, N.D. Inhibition of the hantavirus fusion process by predicted domain III and stem peptides from glycoprotein Gc. *PLoS Negl. Trop. Dis.* **2016**. under review.
156. Qin, Z.L.; Zheng, Y.; Kielian, M. Role of conserved histidine residues in the low-pH dependence of the Semliki Forest virus fusion protein. *J. Virol.* **2009**, *83*, 4670–4677. [CrossRef] [PubMed]
157. Delos, S.E.; La, B.; Gilmartin, A.; White, J.M. Studies of the "chain reversal regions" of the avian sarcoma/leukosis virus (ASLV) and Ebolavirus fusion proteins: analogous residues are important, and a His residue unique to EnvA affects the pH dependence of ASLV entry. *J. Virol.* **2010**, *84*, 5687–5694. [CrossRef] [PubMed]
158. Carneiro, F.A.; Stauffer, F.; Lima, C.S.; Juliano, M.A.; Juliano, L.; Da Poian, A.T. Membrane fusion induced by vesicular stomatitis virus depends on histidine protonation. *J. Biol. Chem.* **2003**, *278*, 13789–13794. [CrossRef] [PubMed]
159. Kampmann, T.; Mueller, D.S.; Mark, A.E.; Young, P.R.; Kobe, B. The Role of histidine residues in low-pH-mediated viral membrane fusion. *Structure* **2006**, *14*, 1481–1487. [CrossRef] [PubMed]
160. Edgcomb, S.P.; Murphy, K.P. Variability in the pKa of histidine side-chains correlates with burial within proteins. *Proteins* **2002**, *49*, 1–6. [CrossRef] [PubMed]
161. Hacker, J.K.; Hardy, J.L. Adsorptive endocytosis of California encephalitis virus into mosquito and mammalian cells: a role for G1. *Virology* **1997**, 23540–23547. [CrossRef] [PubMed]
162. Meier, R.; Franceschini, A.; Horvath, P.; Tetard, M.; Mancini, R.; von Mering, C.; Helenius, A.; Lozach, P.Y. Genome-wide small interfering RNA screens reveal VAMP3 as a novel host factor required for Uukuniemi virus late penetration. *J. Virol.* **2014**, *88*, 8565–8578. [CrossRef] [PubMed]
163. Hopkins, K.C.; McLane, L.M.; Maqbool, T.; Panda, D.; Gordesky-Gold, B.; Cherry, S. A genome-wide RNAi screen reveals that mRNA decapping restricts bunyaviral replication by limiting the pools of DCP2-accessible targets for cap-snatching. *Genes Dev.* **2013**, *27*, 1511–1525. [CrossRef] [PubMed]
164. Eckerle, I.; Lenk, M.; Ulrich, R.G. More novel hantaviruses and diversifying reservoir hosts–time for development of reservoir-derived cell culture models? *Viruses* **2014**, *6*, 951–967. [CrossRef] [PubMed]
165. Sundstrom, K.B.; Nguyen Hoang, A.T.; Gupta, S.; Ahlm, C.; Svensson, M.; Klingstrom, J. Andes hantavirus-infection of a 3D human lung tissue model reveals a late peak in progeny virus production followed by increased levels of proinflammatory cytokines and VEGF-A. *PLoS ONE* **2016**, *11*, e0149354. [CrossRef] [PubMed]
166. Bell-Sakyi, L.; Kohl, A.; Bente, D.A.; Fazakerley, J.K. Tick cell lines for study of Crimean-Congo hemorrhagic fever virus and other arboviruses. *Vector Borne Zoonotic Dis.* **2012**, *12*, 769–781. [CrossRef] [PubMed]
167. Mazelier, M.R.; R.N.; Zumstein, M.; Mancini, R.; Bell-Sakyi, L.; Lozach, P.Y. Uukuniemi virus as a tick-borne virus model. *J. Virol.* **2016**. in press. [CrossRef] [PubMed]

168. Li, Y.; Modis, Y. A novel membrane fusion protein family in *Flaviviridae*? *Trends Microbiol.* **2014**, *22*, 176–182. [CrossRef] [PubMed]
169. Rusu, M.; Bonneau, R.; Holbrook, M.R.; Watowich, S.J.; Birmanns, S.; Wriggers, W.; Freiberg, A.N. An assembly model of rift valley Fever virus. *Front. Microbiol.* **2012**, *3*, 254. [CrossRef] [PubMed]

© 2016 by the authors. Licensee MDPI, Basel, Switzerland. This article is an open access article distributed under the terms and conditions of the Creative Commons Attribution (CC BY) license (http://creativecommons.org/licenses/by/4.0/).

Review

What Do We Know about How Hantaviruses Interact with Their Different Hosts?

Myriam Ermonval *, Florence Baychelier and Noël Tordo

Unité des Stratégies Antivirales, Département de Virologie, Institut Pasteur, 25 Rue du Docteur Roux, 75015 Paris, France; Florence.Baychelier-tine@sanofi.com (F.B.); ntordo@pasteur.fr (N.T.)
* Correspondence: myriam.ermonval@pasteur.fr; Tel.: +33-1-4061-3632

Academic Editors: Jane Tao and Pierre-Yves Lozach
Received: 4 May 2016; Accepted: 5 August 2016; Published: 11 August 2016

Abstract: Hantaviruses, like other members of the *Bunyaviridae* family, are emerging viruses that are able to cause hemorrhagic fevers. Occasional transmission to humans is due to inhalation of contaminated aerosolized excreta from infected rodents. Hantaviruses are asymptomatic in their rodent or insectivore natural hosts with which they have co-evolved for millions of years. In contrast, hantaviruses cause different pathologies in humans with varying mortality rates, depending on the hantavirus species and its geographic origin. Cases of hemorrhagic fever with renal syndrome (HFRS) have been reported in Europe and Asia, while hantavirus cardiopulmonary syndromes (HCPS) are observed in the Americas. In some cases, diseases caused by Old World hantaviruses exhibit HCPS-like symptoms. Although the etiologic agents of HFRS were identified in the early 1980s, the way hantaviruses interact with their different hosts still remains elusive. What are the entry receptors? How do hantaviruses propagate in the organism and how do they cope with the immune system? This review summarizes recent data documenting interactions established by pathogenic and nonpathogenic hantaviruses with their natural or human hosts that could highlight their different outcomes.

Keywords: *Bunyaviridae*; hantavirus; virus-host interaction; rodents; humans; insectivores

---

1. Introduction

Hantaviruses are tri-segmented, enveloped, RNA viruses of negative polarity, belonging to the family *Bunyaviridae*. In contrast to members of the four other genera, viruses from the Hantavirus genus are not arthropod borne. Their human transmission occurs through inhalation of aerosolized virus particles present in dried excreta of naturally infected rodents, among which, hantaviruses circulate without giving any recognized symptoms. Human pathologies linked to hantaviruses have been observed since the 1950s, with the documentation of Korean Hemorrhagic Fever, but the Hantaan virus (HTNV) prototype was only isolated from the striped field mouse (*Apodemus agrarius*) in 1978 [1,2]. To date, only hantaviruses circulating among rodent reservoirs (Murinae, Arvicolinae, and Sigmodontinae) have been found to be pathogenic to humans.

New hantaviruses are now described in many small insectivorous mammals [3], such as moles and shrews (Soricomorpha), as well as in bats (Chiroptera). A recent study performed in Brazilian bats showed the presence of hantaviruses among different bats species, and not solely in insectivorous bats as previously suggested [4]. These discoveries point to the fact that hantaviruses circulate in host reservoirs present worldwide, extending the risk of possible host co-infection by different hantaviruses, and therefore virus reassortment and host spill-over. Moreover, changes in reservoir ecology due to human impacts on climate and biodiversity are additional factors that make hantaviruses a global public health concern [5,6].

Since their initial discovery, more than 20 different species of hantaviruses that are pathogenic for humans have been described, with new viruses regularly found all over the world. They cause hemorrhagic fever with renal syndrome (HFRS) in the Old World (Europe and Asia) while hantavirus cardiopulmonary syndrome (HCPS) is more specifically associated with human diseases in the New World (Americas). The major HFRS symptoms are acute kidney injury (AKI) and hemorrhagic fevers.

The most frequent hantavirus, endemic to European countries, is the Puumala virus (PUUV), causing more than 10,000 cases of nephropathia epidemica each year. This mild form of HFRS has a case fatality of less than 0.2%. However, mortality associated with the Dobrava virus (DOBV) in central Europe, and HTNV or Seoul virus (SEOV) in Asia may reach 15%. It is worth mentioning that SEOV has recently been described in wild and pet rats in several European countries [7–10] and that a few human cases have also been detected in the UK and France [11,12]. The easy adaptation of rats to other territories, in particular urban areas, has led to the extension of SEOV outside of Asia. In the Americas, HCPS presents with pulmonary edema, and although cases are less frequent (hundreds per year), hantaviruses, such as the Andes virus (ANDV) in South America and the Sin Nombre virus (SNV) in the US, can give rise to mortality rates of up to 50% of infected individuals. Despite these variable clinical pictures [13,14], which depend on the hantavirus and its geographic origin, HFRS and HCPS share common characteristics. In particular, thrombocytopenia [15,16] and vascular leakage [17,18] correlate with disease severity. These features have been compared to the different forms of dengue presenting similar clinical manifestations [19].

Many questions concerning the mechanisms of transmission and pathogenesis in humans, and the way hantaviruses propagate in individuals (entry, tropism and targeted organs for diseases) remain to be answered. Understanding how hantaviruses interact with specific host factors may explain: (a) the diverse degrees of pathogenicity observed in humans, from nonpathogenic to nephropathia epidemica, HFRS or HCPS; and (b) how a persistent and asymptomatic state is achieved in natural hosts. Recent specific reviews have addressed hantavirus pathogenesis [14,20], hantavirus interactions with the immune system [21–23], or their persistence in rodents [24,25]. These topics will not be detailed here. Rather, the present review summarizes information on pathogenic factors, as deduced from a comparison of the effects of pathogenic versus nonpathogenic hantaviruses in humans and natural hosts. A better understanding of the mechanisms leading to the different hantavirus outcomes should throw light on the physiopathology of these viruses and help to identify factors of pathogenesis that could represent potential therapeutic targets.

2. Outcomes of Hantaviruses: Persistence in Rodents and Pathogenicity in Human Hosts

Although hantaviruses are asymptomatic in their rodent reservoir, they are highly viremic and seroconversion takes place in this natural host, as indicated by antibody production, including neutralizing antibodies. Recent data using laboratory animals show that viremic phases in wild rodents are longer than previously described [26]. The virus is found in many organs, such as the kidneys, liver, spleen, heart, and in greater amounts, the lungs. However, the type of infected cells in each tissue or organ has not always been determined. The virus is also found in the saliva, urine and feces of rodents [27], and can persist for several weeks in the environment [28], which increases the risk of transmission to humans. Variations in the levels of virus shedding are observed depending on the hantavirus species. Hantaviruses replicate in natural hosts despite an adaptive immune response. This raises the question of the way tolerance and resistance are balanced in the rodent host such that hantaviruses escape the innate and adaptive immunity and establish a persistent state without being eliminated.

In contrast to what is observed in rodents, human hantavirus infection may lead to clinical signs. Similar to a few other human pathogens causing hemorrhagic fevers, endothelial cells are the main targets of hantavirus infection [29]. In HFRS and HCPS patients, endothelial cells are ubiquitously infected throughout the body, although injury is most prominent in the lungs and kidneys. The infection is non-lytic, but the altered function of the infected endothelium accounts for increased vascular permeability, hemorrhage, acute thrombocytopenia and pulmonary edemas or kidney failure as described for hantavirus diseases [30].

The narrow host specificity of hantaviruses, with one virus species adapted to one rodent species, and the fact that natural rodent hosts do not harbor any obvious symptoms, make it difficult to develop animal models to study the physiopathology of these viruses. The few studies that have been conducted using rodent models of hantavirus disease were performed with the Turkish hamster [31] or the Syrian hamster, in which only pathogenic New World hantaviruses induce symptoms resembling HCPS [32,33]. However, a recent report has shown that the Imjin virus (MJNV) hosted by some Asian shrews was able to induce pulmonary syndromes in Syrian hamsters [34]. The best animal model remains the macaque nonhuman primate for both New and Old World hantaviruses [35–37]. Recent data described possible infection of humanized SCID mice with hantaviruses [38]. Today, most data on the physiopathology of hantaviruses come from analyses of human clinical samples, or host reservoir samples, as well as from in vitro studies with cellular models.

The mechanisms underlying the major symptoms of vascular permeability observed in both HFRS and HCPS are not fully understood. In particular, endothelial cells are not directly lysed by hantavirus infection. The current hypothesis to explain the vascular leakage is that an excessive innate immune response could impair the barrier functions of the endothelium [14,39]. It was then proposed that infected endothelial cells from lung or kidney would secrete factors leading to the recruitment of innate immune cells (macrophages, dendritic cells, neutrophils, T cells). These recruited immune cells, in turn, would secrete pro-inflammatory cytokines upon activation, affecting barrier integrity. In addition, macrophages can be infected by hantaviruses, which would also deliver inflammatory factors. In addition, many studies on HCPS have highlighted the role of vascular endothelial growth factor on the regulation of vascular permeability, as is the case following hypoxia [30].

While pathogenic hantaviruses do not always give rise to severe diseases, some serologic evidence has been obtained for human infection by hantaviruses that are considered to be nonpathogenic, like Prospect Hill (PHV) or Tula (TULV) viruses [40,41]. Additional epidemiological studies are needed to clarify whether or not other nonpathogenic rodent- or insectivore-borne hantaviruses may replicate in humans. Understanding how these different viruses interact with human and small mammals such that the endothelium can be infected with or without harboring dysfunction is of great importance. It would be particularly interesting to investigate if the same mechanism operates in asymptomatic rodents and in humans infected with nonpathogenic hantaviruses.

3. Hantavirus Propagation in Different Cell Types

Until now, there has been no observed human–human transmission of hantaviruses [6], except in some HCPS cases caused by ADNV [42]. Although bites between fighting males may be a mode of transmission between rodents, the major portal for hantavirus entry into rodents and humans is inhalation of aerosolized viruses. For subsequent maintenance and transmission between individuals, the virus must replicate in target cells, then disseminate from the lung to other organs, and then must be shed.

It is of interest to evaluate whether or not differences can be found in the way each hantavirus species infects its main target cell. Viral envelope glycoproteins, Gn and Gc, play an important role in this targeting process by interacting with specific entry receptors, which can vary according to the hantavirus. Virus particle budding and interaction with cellular factors may also occur in different intracellular spaces during the virus cycle [43].

3.1. Entry Receptors

As described for some other viruses, evidence strongly suggests that the cellular entry of hantaviruses is mediated by interaction of the viral glycoproteins Gn-Gc with integrins. These cellular membrane proteins promote cell–cell and cell–extracellular matrix adhesion. They also induce signaling cascades regulating cell proliferation, survival, differentiation, activation and migration [44].

Interestingly, pathogenic and nonpathogenic hantaviruses use different integrin receptors at the surface of human cells: α11β3 is used by hantaviruses inducing HCPS, αvβ3 by hantaviruses inducing HFRS, and α5β1 by nonpathogenic hantaviruses [45,46]. Different cell types express different integrins and the β3 chain is an abundant surface receptor of endothelial cells, dendritic cells (DC) and platelets, which are known to be susceptible to hantavirus infection [47,48]. In addition, DC and platelets are involved in the pathogenic process of vascular leakage and thrombocytopenia [49].

All pathogenic hantaviruses use β3 integrins to enter human cells. This is rather intriguing given their diversity and specific association, each to a different natural host. However, little information is available about the integrin receptors and co-receptors used by the different hantaviruses in their rodent reservoirs. Indeed, natural host reservoirs of hantaviruses are very diverse and the lack of genetic markers and specific detection tools make it difficult to identify receptors or other partners of interaction in the reservoir hosts. Interestingly, Sangassou virus, an African hantavirus, has been recently described and shown to interact with β1 rather than β3 integrin in its Murinae reservoir [50]. It would be interesting to understand whether or not the specificity of interaction with an integrin receptor, and therefore with the type of target cells, could impact pathogenesis. In this respect, the question of which cells express the α5β1 receptor for nonpathogenic viruses remains to be addressed. It has been recently shown that hantaviruses could stimulate human neutrophils (PMN) and that the β2 integrin receptor that is highly expressed by PMN could act as a new entry receptor [51].

Other membrane proteins of the complement regulatory system, such as the decay-accelerating factor (DAF) for HTNV and PUUV, and gC1qR for HTNV, were identified as co-receptors for hantavirus entry [52,53]. These co-receptors together with integrins could also participate in hantavirus tropism and organ targeting, differentiating pathogenic from nonpathogenic viruses. Other cellular proteins, such as the 70 kDa [54] and 30 kDa protein [55], have also been described for HTNV entry.

Whether or not pathogenic and nonpathogenic viruses for humans use a similar entry receptor in rodents and how this might impact virus propagation has yet to be defined. Viral entry is required, but insufficient to account for pathogenesis since both pathogenic and nonpathogenic hantaviruses are able to infect endothelial cells and macrophages. Therefore, interaction with intracellular factors involved in virus replication and assembly might also play an important role in the outcomes of hantavirus infections.

3.2. Maturation and Cell Tropism

Envelope glycoproteins are important for entry, and also in many aspects of virus trafficking, maturation and assembly [43]. Interaction of Gn with Gc for the formation of the spike that interacts with entry receptors, or interaction of the cytosolic tail of Gn (GnCT) with the N nucleocapsid [56] that is required for viral assembly must be finely regulated. Interestingly, Old and New World hantaviruses bud on Golgi membranes or the plasma membrane during virus assembly, respectively. Interestingly, different functional domains have been identified on the GnCT of different hantaviruses. A conserved zinc finger domain favoring interaction has been found in the GnCT of both pathogenic, ANDV, and nonpathogenic, PHV, viruses. In contrast, a signal sequence for degradation by autophagy seems specific to pathogenic hantaviruses [57]. Such structures could play important roles in viral assembly and pathogenesis. Finally, ITAM domains known to be important for triggering intracellular signaling in response to receptor activation are present in the GnCT of pathogenic hantaviruses associated with HCPS. This suggests that GnCT could participate in the deregulation of immune and endothelial functions [58].

In vitro, hantaviruses can infect endothelial cells, epithelial cells, and cells from the immune system, such as macrophages, follicular dendritic cells and lymphocytes via attachment of the viral glycoprotein to surface receptors in a cell- and virus-specific way. For instance, pathogenic, but not nonpathogenic hantaviruses may only infect megakaryocytes that have differentiated [59]. Upon infection, human DC could serve both as vehicles of hantaviruses or allow them to evade the immune system. Moreover, infection induces DC maturation and increases expression of β3 integrin, interferon (IFN)-α and tumor necrosis factor (TNF)-α [60]. Macrophages are the second-most important targets of hantaviruses after endothelial cells in both rodents and humans. Despite a low replication rate and release of infectious virus due to inhibition by IFN-α, monocytes/macrophages from peripheral blood are susceptible to PUUV infection [61,62]. Recently, keratinocytes have also been shown to be permissive to hantavirus infection [63], which is of interest considering that hantaviruses can be transmitted by bites between small mammals.

The difficulties in infecting cell cultures with hantaviruses and obtaining virus progeny have led to only fragmentary information on differential mechanisms of virus-entry, maturation and propagation. Indeed, studies are complicated by the great diversity of rodent and insectivore reservoirs, and at the same time the fine specificity of each hantavirus for one host species. Therefore, development of new primary cellular models from natural hosts is of crucial importance [64] for evaluating selective factors in the different host cells. This will be the only way to understand how rodent tissues are persistently infected, and why some organs are predominantly targeted in a virus-specific manner during human pathogenesis.

3.3. *Apoptosis and Cell Survival*

It is well admitted that hantaviruses are not lytic for their endothelial target cells since no signs of cytotoxicity were reported in vitro in efficiently infected primary cells including endothelial, epithelial, dendritic and mast cells [65]. However, cell damage in PUUV patients presenting with nephropathies is manifested by increased levels of perforin, granzyme B, and LDH, as well as markers of epithelial cell apoptosis [66]. There are conflicting reports on whether or not hantaviruses induce apoptosis in vitro. There were no effects from PUUV recorded on the viability of Vero E6 and A549 cells [67], whereas TULV was shown to induce apoptosis in VeroE6 cells [68]. The effect could depend on the virus species and/or the cell type because a cytopathic effect of different hantaviruses (HTNV, SEOV, and ANDV) was detected in human HEK293 cells [69]. More recently, it was shown that the N protein of HTNV has an antiapoptotic effect in A549 and Hela cells that is mediated by a down-regulation of p53 [70]. Although it could be suspected that apoptosis will not be induced in infected cells of the rodent reservoir, the mechanisms of hantavirus persistence in these natural hosts are still to be defined.

Interestingly, recent investigations are in favor of a protective role from cells infected by hantaviruses through two mechanisms that could promote pathogenesis. The first mechanism is to induce survival of cells of the innate immunity. As already mentioned, a high level of immune cell activation could be detrimental. For instance, a subset of natural killer (NK) cells expressing the NKG2C receptor, specific for the human leukocyte antigen (HLA)-E ligand expressed by endothelial cells, expands in patients during acute PUUV infection. Interleukin (IL)-15 and the anti-apoptotic Bcl-2 molecules both play a role in promoting the survival of these proliferating NK cells [71,72]. Similarly, neutrophils that are now thought to play an important role in vascular leakage observed in hantavirus disease [39] are activated in vitro and in vivo and survived longer in response to hantavirus exposure [51,73]. We also have recently observed prolonged survival of neutrophils due to a delayed apoptosis, specifically induced by the pathogenic PUUV, but interestingly, not by the nonpathogenic TULV or PHV hantaviruses (Baychelier, personal communication). The second mechanism is to protect infected cells from the cytotoxic effect of immune cells. For instance, ANDV and HTNV infected endothelial cells are resistant to NK cell-mediated killing. Consequently, infected cells will be protected, but non-infected surrounding cells will not be protected and are killed by NK cells [74].

Taken together, the picture is still unclear about the effect of hantaviruses on cell survival and how different mechanisms could operate depending on whether hantaviruses are pathogenic or not.

4. Interaction of Hantaviruses with the Immune System

The IFN cytokine family is the first line of antiviral defense. Invading viruses are detected by nonimmune cells early during infection. Recognition of viral molecules by pattern recognition receptors (PRRs), such as Toll-like receptors (TLR) or the RIG-I and MDA5 cytoplasmic helicase receptors, modulate signaling pathways, or transcription factors, resulting in the induction of IFN-α/β [75]. IFNs then induce expression of a large pattern of different IFN-stimulated genes (ISGs) to set up an antiviral state. In addition, PRRs directly trigger pro-inflammatory responses that induce host resistance to infection and activate innate immune cells before the establishment of an adaptive immune response. Among the ISGs targeted by viral infection, ISG15, a ubiquitin homolog, MxA monomers which bind and degrade viral components, RNase L which cleaves cellular and viral RNAs, and PKR which inhibits the phosphorylation of the eIF2α translation initiation factor, have been described as directly promoting antiviral states [75].

4.1. Induction of the Different Interferon Responses

Upstream of type I IFN induction, TLR4 expression has been shown to be higher in HTNV-infected human endothelial vein cells (EVC-304) as compared to their uninfected counterparts, leading to an increase of IFN-β, IL-6, and TNF-α secretion [76]. Increased IFN-β production may confer an antiviral state, whereas IL-6 is likely responsible for a pro-inflammatory response. Furthermore, TNF-α, which contributes to endothelial permeability, plays a key role in the pathogenesis of HFRS. Indeed, the permeability of endothelial cells is significantly prolonged upon TNF-α treatment in HTNV-infected cells as compared to uninfected cells [77]. In this same cellular model, the adaptor protein TRIF is up-regulated downstream of TLR4 [78]. Another study, using a human A549 epithelial lung cell line and HuH7 hepatoma cell line, has demonstrated that pathogenic HTNV and nonpathogenic PHV activate innate immune responses in different ways. Both viruses induce IFN signaling via an activation of Mxa, and, in the case of HTNV, but not PHV, this follows the recruitment of TLR3 [79]. These findings are in line with the fact that in endothelial cells, the pathogenic ANDV virus regulates the early interferon response, whereas the nonpathogenic PHV does not. This may explain how PHV can infect EC without being able to replicate [30]. In support of this idea, PHV, but not the pathogenic HTNV, New York 1 virus (NY-1) and ANDV hantaviruses, induces a high level of IFN in human EC early after infection [80]. However, the situation is more complex because another nonpathogenic hantavirus (TULV) replicates as successfully as pathogenic ones in human ECs suggesting that TULV is also able to regulate cellular IFN responses [81].

Type II IFN-γ also exhibits antiviral activities. It has been shown for instance that pre-treatment of Vero E6 and A549 cells with IFN-γ inhibits HTNV infection in a dose dependent way and by a mechanism that is MxA-independent [82]. However, an already-established HTNV infection is insensitive to subsequent addition of IFN-γ stimulation. The same observation has been made for IFN-α/β and IFN-λ [83]. These observations are consistent with the early antiviral effect of IFN on viral replication.

Type III IFNs, which include IFN-λ1/IL-29, IFN-λ2/IL-28A and IFN-λ3/IL-28B, are regulated in a similar manner to type I IFNs [84]. The antiviral effect of IFN-λ has been tested on infected cells. Interestingly, a synergistic inhibitory effect of IFN-λ with IFN-γ, but not with IFN-α/β, was seen on HTNV replication in A549 cells [83]. In line with this finding, a high level of IFN-λ1 is induced in HTNV infected A549 cells, and MRC-5, a human fibroblast cell line lacking the IFN-λ receptor. Expression of IFN-λ1 preceded the induction of MxA and IFN-β. Furthermore, induction of IFN-λ1 and MxA has been observed in Vero-E6 cells [85], which are unable to produce type I IFNs and are used to prepare viral stocks. Three New World hantaviruses (SNV, ANDV and PHV) also induce IFN-λ in Vero E6 cells. Presence of IFN-λ in virus stock prepared from the supernatant of infected Vero E6 cells, has been demonstrated to activate ISG56 and MxA genes in Huh7 and A549 cells. This happens independently of the virus used to infect these cells, as demonstrated by neutralization of the effect using specific IFN-λ antibodies. The situation is different in human umbilical vein endothelial cells (HUVEC) infected by SNV where IFN-λ induction of ISGs is virus-specific, and therefore, not affected by IFN-λ neutralizing antibodies [86]. This is in good agreement with the fact that HUVEC cells, which lack the IL-28Rα chain, are unable to respond to IFN-λ [87]. However, IFN-λ levels decrease in the serum of PUUV-infected patients, whereas IFN-α/β levels remain unchanged and IFN-γ is elevated. This apparent contradiction could be due to an efficient counteraction of IFN-λ by hantaviral proteins [83].

4.2. How Do Hantaviruses Evade the IFN Antiviral Response?

In order to successfully replicate in cells, viruses have evolved many mechanisms to counteract the IFN-induced host defense at almost every step of the signaling pathway [88]. This is supported by the fact that pre-treatment of cells with type I IFN-α/β can block hantavirus replication [82,89].

Transcriptional activation of the IFN-β gene requires assembly of an enhanceosome containing ATF-2/c-Jun, IRF-3/IRF-7, and NF-κB [90]. The N nucleocapsid of HTNV blocks TNF-α induced activation of NF-κB by impairing its nuclear translocation [91]. This effect is shared by the N protein of DOBV and SEOV, but not of PUUV, SNV and ANDV [92]. More recently the N nucleocapsid of ANDV has been shown to carry a virulence domain able to inactivate PKR, and therefore counteract its antiviral effect [93].

The S and M segments of some bunyaviruses encode nonstructural proteins (NSs and NSm), which can block IFN-β transcription as shown for the NSs of the Rift Valley Fever virus [94]. Since hantaviruses do not possess NSs or a NSm proteins, these activities must be carried by one of their four structural proteins (N, Gn, Gc or the L polymerase). It should be noted that a few hantaviruses, hosted by rodents of the Arvicolinae family, contain an evolutionary conserved short open reading frame. This sequence could encode a NSs protein that has been shown to influence the interferon response in PUUV-infected cells. It seems unlikely that this NSs represents a virulence factor for human infection since it is found both in pathogenic, PUUV and nonpathogenic, TULV hantaviruses and is absent from highly pathogenic hantaviruses such as ANDV, NY-1, and HTNV [95–97].

The Gn envelope glycoprotein contains a 142 amino acid long cytoplasmic tail (GnCT) involved in IFN regulation. An interaction of GnCT from the pathogenic NY-1, but not from PHV, with interferon elements inhibits the antiviral action of IFN in endothelial cells [98]. GnCT recruits TRAF3 leading to its dissociation from TBK1. The ubiquitin ligase activity of the complex TRAF3-TBK1 that phosphorylates IRF3 is then abolished along with signal transmission required for the transcription of IFN-β [30,99]. This allows pathogenic NY-1 hantavirus, but not nonpathogenic PHV, to bypass innate immune responses and to successfully replicate within endothelial cells. In contrast to PHV, the GnCT from the nonpathogenic TULV behaves similarly to the pathogenic NY-1 by inhibiting IFN- and ISRE-directed responses upstream of IRF3 at the level of the TBK1 complex. However, unlike pathogenic hantaviruses, TULV GnCT failed to bind TRAF3 [100]. GnCT could be a virulence factor responsible for the delayed IFN responses observed with NY-1, ANDV and HTNV in infected cells. In line with the pivotal role of IRF3 activation in eliciting IFN responses, the nuclear translocation of IRF-3 is impaired upon SNV, HTNV and SEOV infection, whereas nuclear accumulation of IRF-3 is seen with nonpathogenic PHV, TULV and the Thottapalayam virus [101]. Strikingly, the ability of hantaviruses to provoke an early IFN response appears necessary for pathogenesis, but also leads to distinct responses for the two nonpathogenic hantaviruses, PHV and TULV. The fact that the nonpathogenic TULV replicates successfully in EC, suggests the existence of additional viral determinants of pathogenesis [81].

Another way for hantaviruses to interfere with the early innate response, as shown in HTNV infected A549 cells [83], could be at the level of STAT1 transcription which controls the three types of IFNs signaling. Unexpectedly, the glycoproteins of both pathogenic ANDV and nonpathogenic PHV could inhibit STAT-1 nuclear translocation thus impairing IFN signaling. The difference in pathogenicity of these viruses could then be based on different strengths of IRF-3 activation [102] in ANDV and PHV infected human lung microvascular endothelial cells (HMVEC-Ls).

The various outcomes of hantavirus infection could then relate to differential interaction with early steps of the IFN antiviral pathway. Pathogenesis could result, at least in part, from the delayed transcription of IFN-β and other ISGs, allowing pathogenic hantaviruses to rapidly replicate and spread through endothelial cell barriers.

Concerning the role of domains from the GnCT and N viral proteins as a virulence factor, and the fact that hantaviruses are asymptomatic in rodents, one speculation is that interactions of viral proteins with host factors have differently evolved according to the genetic background of their rodent reservoir. This could explain why glycoproteins of hantaviruses behave differently in terms of interaction with elements of the innate immune system leading to different degrees of pathogenesis in humans and to persistence only in specific natural hosts. Altogether, these observations point to the complexity of the mechanisms by which hantaviral proteins interfere with the innate immune system of different hosts, and suggest the involvement of other processes, such as adaptive immunity.

4.3. Differences in B and T Cell Immunity Induced by Hantaviruses

Hantaviruses are viremic in humans and in rodents, and neutralizing antibodies are produced in both cases. Antibodies protect humans from re-infection, but obviously do not impair the virus circulation and persistence in rodents. This phenomenon cannot be easily explained and the differential role of T cells has been put forward.

NK cells are at the frontier between innate and adaptive immunity. Under pathogenic conditions, they rapidly expand and then persist during acute infection in humans, with their number remaining elevated for at least 60 days. A strong T cell response involving CD8+ cytotoxic lymphocytes, and to a lesser extent CD4+ T cells, also accompanies the acute phase of PUUV infection [103]. A mixed pattern of T cells of the Th1 and Th2 phenotypes as well as high levels of pro-inflammatory cytokines, not efficiently suppressed by regulatory cytokines, lead to harmful effects in infected patients [22]. Thereafter, T cells decrease with the decline in viral load and the clearance of viremia that are probably due to the effect of intrinsic negative signals and extrinsic regulation.

It is hypothesized that CD4+ regulatory T cells (Treg) can account for viral persistence in rodents [61], whereas NK and CTL have a role to play in human pathogenesis. Presently, induction of regulatory FoxP3+ CD8+ or CD4+ T cells has not been detected in patients during acute hantavirus infection. However, CD8+ and CD4+ T cells may modulate effector responses through different inhibitory receptors during an acute viral infection [104]. It is thought that hantavirus infection induces DC differentiation, and subsequently antigen presenting cell (APC) transition and T cell stimulation. In particular, induction of memory T cells with long lasting protection is found in infected patients [105]. Different infection outcomes in humans and rodents could be explained by the fact that T-CD8+ attracted by infected endothelium would be damaging in humans, while in rodents, these T cells will be inactivated by the induction of Tregs. However, this picture is an over simplification because increased severity of PUUV-induced nephropathia epidemica correlates with higher levels of FoxP3+ Treg [106].

The importance of T cell responses is also attested to by the fact that the major response to hantaviruses is against immunogenic CD8+ T cell epitopes restricted to HLA-I [104]. In this respect, hantavirus infection in different cell types up-regulates the expression of HLA-I, which is involved in antigen presentation, as well as DC cross-presentation to CD8+ T cells [107].

As already mentioned, little is known about the way hantaviruses can replicate and persist in small mammal reservoir species despite the induction of seroconversion and high titers of neutralizing antibodies in these natural hosts. Of note, epitopes on the N and Gn structural proteins recognized by antibodies induced in rodents are different from those inducing human antibody reactivity. In contrast, three epitopes of the Gc protein are immunogenic in both human and rodents. Each hantavirus species also exhibits a narrow specificity for a specific host reservoir and little is known about the mechanism that limits spill over and adaptation to new mammals. It has been shown that deer mice, the natural hosts of pathogenic SNV, can be infected by ANDV, which is also highly pathogenic to humans and asymptomatic in its host reservoir, the pigmy rice rat. In both cases of virus infection, endothelial cells are infected, but deer mice do not present obvious symptoms. However, ANDV is not maintained in deer mice, whereas SNV persists in its specific host. Interestingly, the deer mice immune reaction to infection is different depending on the hantavirus. Persistence of SNV is linked to a low induction of immune cells. In contrast, heterologous ANDV is cleared after induction of a strong B and T cell immune response [108]. This phenomenon must be a consequence of the long-term evolution and adaptation of hantaviruses with their natural rodent hosts [109].

5. Activation of Host Cell Factors by Hantaviruses

5.1. Differences Associated with HCPS, HFRS, and Nephropathia Epidemica Pathogenesis in Hantavirus Infected Patients

Although differences have been noted in terms of entry receptors, partners for interaction and modulation of the interferon response, a clear representation of the differences leading to the various degrees of pathogenicity associated with hantaviruses has not yet been obtained. This is not surprising considering the complexity of the signaling cascades that are downstream of the IFN, which can be triggered in different innate immune situations. Moreover, the levels of interferon and their receptors will depend on the cell type as for instance the IFN-λ receptor that is only expressed on epithelial cells. Additionally, IFN signaling is connected to other cellular processes such that during the establishment of hantavirus diseases, pro-inflammatory events also impact coagulation and vascular permeability processes as well as on the recruitment of cells of both innate and adaptive immunity.

Studies using high throughput or multiplex technology reveal further complexity. For instance, in one study using the serum of SNV-infected patients presenting with HCPS, the levels of cytokines specific of the Th1 and Th2 responses vary, revealing increases in IL-6, IFN-γ, sIL2R and TNF-α accompanied by a decrease in IL-10 [110]. In another analysis, 68 different cytokines, including chemokines, angiogenic and growth factors, were tested. The most important changes were an up-regulation of IL-6, CXCL10, CX3CL11, MIF and MIG, and a down-regulation of CXCL12 and to a lesser extent CCL21, 22, 27 and sCD40L [111]. This up-regulation of cytokine expression, could promote tissue migration of mononuclear cells (T lymphocytes, NK, and DC), with leukocytes playing a role in the repair of lung tissue, as well as in increasing endothelial monolayer permeability. Conversely, the down-regulation of cytokines associated with platelet homeostasis is consistent with the thrombocytopenia observed in patients.

For HFRS patients, the serum levels of IFN-γ, IL-10, CCL2, and IL-12 have been shown to be up-regulated as compared to healthy controls. Variations then depends on the phase and severity of the disease, with IFN-γ and IL-12 relating to mild forms [112]. An up-regulation of uPAR has also been described in PUUV patients. This receptor binds the β3-integrin and can cause proteinuria by acting on the glomerular endothelium of the kidney [113]. As for HCPS, a significant elevation of IL-6 and TNF-α plasma levels and also of IL-10 has been detected at the onset of the HFRS acute phase [114]. In addition to factors involved in kidney failure (creatinine, C reactive protein and NO), an up-regulation of these inflammatory cytokines has been confirmed in a macaque model of PUUV infection [115]. An increase in CXCL10 has also been described in the serum of patients infected by HTNV depending on the disease severity [116]. More recently, factors linked to inflammation or coagulation such as tPA and PAI-1 have also been found in the serum of both HFRS patients and macaques experimentally infected with PUUV [117].

5.2. Regulation of Cellular Factors Induced by Hantaviruses in Vitro

There are striking differences in the early induction of ISGs in HUVEC that have been infected with either pathogenic hantaviruses that cause HCPS (NY-1) and HFRS (HTNV) or nonpathogenic PHV hantavirus. Pathogenic viruses suppress the IFN response whereas nonpathogenic PHV activates the response. Induction of IL-6, IL-8, and adhesion molecules involved in the recruitment of leukocytes appears specific to the pathogenic viruses [80]. A similar transcriptional analysis revealed different impacts of the pathogenic SNV and nonpathogenic PHV on vascular endothelial cells [118]. In particular, among other genes, CCL5 and CXCL10 appear to be specifically up-regulated by SNV. Human HUVEC cells infected by HTNV, also show an increase in CCL5 and IL-6 as well as an induction of ICAM-1 and VCAM-1 adhesion molecules [119]. A cytokine and receptor screen has been performed with primary DC infected by ANDV. It revealed the induction of pro-inflammatory cytokines amongst which IL-10 and the matrix metalloprotease, MMP9, are able to indirectly affect the permeability of HUVEC endothelial cells [120]. HUVEC endothelial cells and THP1 macrophages have been used for microarray analysis. The effect of nonpathogenic shrew borne (MJNV and Thottapalaiam virus) and rodent-borne (PHV) hantaviruses was compared to pathogenic HTNV. Currently, no human diseases have been associated with hantaviruses circulating in insectivores. It is worth considering that MJNV and Thottapalaiam virus, thought to be nonpathogenic for humans, might induce pro-inflammatory cytokines as does HTNV, but not PHV [121]. Also as already mentioned, MJNV causes diseases in the Syrian hamster model [34]. This could represent a risk for human emergence of hantaviruses from reservoirs other than rodents and requires further attention.

The differential regulation of the innate immune response by pathogenic and nonpathogenic hantaviruses has been evaluated on other cell models. The fact that the chemokine CCL5 is activated in A549 and HuH7 cells infected by HTNV but not by PHV, suggests the involvement of different downstream signaling cascades [79]. Proteomics has revealed a high degree of CXCL10 activation in HuH7 cells infected with PUUV, as compared to non-infected control cells or cells infected with nonpathogenic TULV or PHV (our unpublished data). CCL5 and CXCL10 were also up-regulated in keratinocytes derived cells that are susceptible to HTNV [63].

5.3. *Differential miRNA Signatures in Hantavirus Infections*

It has been reported that the expression pattern of single-stranded non coding miRNA gene regulators, differ in response to pathogenic (HTNV) versus nonpathogenic (PHV) hantaviruses [122]. This differential expression occurs in HUVEC endothelial, A549 epithelial or THP1 macrophage human cell lines, in a cell-type-specific manner. Moreover, the expression of several miRNAs involved in the regulation of proteins activated during the innate immune response including Mxa, IP-10 (CXCL10), INF-β or RANTES (CCL5), varies inversely in HTNV and PHV-infected cells. Differentially expressed miRNAs also target immune receptor signaling through RIG-I-like, NOD-like, Toll-like receptors and inflammatory pathways including JAK-STAT, PI3K-Akt and MAPK signaling pathways. It has also been shown that ANDV infection alters the expression of several EC specific miRNAs involved in vascular integrity, adherence and angiogenesis [123].

5.4. *Persistence of Hantaviruses in Host Reservoirs*

It is important to understand the mechanisms that explain: (a) the persistence and specificity of one hantavirus for a given host species; (b) whether or not spill-over of hantaviruses can occur between species; and (c) if spill over can result in host switching, i.e., in adaptation to a new host.

As already mentioned, SNV is persistent in its natural host, the deer mouse, which is also susceptible to an experimental infection by ANDV. Gene expression analyses by quantitative real time PCR arrays have shown that these two viruses interact differently with cell factors of lymph node cells from infected deer mice [124]. The expression profiles are coherent in the way B and T cells induce persistence of SNV or clearance of ANDV [108]. For instance, Th2 and IL-4 signaling factors are upregulated in ANDV infected mice, whereas cytokines corresponding to the Th1 and Treg phenotypes are predominant with SNV. Treg activation has also been reported in Norway rats, a natural host reservoir of SEOV, and could contribute to its persistence. In such a situation, pro-inflammatory mediators are not activated [125].

Together, data highlight the importance of the innate immune response in the different outcomes of hantavirus infections (pathogenic versus non-pathogenic or persistent). The complexity of the balance between different signaling networks regulating cell homeostasis and the time post infection of the study, might contribute to the variability of results obtained in different cellular models.

6. Conclusions

The mechanisms sustaining hantavirus pathogenesis in humans or leading to persistence in the animal reservoir most likely result from an interaction of hantaviruses with the immune system. Many data have accumulated, however, much remains to be understood before a comprehensive synthesis can be established. Arguments converge on the role of inflammatory mediators and immune cells targeting the endothelium during pathogenesis. However, some discrepancies exist and may be due to limitations of cellular models, the way hantaviruses are produced in vitro, or the time post-infection of the observation. Transcriptome or proteome analyses support the existence of differences in the relationships of pathogenic and nonpathogenic hantaviruses with their different hosts. The data obtained so far correlate well with some clinical pathogenic markers. These large-scale approaches will allow for comparative experiments using different viruses in different cell models. These technologies will be fundamental to the discovery of mechanisms that underlie the different ways these emerging viruses manipulate antiviral responses and perturb cellular function. Additionally, these technologies will also help to identify specific factors that could be targeted to counteract pathogenesis and to design specific therapies.

Acknowledgments: This work was supported by the European FP7 ANTIGONE Program n°278976. The contents of this publication are the sole responsibility of the authors and do not necessarily reflect the views of the European Commission. The authors are grateful to Stuart Moore for careful reading of the manuscript and for helpful suggestions.

Author Contributions: M.E. and F.B. compiled the bibliographic references and wrote the review and all authors contributed to the discussion about the review content. All authors have read and approved the final manuscript.

Conflicts of Interest: The authors declare no conflict of interest.

References

1. Lee, H.W.; Lee, P.W.; Baek, L.J.; Song, C.K.; Seong, I.W. Intraspecific transmission of Hantaan virus, etiologic agent of Korean hemorrhagic fever, in the rodent *Apodemus agrarius*. *Am. J. Trop Med. Hyg.* **1981**, *30*, 1106–1112. [PubMed]
2. Mir, M.A. Hantaviruses. *Clin. Lab. Med.* **2010**, *30*, 67–91. [CrossRef] [PubMed]
3. Guo, W.P.; Lin, X.D.; Wang, W.; Tian, J.H.; Cong, M.L.; Zhang, H.L.; Wang, M.R.; Zhou, R.H.; Wang, J.B.; Li, M.H.; et al. Phylogeny and origins of hantaviruses harbored by bats, insectivores, and rodents. *PLoS Pathog.* **2013**, *9*, e1003159. [CrossRef] [PubMed]
4. Sabino-Santos, G., Jr.; Maia, F.G.; Vieira, T.M.; de Lara Muylaert, R.; Lima, S.M.; Goncalves, C.B.; Barroso, P.D.; Melo, M.N.; Jonsson, C.B.; Goodin, D.; et al. Evidence of hantavirus infection among bats in Brazil. *Am. J. Trop Med. Hyg.* **2015**, *93*, 404–406. [CrossRef] [PubMed]
5. Heyman, P.; Thoma, B.R.; Marie, J.L.; Cochez, C.; Essbauer, S.S. In search for factors that drive hantavirus epidemics. *Front. Physiol.* **2012**, *3*, 237. [CrossRef] [PubMed]
6. Jonsson, C.B.; Figueiredo, L.T.; Vapalahti, O. A global perspective on hantavirus ecology, epidemiology, and disease. *Clin. Microbiol. Rev.* **2010**, *23*, 412–441. [CrossRef] [PubMed]
7. Dupinay, T.; Pounder, K.C.; Ayral, F.; Laaberki, M.H.; Marston, D.A.; Lacote, S.; Rey, C.; Barbet, F.; Voller, K.; Nazaret, N.; et al. Detection and genetic characterization of Seoul virus from commensal brown rats in France. *Virol. J.* **2014**, *11*, 32. [CrossRef] [PubMed]
8. Goeijenbier, M.; Verner-Carlsson, J.; van Gorp, E.C.; Rockx, B.; Koopmans, M.P.; Lundkvist, A.; van der Giessen, J.W.; Reusken, C.B. Seoul hantavirus in brown rats in The Netherlands: Implications for physicians–epidemiology, clinical aspects, treatment and diagnostics. *Neth. J. Med.* **2015**, *73*, 155–160. [PubMed]
9. Heyman, P.; Baert, K.; Plyusnina, A.; Cochez, C.; Lundkvist, A.; Esbroeck, M.V.; Goossens, E.; Vandenvelde, C.; Plyusnin, A.; Stuyck, J. Serological and genetic evidence for the presence of Seoul hantavirus in *Rattus norvegicus* in Flanders, Belgium. *Scand. J. Infect. Dis.* **2009**, *41*, 51–56. [CrossRef] [PubMed]
10. McElhinney, L.; Fooks, A.R.; Featherstone, C.; Smith, R.; Morgan, D. Hantavirus (Seoul virus) in pet rats: A zoonotic viral threat. *Vet. Rec.* **2016**, *178*, 171–172. [CrossRef] [PubMed]

11. Jameson, L.J.; Logue, C.H.; Atkinson, B.; Baker, N.; Galbraith, S.E.; Carroll, M.W.; Brooks, T.; Hewson, R. The continued emergence of hantaviruses: Isolation of a Seoul virus implicated in human disease, United Kingdom, October 2012. *Euro Surveill.* **2013**, *18*, 4–7. [PubMed]
12. Mace, G.; Feyeux, C.; Mollard, N.; Chantegret, C.; Audia, S.; Rebibou, J.M.; Spagnolo, G.; Bour, J.B.; Denoyel, G.A.; Sagot, P.; et al. Severe Seoul hantavirus infection in a pregnant woman, France, October 2012. *Euro Surveill.* **2013**, *18*, 20464. [PubMed]
13. Borges, A.A.; Campos, G.M.; Moreli, M.L.; Souza, R.L.; Aquino, V.H.; Saggioro, F.P.; Figueiredo, L.T. Hantavirus cardiopulmonary syndrome: Immune response and pathogenesis. *Microbes Infect.* **2006**, *8*, 2324–2330. [CrossRef] [PubMed]
14. Hepojoki, J.; Vaheri, A.; Strandin, T. The fundamental role of endothelial cells in hantavirus pathogenesis. *Front. Microbiol.* **2014**, *5*, 727. [CrossRef] [PubMed]
15. Connolly-Andersen, A.M.; Sundberg, E.; Ahlm, C.; Hultdin, J.; Baudin, M.; Larsson, J.; Dunne, E.; Kenny, D.; Lindahl, T.L.; Ramstrom, S.; et al. Increased thrombopoiesis and platelet activation in hantavirus-infected patients. *J. Infect. Dis.* **2015**, *212*, 1061–1069. [CrossRef] [PubMed]
16. Latus, J.; Kitterer, D.; Segerer, S.; Artunc, F.; Alscher, M.D.; Braun, N. Severe thrombocytopenia in hantavirus-induced nephropathia epidemica. *Infection* **2014**, *43*, 83–87. [CrossRef] [PubMed]
17. Connolly-Andersen, A.M.; Thunberg, T.; Ahlm, C. Endothelial activation and repair during hantavirus infection: Association with disease outcome. *Open Forum Infect. Dis.* **2015**, *1*. [CrossRef] [PubMed]
18. Gorbunova, E.; Gavrilovskaya, I.N.; Mackow, E.R. Pathogenic hantaviruses Andes virus and Hantaan virus induce adherens junction disassembly by directing vascular endothelial cadherin internalization in human endothelial cells. *J. Virol.* **2010**, *84*, 7405–7411. [CrossRef] [PubMed]
19. Spiropoulou, C.F.; Srikiatkhachorn, A. The role of endothelial activation in dengue hemorrhagic fever and hantavirus pulmonary syndrome. *Virulence* **2013**, *4*, 525–536. [CrossRef] [PubMed]
20. Vaheri, A.; Strandin, T.; Hepojoki, J.; Sironen, T.; Henttonen, H.; Makela, S.; Mustonen, J. Uncovering the mysteries of hantavirus infections. *Nat. Rev. Microbiol.* **2013**, *11*, 539–550. [CrossRef] [PubMed]
21. Khaiboullina, S.F.; St Jeor, S.C. Hantavirus immunology. *Viral Immunol.* **2002**, *15*, 609–625. [CrossRef] [PubMed]
22. Schonrich, G.; Rang, A.; Lutteke, N.; Raftery, M.J.; Charbonnel, N.; Ulrich, R.G. Hantavirus-induced immunity in rodent reservoirs and humans. *Immunol. Rev.* **2008**, *225*, 163–189. [CrossRef] [PubMed]
23. Terajima, M.; Ennis, F.A. T cells and pathogenesis of hantavirus cardiopulmonary syndrome and hemorrhagic fever with renal syndrome. *Viruses* **2011**, *3*, 1059–1073. [CrossRef] [PubMed]
24. Easterbrook, J.D.; Klein, S.L. Immunological mechanisms mediating hantavirus persistence in rodent reservoirs. *PLoS Pathog.* **2008**, *4*, e1000172. [CrossRef] [PubMed]
25. Meyer, B.J.; Schmaljohn, C.S. Persistent hantavirus infections: Characteristics and mechanisms. *Trends Microbiol.* **2000**, *8*, 61–67. [CrossRef]
26. Voutilainen, L.; Sironen, T.; Tonteri, E.; Back, A.T.; Razzauti, M.; Karlsson, M.; Wahlstrom, M.; Niemimaa, J.; Henttonen, H.; Lundkvist, A. Life-long shedding of Puumala hantavirus in wild bank voles (*Myodes glareolus*). *J. Gen. Virol.* **2015**, *96*, 1238–1247. [CrossRef] [PubMed]
27. Hardestam, J.; Karlsson, M.; Falk, K.I.; Olsson, G.; Klingstrom, J.; Lundkvist, A. Puumala hantavirus excretion kinetics in bank voles (*Myodes glareolus*). *Emerg. Infect. Dis.* **2008**, *14*, 1209–1215. [CrossRef] [PubMed]
28. Kallio, E.R.; Klingstrom, J.; Gustafsson, E.; Manni, T.; Vaheri, A.; Henttonen, H.; Vapalahti, O.; Lundkvist, A. Prolonged survival of Puumala hantavirus outside the host: Evidence for indirect transmission via the environment. *J. Gen. Virol.* **2006**, *87*, 2127–2134. [CrossRef] [PubMed]
29. Valbuena, G.; Walker, D.H. The endothelium as a target for infections. *Annu. Rev. Pathol.* **2006**, *1*, 171–198. [CrossRef] [PubMed]
30. Mackow, E.R.; Dalrymple, N.A.; Cimica, V.; Matthys, V.; Gorbunova, E.; Gavrilovskaya, I. Hantavirus interferon regulation and virulence determinants. *Virus Res.* **2014**, *187*, 65–71. [CrossRef] [PubMed]
31. Hardcastle, K.; Scott, D.; Safronetz, D.; Brining, D.L.; Ebihara, H.; Feldmann, H.; LaCasse, R.A. Laguna Negra virus infection causes hantavirus pulmonary syndrome in Turkish hamsters (*Mesocricetus brandti*). *Vet. Pathol.* **2016**, *53*, 182–189. [CrossRef] [PubMed]
32. Ogg, M.; Jonsson, C.B.; Camp, J.V.; Hooper, J.W. Ribavirin protects Syrian hamsters against lethal hantavirus pulmonary syndrome–after intranasal exposure to Andes virus. *Viruses* **2013**, *5*, 2704–2720. [CrossRef] [PubMed]

33. Safronetz, D.; Ebihara, H.; Feldmann, H.; Hooper, J.W. The Syrian hamster model of hantavirus pulmonary syndrome. *Antivir. Res.* **2012**, *95*, 282–292. [CrossRef] [PubMed]
34. Gu, S.H.; Kim, Y.S.; Baek, L.J.; Kurata, T.; Yanagihara, R.; Song, J.W. Lethal disease in infant and juvenile Syrian hamsters experimentally infected with Imjin virus, a newfound crocidurine shrew-borne hantavirus. *Infect. Genet. Evol.* **2015**, *36*, 231–239. [CrossRef] [PubMed]
35. Klingstrom, J.; Falk, K.I.; Lundkvist, A. Delayed viremia and antibody responses in Puumala hantavirus challenged passively immunized *Cynomolgus macaques*. *Arch. Virol.* **2005**, *150*, 79–92. [CrossRef] [PubMed]
36. Safronetz, D.; Prescott, J.; Feldmann, F.; Haddock, E.; Rosenke, R.; Okumura, A.; Brining, D.; Dahlstrom, E.; Porcella, S.F.; Ebihara, H.; et al. Pathophysiology of hantavirus pulmonary syndrome in *Rhesus macaques*. *Proc. Natl. Acad. Sci. USA* **2014**, *111*, 7114–7119. [CrossRef] [PubMed]
37. Sironen, T.; Klingstrom, J.; Vaheri, A.; Andersson, L.C.; Lundkvist, A.; Plyusnin, A. Pathology of Puumala hantavirus infection in macaques. *PLoS ONE* **2008**, *3*, e3035. [CrossRef] [PubMed]
38. Kobak, L.; Raftery, M.J.; Voigt, S.; Kuhl, A.A.; Kilic, E.; Kurth, A.; Witkowski, P.; Hofmann, J.; Nitsche, A.; Schaade, L.; et al. Hantavirus-induced pathogenesis in mice with a humanized immune system. *J. Gen. Virol.* **2015**, *96*, 1258–1263. [CrossRef] [PubMed]
39. Schonrich, G.; Kruger, D.H.; Raftery, M.J. Hantavirus-induced disruption of the endothelial barrier: Neutrophils are on the payroll. *Front. Microbiol.* **2015**, *6*, 222. [PubMed]
40. Mertens, M.; Hofmann, J.; Petraityte-Burneikiene, R.; Ziller, M.; Sasnauskas, K.; Friedrich, R.; Niederstrasser, O.; Kruger, D.H.; Groschup, M.H.; Petri, E.; et al. Seroprevalence study in forestry workers of a non-endemic region in Eastern Germany reveals infections by Tula and Dobrava-Belgrade hantaviruses. *Med. Microbiol. Immunol.* **2011**, *200*, 263–268. [CrossRef] [PubMed]
41. Yanagihara, R.; Gajdusek, D.C.; Gibbs, C.J., Jr.; Traub, R. Prospect Hill virus: Serologic evidence for infection in mammologists. *N. Engl. J. Med.* **1984**, *310*, 1325–1326. [PubMed]
42. Martinez, V.P.; Bellomo, C.; San Juan, J.; Pinna, D.; Forlenza, R.; Elder, M.; Padula, P.J. Person-to-person transmission of Andes virus. *Emerg. Infect. Dis.* **2005**, *11*, 1848–1853. [CrossRef] [PubMed]
43. Cifuentes-Munoz, N.; Salazar-Quiroz, N.; Tischler, N.D. Hantavirus Gn and Gc envelope glycoproteins: Key structural units for virus cell entry and virus assembly. *Viruses* **2014**, *6*, 1801–1822. [CrossRef] [PubMed]
44. Hynes, R.O. Integrins: Bidirectional, allosteric signaling machines. *Cell* **2002**, *110*, 673–687. [CrossRef]
45. Gavrilovskaya, I.N.; Brown, E.J.; Ginsberg, M.H.; Mackow, E.R. Cellular entry of hantaviruses which cause hemorrhagic fever with renal syndrome is mediated by beta3 integrins. *J. Virol.* **1999**, *73*, 3951–3959. [PubMed]
46. Gavrilovskaya, I.N.; Shepley, M.; Shaw, R.; Ginsberg, M.H.; Mackow, E.R. Beta3 integrins mediate the cellular entry of hantaviruses that cause respiratory failure. *Proc. Natl. Acad. Sci. USA* **1998**, *95*, 7074–7079. [CrossRef] [PubMed]
47. Dalrymple, N.A.; Mackow, E.R. Virus interactions with endothelial cell receptors: Implications for viral pathogenesis. *Curr. Opin. Virol.* **2014**, *7*, 134–140. [CrossRef] [PubMed]
48. Mackow, E.R.; Gavrilovskaya, I.N. Cellular receptors and hantavirus pathogenesis. *Curr. Top. Microbiol. Immunol.* **2001**, *256*, 91–115. [PubMed]
49. Gavrilovskaya, I.N.; Gorbunova, E.E.; Mackow, N.A.; Mackow, E.R. Hantaviruses direct endothelial cell permeability by sensitizing cells to the vascular permeability factor VEGF, while angiopoietin 1 and sphingosine 1-phosphate inhibit hantavirus-directed permeability. *J. Virol.* **2008**, *82*, 5797–5806. [CrossRef] [PubMed]
50. Klempa, B.; Witkowski, P.T.; Popugaeva, E.; Auste, B.; Koivogui, L.; Fichet-Calvet, E.; Strecker, T.; Ter Meulen, J.; Kruger, D.H. Sangassou virus, the first hantavirus isolate from Africa, displays genetic and functional properties distinct from those of other murinae-associated hantaviruses. *J. Virol.* **2012**, *86*, 3819–3827. [CrossRef] [PubMed]
51. Raftery, M.J.; Lalwani, P.; Krautkrmer, E.; Peters, T.; Scharffetter-Kochanek, K.; Kruger, R.; Hofmann, J.; Seeger, K.; Kruger, D.H.; Schonrich, G. Beta2 integrin mediates hantavirus-induced release of neutrophil extracellular traps. *J. Exp. Med.* **2014**, *211*, 1485–1497. [CrossRef] [PubMed]
52. Krautkramer, E.; Zeier, M. Hantavirus causing hemorrhagic fever with renal syndrome enters from the apical surface and requires decay-accelerating factor (DAF/CD55). *J. Virol.* **2008**, *82*, 4257–4264. [CrossRef] [PubMed]

53. Choi, Y.; Kwon, Y.C.; Kim, S.I.; Park, J.M.; Lee, K.H.; Ahn, B.Y. A hantavirus causing hemorrhagic fever with renal syndrome requires gC1qR/p32 for efficient cell binding and infection. *Virology* **2008**, *381*, 178–183. [CrossRef] [PubMed]
54. Mou, D.L.; Wang, Y.P.; Huang, C.X.; Li, G.Y.; Pan, L.; Yang, W.S.; Bai, X.F. Cellular entry of Hantaan virus A9 strain: Specific interactions with beta3 integrins and a novel 70kda protein. *Biochem. Biophys. Res. Commun.* **2006**, *339*, 611–617. [CrossRef] [PubMed]
55. Kim, T.Y.; Choi, Y.; Cheong, H.S.; Choe, J. Identification of a cell surface 30 kda protein as a candidate receptor for Hantaan virus. *J. Gen. Virol.* **2002**, *83*, 767–773. [CrossRef] [PubMed]
56. Wang, H.; Alminaite, A.; Vaheri, A.; Plyusnin, A. Interaction between hantaviral nucleocapsid protein and the cytoplasmic tail of surface glycoprotein Gn. *Virus Res.* **2010**, *151*, 205–212. [CrossRef] [PubMed]
57. Ganaie, S.S.; Mir, M.A. The role of viral genomic RNA and nucleocapsid protein in the autophagic clearance of hantavirus glycoprotein Gn. *Virus Res.* **2014**, *187*, 72–76. [CrossRef] [PubMed]
58. Geimonen, E.; LaMonica, R.; Springer, K.; Farooqui, Y.; Gavrilovskaya, I.N.; Mackow, E.R. Hantavirus pulmonary syndrome-associated hantaviruses contain conserved and functional itam signaling elements. *J. Virol.* **2003**, *77*, 1638–1643. [CrossRef] [PubMed]
59. Lutteke, N.; Raftery, M.J.; Lalwani, P.; Lee, M.H.; Giese, T.; Voigt, S.; Bannert, N.; Schulze, H.; Kruger, D.H.; Schonrich, G. Switch to high-level virus replication and HLA class i upregulation in differentiating megakaryocytic cells after infection with pathogenic hantavirus. *Virology* **2010**, *405*, 70–80. [CrossRef] [PubMed]
60. Raftery, M.J.; Kraus, A.A.; Ulrich, R.; Kruger, D.H.; Schonrich, G. Hantavirus infection of dendritic cells. *J. Virol.* **2002**, *76*, 10724–10733. [CrossRef] [PubMed]
61. Li, W.; Klein, S.L. Seoul virus-infected rat lung endothelial cells and alveolar macrophages differ in their ability to support virus replication and induce regulatory t cell phenotypes. *J. Virol.* **2012**, *86*, 11845–11855. [CrossRef] [PubMed]
62. Temonen, M.; Lankinen, H.; Vapalahti, O.; Ronni, T.; Julkunen, I.; Vaheri, A. Effect of interferon-alpha and cell differentiation on Puumala virus infection in human monocyte/macrophages. *Virology* **1995**, *206*, 8–15. [CrossRef]
63. Ye, W.; Xu, Y.; Wang, Y.; Dong, Y.; Xi, Q.; Cao, M.; Yu, L.; Zhang, L.; Cheng, L.; Wu, X.; et al. Hantaan virus can infect human keratinocytes and activate an interferon response through the nuclear translocation of IRF-3. *Infect. Genet. Evol.* **2015**, *29*, 146–155. [CrossRef] [PubMed]
64. Eckerle, I.; Lenk, M.; Ulrich, R.G. More novel hantaviruses and diversifying reservoir hosts—Time for development of reservoir-derived cell culture models? *Viruses* **2014**, *6*, 951–967. [CrossRef] [PubMed]
65. Temonen, M.; Vapalahti, O.; Holthofer, H.; Brummer-Korvenkontio, M.; Vaheri, A.; Lankinen, H. Susceptibility of human cells to Puumala virus infection. *J. Gen. Virol.* **1993**, *74 Pt 3*, 515–518. [CrossRef] [PubMed]
66. Klingstrom, J.; Hardestam, J.; Stoltz, M.; Zuber, B.; Lundkvist, A.; Linder, S.; Ahlm, C. Loss of cell membrane integrity in Puumala hantavirus-infected patients correlates with levels of epithelial cell apoptosis and perforin. *J. Virol.* **2006**, *80*, 8279–8282. [CrossRef] [PubMed]
67. Hardestam, J.; Klingstrom, J.; Mattsson, K.; Lundkvist, A. Hfrs causing hantaviruses do not induce apoptosis in confluent Vero E6 and A-549 cells. *J. Med. Virol.* **2005**, *76*, 234–240. [CrossRef] [PubMed]
68. Li, X.D.; Kukkonen, S.; Vapalahti, O.; Plyusnin, A.; Lankinen, H.; Vaheri, A. Tula hantavirus infection of Vero E6 cells induces apoptosis involving caspase 8 activation. *J. Gen. Virol.* **2004**, *85*, 3261–3268. [CrossRef] [PubMed]
69. Markotic, A.; Hensley, L.; Geisbert, T.; Spik, K.; Schmaljohn, C. Hantaviruses induce cytopathic effects and apoptosis in continuous human embryonic kidney cells. *J. Gen. Virol.* **2003**, *84*, 2197–2202. [CrossRef] [PubMed]
70. Park, S.W.; Han, M.G.; Park, C.; Ju, Y.R.; Ahn, B.Y.; Ryou, J. Hantaan virus nucleocapsid protein stimulates MDM2-dependent p53 degradation. *J. Gen. Virol.* **2013**, *94*, 2424–2428. [CrossRef] [PubMed]
71. Bjorkstrom, N.K.; Lindgren, T.; Stoltz, M.; Fauriat, C.; Braun, M.; Evander, M.; Michaelsson, J.; Malmberg, K.J.; Klingstrom, J.; Ahlm, C.; et al. Rapid expansion and long-term persistence of elevated NK cell numbers in humans infected with hantavirus. *J. Exp. Med.* **2011**, *208*, 13–21. [CrossRef] [PubMed]

72. Braun, M.; Bjorkstrom, N.K.; Gupta, S.; Sundstrom, K.; Ahlm, C.; Klingstrom, J.; Ljunggren, H.G. NK cell activation in human hantavirus infection explained by virus-induced IL-15/IL15Ralpha expression. *PLoS Pathog.* **2014**, *10*, e1004521. [CrossRef] [PubMed]
73. Koma, T.; Yoshimatsu, K.; Nagata, N.; Sato, Y.; Shimizu, K.; Yasuda, S.P.; Amada, T.; Nishio, S.; Hasegawa, H.; Arikawa, J. Neutrophil depletion suppresses pulmonary vascular hyperpermeability and occurrence of pulmonary edema caused by hantavirus infection in C.B-17 SCID mice. *J. Virol.* **2014**, *88*, 7178–7188. [CrossRef] [PubMed]
74. Gupta, S.; Braun, M.; Tischler, N.D.; Stoltz, M.; Sundstrom, K.B.; Bjorkstrom, N.K.; Ljunggren, H.G.; Klingstrom, J. Hantavirus-infection confers resistance to cytotoxic lymphocyte-mediated apoptosis. *PLoS Pathog.* **2013**, *9*, e1003272. [CrossRef] [PubMed]
75. Borden, E.C.; Sen, G.C.; Uze, G.; Silverman, R.H.; Ransohoff, R.M.; Foster, G.R.; Stark, G.R. Interferons at age 50: Past, current and future impact on biomedicine. *Nat. Rev. Drug Discov.* **2007**, *6*, 975–990. [CrossRef] [PubMed]
76. Jiang, H.; Wang, P.Z.; Zhang, Y.; Xu, Z.; Sun, L.; Wang, L.M.; Huang, C.X.; Lian, J.Q.; Jia, Z.S.; Li, Z.D.; et al. Hantaan virus induces toll-like receptor 4 expression, leading to enhanced production of beta interferon, interleukin-6 and tumor necrosis factor-alpha. *Virology* **2008**, *380*, 52–59. [CrossRef] [PubMed]
77. Niikura, M.; Maeda, A.; Ikegami, T.; Saijo, M.; Kurane, I.; Morikawa, S. Modification of endothelial cell functions by Hantaan virus infection: Prolonged hyper-permeability induced by TNF-alpha of Hantaan virus-infected endothelial cell monolayers. *Arch. Virol.* **2004**, *149*, 1279–1292. [CrossRef] [PubMed]
78. Yu, H.T.; Jiang, H.; Zhang, Y.; Nan, X.P.; Li, Y.; Wang, W.; Jiang, W.; Yang, D.Q.; Su, W.J.; Wang, J.P.; et al. Hantaan virus triggers TLR4-dependent innate immune responses. *Viral Immunol.* **2012**, *25*, 387–393. [CrossRef] [PubMed]
79. Handke, W.; Oelschlegel, R.; Franke, R.; Kruger, D.H.; Rang, A. Hantaan virus triggers TLR3-dependent innate immune responses. *J. Immunol.* **2009**, *182*, 2849–2858. [CrossRef] [PubMed]
80. Geimonen, E.; Neff, S.; Raymond, T.; Kocer, S.S.; Gavrilovskaya, I.N.; Mackow, E.R. Pathogenic and nonpathogenic hantaviruses differentially regulate endothelial cell responses. *Proc. Natl. Acad. Sci. USA* **2002**, *99*, 13837–13842. [CrossRef] [PubMed]
81. Matthys, V.; Mackow, E.R. Hantavirus regulation of type i interferon responses. *Adv. Virol.* **2012**, *2012*, 524024. [CrossRef] [PubMed]
82. Oelschlegel, R.; Kruger, D.H.; Rang, A. MxA-independent inhibition of hantaan virus replication induced by type I and type II interferon in vitro. *Virus Res.* **2007**, *127*, 100–105. [CrossRef] [PubMed]
83. Stoltz, M.; Ahlm, C.; Lundkvist, A.; Klingstrom, J. Lambda interferon (IFN-lambda) in serum is decreased in hantavirus-infected patients, and in vitro-established infection is insensitive to treatment with all IFNs and inhibits IFN-gamma-induced nitric oxide production. *J. Virol.* **2007**, *81*, 8685–8691. [CrossRef] [PubMed]
84. Ank, N.; West, H.; Bartholdy, C.; Eriksson, K.; Thomsen, A.R.; Paludan, S.R. Lambda interferon (IFN-lambda), a type III IFN, is induced by viruses and IFNs and displays potent antiviral activity against select virus infections in vivo. *J. Virol.* **2006**, *80*, 4501–4509. [CrossRef] [PubMed]
85. Stoltz, M.; Klingstrom, J. Alpha/beta interferon (IFN-alpha/beta)-independent induction of IFN-lambda 1 (interleukin-29) in response to Hantaan virus infection. *J. Virol.* **2010**, *84*, 9140–9148. [CrossRef] [PubMed]
86. Prescott, J.; Hall, P.; Acuna-Retamar, M.; Ye, C.; Wathelet, M.G.; Ebihara, H.; Feldmann, H.; Hjelle, B. New World hantaviruses activate IFN-lambda production in type I IFN-deficient Vero E6 cells. *PLoS ONE* **2010**, *5*, e11159. [CrossRef] [PubMed]
87. Zhou, Z.; Hamming, O.J.; Ank, N.; Paludan, S.R.; Nielsen, A.L.; Hartmann, R. Type III interferon (IFN) induces a type I IFN-like response in a restricted subset of cells through signaling pathways involving both the JAK-STAT pathway and the mitogen-activated protein kinases. *J. Virol.* **2007**, *81*, 7749–7758. [CrossRef] [PubMed]
88. Seth, R.B.; Sun, L.; Chen, Z.J. Antiviral innate immunity pathways. *Cell Res.* **2006**, *16*, 141–147. [CrossRef] [PubMed]
89. Nam, J.H.; Hwang, K.A.; Yu, C.H.; Kang, T.H.; Shin, J.Y.; Choi, W.Y.; Kim, I.B.; Joo, Y.R.; Cho, H.W.; Park, K.Y. Expression of interferon inducible genes following Hantaan virus infection as a mechanism of resistance in A549 cells. *Virus Genes* **2003**, *26*, 31–38. [CrossRef] [PubMed]
90. Yie, J.; Senger, K.; Thanos, D. Mechanism by which the IFN-beta enhanceosome activates transcription. *Proc. Natl. Acad. Sci. USA* **1999**, *96*, 13108–13113. [CrossRef] [PubMed]

91. Taylor, S.L.; Frias-Staheli, N.; Garcia-Sastre, A.; Schmaljohn, C.S. Hantaan virus nucleocapsid protein binds to importin alpha proteins and inhibits tumor necrosis factor alpha-induced activation of Nuclear Factor kappa B. *J. Virol.* **2009**, *83*, 1271–1279. [CrossRef] [PubMed]
92. Taylor, S.L.; Krempel, R.L.; Schmaljohn, C.S. Inhibition of TNF-alpha-induced activation of NF-kappa B by hantavirus nucleocapsid proteins. *Ann. N. Y. Acad. Sci.* **2009**, *1171* (Suppl. 1), E86–E93. [CrossRef] [PubMed]
93. Wang, Z.; Mir, M.A. Andes virus nucleocapsid protein interrupts protein kinase R dimerization to counteract host interference in viral protein synthesis. *J. Virol.* **2015**, *89*, 1628–1639. [CrossRef] [PubMed]
94. Billecocq, A.; Spiegel, M.; Vialat, P.; Kohl, A.; Weber, F.; Bouloy, M.; Haller, O. Nss protein of Rift Valley Fever virus blocks interferon production by inhibiting host gene transcription. *J. Virol.* **2004**, *78*, 9798–9806. [CrossRef] [PubMed]
95. Jaaskelainen, K.M.; Kaukinen, P.; Minskaya, E.S.; Plyusnina, A.; Vapalahti, O.; Elliott, R.M.; Weber, F.; Vaheri, A.; Plyusnin, A. Tula and Puumala hantavirus NSs Orfs are functional and the products inhibit activation of the interferon-beta promoter. *J. Med. Virol.* **2007**, *79*, 1527–1536. [CrossRef] [PubMed]
96. Jaaskelainen, K.M.; Plyusnina, A.; Lundkvist, A.; Vaheri, A.; Plyusnin, A. Tula hantavirus isolate with the full-length orf for nonstructural protein NSs survives for more consequent passages in interferon-competent cells than the isolate having truncated NSs Orf. *Virol. J.* **2008**, *5*, 3. [CrossRef] [PubMed]
97. Kaukinen, P.; Vaheri, A.; Plyusnin, A. Hantavirus nucleocapsid protein: A multifunctional molecule with both housekeeping and ambassadorial duties. *Arch. Virol.* **2005**, *150*, 1693–1713. [CrossRef] [PubMed]
98. Alff, P.J.; Gavrilovskaya, I.N.; Gorbunova, E.; Endriss, K.; Chong, Y.; Geimonen, E.; Sen, N.; Reich, N.C.; Mackow, E.R. The pathogenic nNY-1 hantavirus G1 cytoplasmic tail inhibits RIG-I- and TBK-1-directed interferon responses. *J. Virol.* **2006**, *80*, 9676–9686. [CrossRef] [PubMed]
99. Alff, P.J.; Sen, N.; Gorbunova, E.; Gavrilovskaya, I.N.; Mackow, E.R. The NY-1 hantavirus Gn cytoplasmic tail coprecipitates TRAF3 and inhibits cellular interferon responses by disrupting TBK1-TRAF3 complex formation. *J. Virol.* **2008**, *82*, 9115–9122. [CrossRef] [PubMed]
100. Matthys, V.; Gorbunova, E.E.; Gavrilovskaya, I.N.; Pepini, T.; Mackow, E.R. The C-terminal 42 residues of the Tula virus Gn protein regulate interferon induction. *J. Virol.* **2011**, *85*, 4752–4760. [CrossRef] [PubMed]
101. Shim, S.H.; Park, M.S.; Moon, S.; Park, K.S.; Song, J.W.; Song, K.J.; Baek, L.J. Comparison of innate immune responses to pathogenic and putative non-pathogenic hantaviruses in vitro. *Virus Res.* **2011**, *160*, 367–373. [CrossRef] [PubMed]
102. Spiropoulou, C.F.; Albarino, C.G.; Ksiazek, T.G.; Rollin, P.E. Andes and Prospect Hill hantaviruses differ in early induction of interferon although both can downregulate interferon signaling. *J. Virol.* **2007**, *81*, 2769–2776. [CrossRef] [PubMed]
103. Rasmuson, J.; Pourazar, J.; Linderholm, M.; Sandstrom, T.; Blomberg, A.; Ahlm, C. Presence of activated airway t lymphocytes in human puumala hantavirus disease. *Chest* **2011**, *140*, 715–722. [CrossRef] [PubMed]
104. Lindgren, T.; Ahlm, C.; Mohamed, N.; Evander, M.; Ljunggren, H.G.; Bjorkstrom, N.K. Longitudinal analysis of the human T cell response during acute hantavirus infection. *J. Virol.* **2011**, *85*, 10252–10260. [CrossRef] [PubMed]
105. Manigold, T.; Mori, A.; Graumann, R.; Llop, E.; Simon, V.; Ferres, M.; Valdivieso, F.; Castillo, C.; Hjelle, B.; Vial, P. Highly differentiated, resting Gn-specific memory CD8+ T cells persist years after infection by Andes hantavirus. *PLoS Pathog.* **2010**, *6*, e1000779. [CrossRef] [PubMed]
106. Koivula, T.T.; Tuulasvaara, A.; Hetemaki, I.; Makela, S.M.; Mustonen, J.; Sironen, T.; Vaheri, A.; Arstila, T.P. Regulatory T cell response correlates with the severity of human hantavirus infection. *J. Infect.* **2014**, *68*, 387–394. [CrossRef] [PubMed]
107. Lalwani, P.; Raftery, M.J.; Kobak, L.; Rang, A.; Giese, T.; Matthaei, M.; van den Elsen, P.J.; Wolff, T.; Kruger, D.H.; Schonrich, G. Hantaviral mechanisms driving HLA class I antigen presentation require both RIG-I and TRIF. *Eur. J. Immunol.* **2013**, *43*, 2566–2576. [CrossRef] [PubMed]
108. Schountz, T.; Quackenbush, S.; Rovnak, J.; Haddock, E.; Black, W.C.T.; Feldmann, H.; Prescott, J. Differential lymphocyte and antibody responses in deer mice infected with Sin Nombre hantavirus or Andes hantavirus. *J. Virol.* **2014**, *88*, 8319–8331. [CrossRef] [PubMed]
109. Spengler, J.R.; Haddock, E.; Gardner, D.; Hjelle, B.; Feldmann, H.; Prescott, J. Experimental Andes virus infection in deer mice: Characteristics of infection and clearance in a heterologous rodent host. *PLoS ONE* **2013**, *8*, e55310. [CrossRef] [PubMed]

110. Borges, A.A.; Campos, G.M.; Moreli, M.L.; Moro Souza, R.L.; Saggioro, F.P.; Figueiredo, G.G.; Livonesi, M.C.; Moraes Figueiredo, L.T. Role of mixed Th1 and Th2 serum cytokines on pathogenesis and prognosis of hantavirus pulmonary syndrome. *Microbes Infect.* **2008**, *10*, 1150–1157. [CrossRef] [PubMed]
111. Morzunov, S.P.; Khaiboullina, S.F.; St Jeor, S.; Rizvanov, A.A.; Lombardi, V.C. Multiplex analysis of serum cytokines in humans with hantavirus pulmonary syndrome. *Front Immunol.* **2015**, *6*, 432. [CrossRef] [PubMed]
112. Khaiboullina, S.F.; Martynova, E.V.; Khamidullina, Z.L.; Lapteva, E.V.; Nikolaeva, I.V.; Anokhin, V.V.; Lombardi, V.C.; Rizvanov, A.A. Upregulation of IFN-gamma and IL-12 is associated with a milder form of hantavirus hemorrhagic fever with renal syndrome. *Eur. J. Clin. Microbiol. Infect. Dis.* **2014**, *33*, 2149–2156. [CrossRef] [PubMed]
113. Outinen, T.K.; Makela, S.; Huttunen, R.; Maenpaa, N.; Libraty, D.; Vaheri, A.; Mustonen, J.; Aittoniemi, J. Urine soluble urokinase-type plasminogen activator receptor levels correlate with proteinuria in Puumala hantavirus infection. *J. Intern. Med.* **2014**, *276*, 387–395. [CrossRef] [PubMed]
114. Linderholm, M.; Ahlm, C.; Settergren, B.; Waage, A.; Tarnvik, A. Elevated plasma levels of tumor necrosis factor (TNF)-alpha, soluble TNF receptors, interleukin (IL)-6, and IL-10 in patients with hemorrhagic fever with renal syndrome. *J. Infect. Dis.* **1996**, *173*, 38–43. [CrossRef] [PubMed]
115. Klingstrom, J.; Plyusnin, A.; Vaheri, A.; Lundkvist, A. Wild-type Puumala hantavirus infection induces cytokines, C-reactive protein, creatinine, and nitric oxide in *Cynomolgus macaques*. *J. Virol.* **2002**, *76*, 444–449. [CrossRef] [PubMed]
116. Zhang, Y.; Liu, B.; Ma, Y.; Yi, J.; Zhang, C.; Xu, Z.; Wang, J.; Yang, K.; Yang, A.; Zhuang, R.; et al. Hantaan virus infection induces CXCL10 expression through TLR3, RIG-I, and MDA-5 pathways correlated with the disease severity. *Mediat. Inflamm.* **2014**, *2014*, 697837. [CrossRef] [PubMed]
117. Strandin, T.; Hepojoki, J.; Laine, O.; Makela, S.; Klingstrom, J.; Lundkvist, A.; Julkunen, I.; Mustonen, J.; Vaheri, A. Interferons induce STAT1-dependent expression of tissue plasminogen activator, a pathogenicity factor in Puumala hantavirus disease. *J. Infect. Dis.* **2016**, *213*, 1632–1641. [CrossRef] [PubMed]
118. Khaiboullina, S.F.; Rizvanov, A.A.; Otteson, E.; Miyazato, A.; Maciejewski, J.; St Jeor, S. Regulation of cellular gene expression in endothelial cells by Sin Nombre and Prospect Hill viruses. *Viral Immunol.* **2004**, *17*, 234–251. [CrossRef] [PubMed]
119. Yu, H.; Jiang, W.; Du, H.; Xing, Y.; Bai, G.; Zhang, Y.; Li, Y.; Jiang, H.; Wang, J.; Wang, P.; et al. Involvement of the AKT/NF-kappaB pathways in the HTNV-mediated increase of IL-6, CCL5, ICAM-1, and VCAM-1 in HUVEC. *PLoS ONE* **2014**, *9*, e93810.
120. Marsac, D.; Garcia, S.; Fournet, A.; Aguirre, A.; Pino, K.; Ferres, M.; Kalergis, A.M.; Lopez-Lastra, M.; Veas, F. Infection of human monocyte-derived dendritic cells by Andes hantavirus enhances pro-inflammatory state, the secretion of active MMP-9 and indirectly enhances endothelial permeability. *Virol. J.* **2011**, *8*, 223. [CrossRef] [PubMed]
121. Shin, O.S.; Yanagihara, R.; Song, J.W. Distinct innate immune responses in human macrophages and endothelial cells infected with shrew-borne hantaviruses. *Virology* **2012**, *434*, 43–49. [CrossRef] [PubMed]
122. Shin, O.S.; Kumar, M.; Yanagihara, R.; Song, J.W. Hantaviruses induce cell type- and viral species-specific host micro-RNA expression signatures. *Virology* **2013**, *446*, 217–224. [CrossRef] [PubMed]
123. Pepini, T.; Gorbunova, E.E.; Gavrilovskaya, I.N.; Mackow, J.E.; Mackow, E.R. Andes virus regulation of cellular micro-RNAs contributes to hantavirus-induced endothelial cell permeability. *J. Virol.* **2010**, *84*, 11929–11936. [CrossRef] [PubMed]
124. Schountz, T.; Shaw, T.I.; Glenn, T.C.; Feldmann, H.; Prescott, J. Expression profiling of lymph node cells from deer mice infected with Andes virus. *BMC Immunol.* **2013**, *14*, 18. [CrossRef] [PubMed]
125. Easterbrook, J.D.; Klein, S.L. Seoul virus enhances regulatory and reduces proinflammatory responses in male Norway rats. *J. Med. Virol.* **2008**, *80*, 1308–1318. [CrossRef] [PubMed]

© 2016 by the authors. Licensee MDPI, Basel, Switzerland. This article is an open access article distributed under the terms and conditions of the Creative Commons Attribution (CC BY) license (http://creativecommons.org/licenses/by/4.0/).

Review

Molecular Insights into Crimean-Congo Hemorrhagic Fever Virus

Marko Zivcec, Florine E. M. Scholte, Christina F. Spiropoulou, Jessica R. Spengler and Éric Bergeron *

Viral Special Pathogens Branch, Division of High Consequence Pathogens and Pathology, Centers for Disease Control and Prevention, Atlanta, GA 30333, USA; mzivcec@cdc.gov (M.Z.); kyj7@cdc.gov (F.E.M.S.); ccs8@cdc.gov (C.F.S.); wsk7@cdc.gov (J.R.S.)

* Correspondence: ebergeron@cdc.gov; Tel.: +1-404-639-1724

Academic Editors: Jane Tao and Pierre-Yves Lozach
Received: 8 March 2016; Accepted: 18 April 2016; Published: 21 April 2016

Abstract: Crimean-Congo hemorrhagic fever virus (CCHFV) is a tick-borne pathogen that causes high morbidity and mortality. Efficacy of vaccines and antivirals to treat human CCHFV infections remains limited and controversial. Research into pathology and underlying molecular mechanisms of CCHFV and other nairoviruses is limited. Significant progress has been made in our understanding of CCHFV replication and pathogenesis in the past decade. Here we review the most recent molecular advances in CCHFV-related research, and provide perspectives on future research.

Keywords: Crimean-Congo hemorrhagic fever; viral hemorrhagic fever; reverse genetics; pathogenesis; tick-borne virus

---

1. Introduction

Crimean-Congo hemorrhagic fever virus (CCHFV) causes a mild to severe hemorrhagic disease (CCHF) exclusively in humans, with case fatality rates of 5%–30%. Presently, efficacy of therapeutic options in controlled clinical trials remains unproven, and supportive care remains the mainstay of treatment. CCHF endemic foci are present over a wide geographic range, including areas in Western and Central Asia, the Middle East, South-Eastern Europe, and Africa [1–4]. CCHFV exists in an enzootic cycle between ticks and mammals; and geographic distribution of the virus mirrors the distribution of the primary tick vector species that include members of the *Hyalomma* genus (*H. marginatum*, *H. anatolicum*, *H. truncatum*, *H. impeltatum*, and *H. impressum*) [5]. Viral transmission to humans can occur via tick bite, or via exposure to body fluids from viremic animals or humans [3].

Following a brief incubation period (usually <7 days), CCHFV infection initially causes non-specific symptoms, including a rapid onset high-grade fever, fatigue, and myalgia, frequently accompanied by vomiting and diarrhea. Progression to severe disease is characterized by thrombocytopenia, elevated circulating liver enzymes, and hemorrhagic manifestation (petechiae, ecchymosis, and epistaxis, as well as gingival, gastrointestinal, and cerebral hemorrhages) [1–4]. Severity of CCHF correlates with increased viral load and dissemination, low anti-CCHFV antibody titers, severity of thrombocytopenia, increased clotting times, hemorrhage, high levels of pro-inflammatory cytokines (e.g., tumor necrosis factor α (TNFα), interleukin 6 (IL-6)), and elevated aspartate aminotransferase and alanine aminotransferase [6–18]. Fatal outcome is typically the result of disseminated intravascular coagulopathy, shock, and/or multi-organ failure [1–4,19]. During the course of disease, CCHFV is widely distributed throughout the body, and has been detected in spleen, lung, heart, and intestinal tissues in fatal human cases [19]. The main cellular targets of infection are mononuclear phagocytes, endothelial cells, and hepatocytes [19]. Infection of monocyte-derived macrophages, endothelial cells, and dendritic cells are confirmed *in vitro* [20–22].

Continued research into CCHF and the development of medical countermeasures is needed based on severity of disease, human-to-human transmission, the absence of vaccines or treatments with proven efficacy and the potential for a severe outbreak in the future. Although research on CCHFV is limited, there have been significant recent advances in CCHFV research. These include modern molecular tools and *in vivo* disease models that offer opportunities for substantial progress in the field, and further development of therapeutics and vaccines. In this article, we summarize current knowledge of CCHFV genome replication, viral protein processing and function, and the development of reverses genetic systems. We also highlight areas of importance for future research.

2. CCHFV Genome and Replication Cycle

2.1. CCHFV Genome

CCHFV is a member of the genus *Nairovirus* in the family *Bunyaviridae*, which includes five genera and over 350 virus species [23]. CCHFV is characterized by a tripartite RNA genome of negative polarity (Figure 1), and by tick-borne maintenance and transmission. Complementary non-coding regions (NCRs) are present at the 5′ and 3′ termini of the small (S), medium (M), and large (L) segments of CCHFV and contribute to the circular appearance of the bunyavirus genomes [24]. The nine terminal nucleotides (5′-UCUCAAAGA and 3′-AGAGUUUCU) are conserved between nairoviruses, and serve as viral promoter regions. The NCRs are necessary for the viral RNA-dependent RNA polymerase (RdRp or L protein) to bind and initiate transcription and/or replication of the viral genome [25–27]. Although the complete NCRs sequences differ between viral segments, each is capable of initiating encapsidation, transcription, replication and packaging of the genomes into nascent virions.

Until recently, all three segments of CCHFV were believed to each encode a single protein. However, a second protein, the non-structural S (NS_S), is encoded in the S segment in the opposite orientation relative to the nucleoprotein (NP) gene, indicating that CCHFV might be considered an ambisense virus (Figure 2) [28]. However, the ambisense coding in CCHFV involves overlapping coding regions. This differs from ambisense coding in bunyaviruses and arenaviruses where the viral proteins are encoded in opposite orientation, and separated by an intergenic region that serves as a transcription termination signal [29,30]. In contrast to the S segment (~1.6 kb), that is comparable in size to other bunyaviruses, the M (~5.4 kb) and L (~12.1 kb) segments of CCHFV are significantly larger than those of other bunyaviruses, and contain a single gene encoding the glycoprotein precursor (GPC) and the L protein, respectively.

The genome segments or viral RNA (vRNA) are encapsidated with NP and L protein to form genomic ribonucleoprotein complexes (RNP). The genomic RNPs are packaged into viral particles by acquiring a lipid envelope and the surface glycoproteins Gn and Gc [31] (Figure 1). Electron microscopy studies of nairoviruses report spherical particles of relatively uniform size with a diameter of ~90–100 nm with small (<10 nm) spike-like projections from the surface of the particle [27,32–36]. Further characterization of the precise arrangement of the surface glycoproteins will require more refined methods such as cryo-electron microscopy.

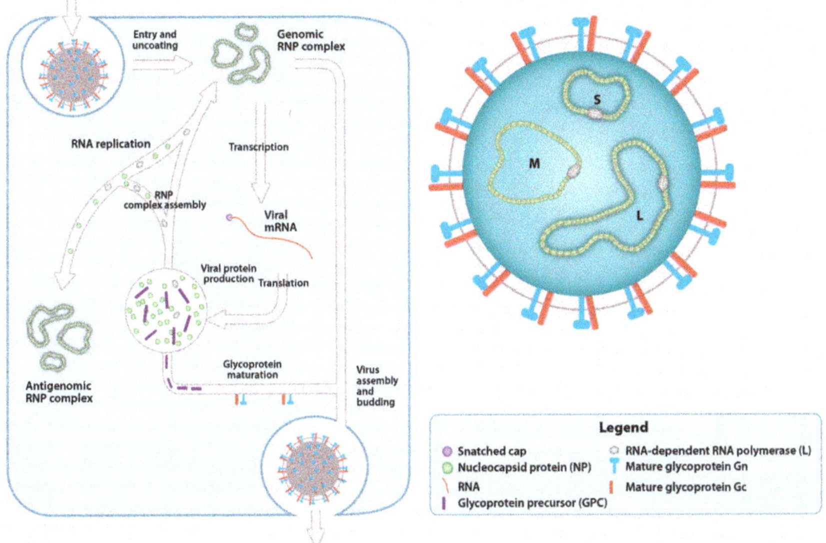

Figure 1. Crimean-Congo hemorrhagic fever virus (CCHFV) virion and replication cycle. The CCHFV virion contains a tri-segmented, negative-sense, single-stranded RNA (vRNA) genome encapsidated by the nucleoprotein (NP) and the RNA-dependent RNA polymerase (RdRp; L protein). Together, vRNA, NP, and RdRp form the genomic ribonucleoprotein complexes (RNP) inside a cellular membrane-derived envelope coated with the mature glycoproteins Gn and Gc. CCHFV attaches to an unidentified cellular receptor and enters the cells in a clathrin-dependent manner. After fusion with the cellular membrane, the viral genomic segments are uncoated and transcribed by L protein into viral mRNA that gain host cell-derived 5′ caps by cap snatching. The viral mRNAs are translated into the NP and L proteins by cytoplasmic ribosomes, while the glycoprotein precursor (GPC) appears to be translated by endoplasmic reticulum (ER)-associated ribosomes. A portion of the newly synthesized NP and L protein are used to replicate the genomic RNA by forming an RNP containing antigenomic RNA (cRNA). The GPC undergoes processing and maturation in the ER and the Golgi, and yields the Gn and Gc. Upon the accumulation of nascent mature glycoproteins and genomic RNPs, new CCHFV particles assembly is believed to occur in the Golgi followed by virion release in Golgi-derived vesicles.

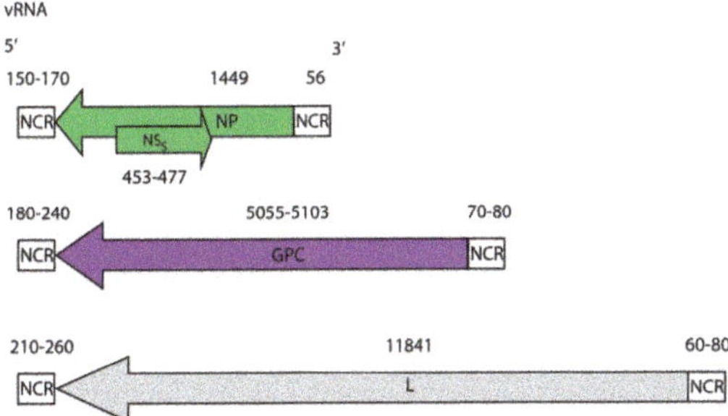

Figure 2. CCHFV genome. CCHFV possess a tri-segmented negative sense RNA genome. The small (~1.6 kb), medium (~5.4 kb) and large (~12.1 kb) segments, code for the NP, the GPC and the L protein, respectively. The small segment also codes for a non-structural S protein (NS_S) in the positive sense. The coding regions are flanked by non-coding regions (NCRs). The nucleotide lengths of the regions (both coding and non-coding) are displayed and based on full-length sequences available in GenBank.

2.2. Cell Entry

The initial binding of CCHFV to the cell surface is mediated by the glycoproteins Gn and/or Gc. However, the details of specific glycoprotein involvement in viral attachment, internalization, and fusion remain unknown. It is suspected that Gc is responsible for binding to the cellular receptors, as monoclonal antibodies targeting Gc can neutralize CCHFV infection of mammalian cells [37]. The Gc of CCHFV is also thought to mediate fusion; the predicted fusion loop of CCHFV [38] shares significant homology with Rift Valley fever virus (RVFV) Gc fusion loop [39]. The cellular receptors required for CCHFV entry have not been identified. A functional interaction has been suggested between CCHFV Gc and cell surface nucleolin, a protein found predominantly within nucleoli [40]. Nucleolin serves as a receptor for the respiratory syncytial virus, functions as an entry factor for Japanese encephalitis virus, and enhances entry of human immunodeficiency virus (HIV) [41–44]. However, further investigations in the context of CCHFV infection are needed to support the involvement of nucleolin in CCHFV cell binding and/or fusion.

After attaching to the cell surface, CCHFV is endocytosed through a clathrin-mediated endocytosis mechanism. Entry requires clathrin and the clathrin pit adaptor protein-2 complex, but not caveolin-1 [45–47]. CCHFV entry is also dependent on cholesterol and a low pH [45,46]. Following endocytosis CCHFV particles are transported to early endosomes and subsequently to multivesicular bodies (MVB) in a process that is dependent on Rab5 [46,47].

In contrast, blocking Rab7-dependent trafficking (from early endosomes to the late endosomes, or transport out of the MVB) has no effect on infection or CCHFV association with MVB [47]. Interfering with the formation of functional MVBs, for example, by depleting components of the endosomal sorting complex required for transport (ESCRT) pathway decreased CCHFV infection levels [47]. This indicates that the MVB is likely the main organelle where the CCHFV envelope fuses with the host membranes.

2.3. Transcription and Replication

Following cell entry and fusion, the genomic RNPs are released into the cytosol and the encapsidated vRNA serves as a template for the L protein to synthesize viral mRNA (Figure 1). No studies have described the 3′ termini of nairovirus mRNA or elements involved in terminating transcription. However, the 5′ terminal regions of Dugbe virus, a related nairovirus, contain a 7-methylguanylate (m7G) cap with sequences derived from cellular mRNA. To initiate viral mRNA synthesis, the L protein uses m7G capped primers that are snatched from cellular mRNA by an endonuclease domain located in the L protein. CCHFV L protein contains a residue (D693) [25] that is predicted to coordinate a Mn^{2+} critical for the cap snatching activity, as demonstrated for endonucleases of other L proteins [48–50]. Mutating D693 selectively abolishes L protein transcription activity, but does not impair its ability to replicate CCHFV genome analogues [26]. This suggests that capped primers are not used to initiate CCHFV replication.

The replication of genomic RNPs is a process requiring the replication and encapsidation of uncapped, negative sense vRNA and positive sense complementary RNA (cRNA). The replication of the RNPs minimally require the L protein and NP [26]. The encapsidated forms of the cRNA and vRNA are respectively defined as antigenomic and genomic RNPs (Figure 1). Since CCHFV is a negative strand RNA virus, genomic RNPs are used as template to synthesize capped mRNA and produce antigenomic RNPs. During the replication of the genomic RNPs, a cRNA is synthesized by the L protein and NP subunits are added to the elongating strands to obtain antigenomic RNPs, and in turn the cRNA of the antigenomic RNPs is used a template to obtain genomic RNPs.

2.4. Glycoprotein Maturation, Viral Assembly, and Egress

RNP replication is followed, and later accompanied, by viral protein processing and subsequent maturation. The maturation of CCHFV GPC is unusually complex and shares little similarity with glycoprotein processing in other bunyaviruses [31]. GPC maturation yields the structural glycoproteins Gn and Gc; secreted non-structural proteins GP160, GP85, and GP38 [31,51]; and the non-structural M protein (NS_M) [52] (Figure 3). Nairoviruses are the only bunyaviruses known to encode secreted non-structural glycoproteins [31,37,52–57]. To yield the complete set of glycoproteins, the GPC is heavily glycosylated [31] and subsequently cleaved by host proteases including the proprotein convertases [51,54,56,58], a family of mammalian serine proteases known to process a variety of cellular proteins, viral glycoproteins, and bacterial toxins [59].

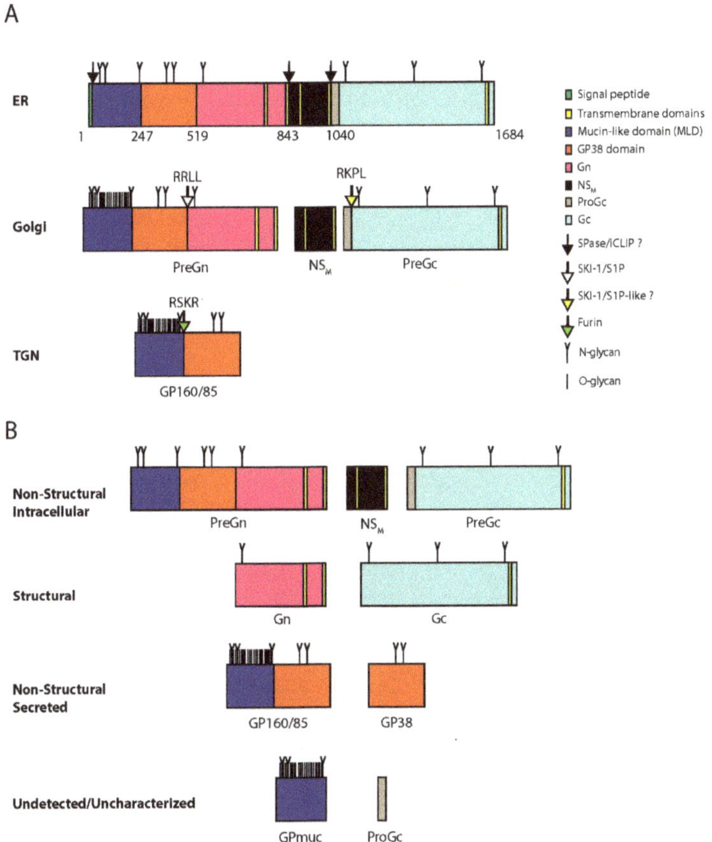

Figure 3. CCHFV glycoprotein processing and products. (**A**) CCHFV glycoprotein processing. The GPC is synthesized in the ER where *N*-glycosylation occurs (*N*-glycan). Numbers indicate amino acids positions. The signal peptidase and/or the intramembrane cleaving proteases (iCLIPs) co-translationally cleave GPC close to or within transmembrane domain-2 and -4. These cleavages yield PreGn, non-structural M protein (NS_M) and PreGc. These proteins traffic to the Golgi where the mucin-like domain of PreGn is *O*-glycosylated (*O*-glycan), and is cleaved by subtilisin kexin isozyme-1/site-1 protease (SKI-1/S1P) at the RRLL motif. PreGc at the RKPL motif by a protease with similar specificity to SKI-1/SIP (SKI-1/S1P-like?). PreGn cleavage liberates an N-terminal fragment with an apparent total molecular weight on SDS-PAGE of 160 kDa (GP160) and 85 kDa (GP85). GP160/85 is later cleaved by furin in the *trans*-Golgi network (TGN). (**B**) GPC products. Processing of GPC yields non-structural products associated with the cell (Non-structural intracellular), associated with the virions (Structural), secreted but not part of the virions (Non-structural secreted), and inferred products that remain uncharacterized or have yet to be detected Undectected/uncharacterized).

The GPC contains an N-terminal signal peptide directing its synthesis to the secretory pathway [31,52]. Upon elongation and translocation into the endoplasmic reticulum (ER), the signal peptide is removed, the GPC is *N*-glycosylated, folded, intra-molecular disulfide bridges are formed, and the transmembrane domains of the precursor protein span the ER membrane five times [31]. Before GPC is completely translated, it is cleaved into the Gn precursor (PreGn), the Gc precursor (PreGc), and the NS_M. This step is thought to require the signal peptidase and the intramembrane cleaving proteases (I-CLiPs), as generation of these three proteins requires cleavage after the signal peptide and near or within the transmembrane domains-2 and -4 (Figure 3) [31,52].

The productive maturation in the ER is followed by PreGn and PreGc transport to the Golgi complex [52,54], where the PreGn mucin-like domain (MLD) acquire numerous *O*-linked glycans [51]. PreGn can traffic to the Golgi complex, the proposed site of CCHFV assembly, in the absence of PreGc [37], but PreGc requires PreGn to exit the ER [37]. Mutating the N577 glycosylation site of PreGn, located in the mature Gn region, blocks PreGn from exiting the ER and prevents secretion of GP160, GP85, and GP38, suggesting a critical role for the glycosylation of N577 in folding and normal trafficking to the Golgi complex [55].

Limited endoproteolysis PreGn and PreGc is required to complete GPC maturation. The host protease required to convert PreGc to Gc remains unidentified. Given that the PreGc cleavage motif of CCHFV (RKPL) is identical to the GPC cleavage motif of Guanarito virus, an arenavirus [60] that is cleaved by the subtilisin kexin isozyme-1/site-1 protease (SKI-1/S1P), this same protease or a protease with similar specificity (SKI/S1P-like) may cleave CCHFV PreGc. The cleavage of PreGn by SKI-1/S1P occurs early in the secretory pathway, either in the ER or after its exit to the *cis*-Golgi apparatus [58]. PreGn cleavage at the RRLL motif liberates N-terminal products (GP160 and GP85), containing the MLD and GP38 domain [58]. GP160/85 can be further cleaved in the *trans*-Golgi network (TGN) by furin at a well-conserved RSKR motif located at the junction of the MLD and the GP38 domain [51]. This cleavage is predicted to free the GP38 domain from the MLD, although available antibodies can only detect GP38 and the uncleaved MLD-containing glycoproteins (GP160/85), but not the cleaved MLD.

Initial speculation was that GP160 was a dimer of GP85 because epitopes located within the MLD and GP38 domains are present in both glycoproteins, but denaturation with urea and reducing agents did not alter the mobility of GP160 or GP85. Additionally, the glycan composition of the non-structural glycoproteins (GP160, GP85 and GP38) is likely similar based on comparable sensitivity to glycosidases specific for *N*- and *O*-linked glycans [31,51]. Therefore, additional experiments are needed to explain the difference in the mobility of GP85 and GP160 on SDS-PAGE.

Our understanding of CCHFV assembly and egress is limited. Like other bunyaviruses, CCHFV RNPs are most likely found in the cytoplasm, and NP is localized in the perinuclear region close to the Golgi complex [61]. The subcellular localization of the viral glycoproteins often dictates the location of virus budding; the accumulation of the glycoproteins in the Golgi complex and TGN [37,54] and NP in perinuclear regions is compatible with budding and assembly of CCHFV particles in the Golgi complex and/or TGN. [61,62]. Following assembly, CCHFV particles are released by exocytosis often in the absence of discernable cytopathology and egress occurs from the basolateral membrane in polarized epithelial cells [63].

3. Viral Protein Function

3.1. S Segment: NP and NS_S

The primary function of NP is to encapsidate the vRNA and cRNA to form RNPs. Consistent with its role in RNP formation, NP has been shown to interact with the N- and C-terminal regions of the L protein and to co-localize with tagged L protein in transfected cells [26,64]. L protein antibodies are currently unavailable for CCHFV, preventing the study of L protein in infected cells. In CCHFV-infected cells, NP is localized to the perinuclear region in an actin filament-dependent manner, and interfering with actin polymerization reduces CCHFV infectivity [63]. NP likely performs multiple other functions. For example, NP can be released from cells even in the absence of other viral particles and leads to the formation of spherical particles resembling bunyaviruses, suggesting a possible role in viral budding [32].

Important insights into the roles of NP were obtained from the crystal structures of CCHFV NP using CCHFV isolates from Iraq [65] and China [66,67]. The protein structure is more closely related to the NP of arenaviruses [68,69] than to that of NP of other bunyaviruses [70–72]. The NP possesses two major domains: a globular head and a flexible arm. The arm domain protrudes from the globular head domain of CCHFV NP. The globular head domain of CCHFV NP comprises 23 α-helices with an overall structure that is similar to the NP of Lassa virus (LASV), an arenavirus. Crystallization of NP as an oligomer in the presence of RNA demonstrated that NP subunits organize in a head-to-tail orientation, with the arm of one subunit interacting with the head domain of the neighboring subunit, that are further assembled into double antiparallel superhelical polymers of NP; this may represent the organization of NP in the RNPs [73].

Oligomerization appears to regulate NP function. In the absence of RNA, NP appears to exist exclusively as a monomer; in this state, NP binds RNA weakly [66,67], suggesting that only NP oligomers effectively bind RNA. Additionally, comparing monomeric and oligomeric NP organization shows that the arm domain changes conformation upon oligomerization allowing stronger binding to RNA [73]. The nature of the chemical bonds between RNA and NP remain unknown, as the RNA structure has yet to be resolved in any of the NP models. However, mutating three CCHFV NP residues (K132, Q300, K411) predicted to bind RNA blocked the transcription and replication of CCHFV minigenomes [65,67].

NP may contribute to viral-mediated immune evasion. Multiple aspects of the mammalian antiviral response appear to target the NP, like the interferon (IFN)-stimulated antiviral gene MxA, which inhibits CCHFV replication [62]. Furthermore, NP is highly immunogenic and is the major target of both B and T cells in mammals [37,74–77]. The NP of LASV and related hantaviruses have been shown to function as IFN antagonists [69,78]. However, CCHFV NP does not suppress the IFN response to Sendai virus infection [67], a model virus that strongly activates the innate immune responses to double stranded RNA (dsRNA). Structurally related arenavirus NPs possess a conserved $3'$ to $5'$ exonuclease activity specific to RNA, and this RNase activity suppresses the innate immune response to dsRNA [69]. *In vitro*, CCHFV NP has endonuclease activity, but appears to be restricted to DNA instead of RNA [67].

Early in infection CCHFV can inhibit apoptosis [79], but later in infection it induces apoptosis [80]. During apotosis, a fraction of NP is cleaved by caspase-3 at a putative DEVD motif [80] located in the apex of the arm domain. Since the arm domain interacts with the head domain in NP oligomers, steric hindrance associated with NP oligomerization would prevent cleavage of NP by caspase-3 [73]. NP monomers may prevent apoptosis by acting as decoy substrate for caspase-3 to delay apoptosis as suggested for a structurally related arenavirus, Junín virus [81].

In transfected cells, the recently described CCHFV NS_S localizes to the mitochondria and induces apoptosis by disrupting the mitochondrial membrane potential [28]. Although NS_S has been detected in CCHFV-infected cells, the proposed apoptotic functions are based exclusively on the overexpression of NS_S, and further investigations of NS_S are warranted.

3.2. M Segment: Glycoproteins

Overall, CCHFV glycoproteins appear to be involved in entry and fusion, virion formation and immune evasion. Gn and Gc are believed to mediate entry and fusion. Their functions are for the most part extrapolated from those of glycoproteins from distantly related bunyaviruses. The cytoplasmic tail of Gn contains a zinc finger domain that binds RNA in vitro, suggesting that Gn may interact with RNA and perform matrix protein-like functions [53]. However, as RNA [3] is encapsidated in the RNP, it is not clear how Gn might bind and incorporate RNP into the nascent virions. Functional studies of the CCHFV GPC highlighted the essential role of the PreGn convertase SKI-1/S1P in the production of infectious particles [54]. Importantly, PreGn and PreGc normally localize in the Golgi complex, which implies that correct processing of these proteins regulates the production of infectious CCHFV particles [54]. Determining the function of GP38 is particularly intriguing as it is nairovirus-specific and does not share sequence homology with other cellular or viral proteins. Furin cleavage was shown to selectively block GP38 production without affecting the secretion of the cleavage products of SKI-1/S1P: GP160 and GP85 [51,56]. Blocking furin cleavage resulted in a transient reduction in viral titers, implying that mature GP38 is not important for CCHFV replication, at least in cell culture. In addition, blocking furin cleavage also results in a slight reduction in PreGn processing, showing that furin cleavage might indirectly regulate SKI-1/S1P-dependent GPC processing [56].

The MLD is found in PreGn, GP85, and GP160. Interestingly, the glycoprotein precursors of filoviruses (e.g., Ebola virus and Marburg virus) also contain a MLD and a furin cleavage site [82,83]. In filoviruses, furin cleavage products, GP1 and GP2, are structural components of virions, and the O-linked glycans of the GP1 MLD can shield and protect GP2 epitopes targeted by neutralizing antibodies [84]. The Ebola virus MLD can be replaced with the CCHFV MLD without affecting protein function, suggesting that the MLD of CCHFV may also shield exposed epitopes [84]. However, the function of the CCHFV MLD may differ significantly from that of filoviruses, as the CCHFV MLD is not incorporated into the viral particles [31]. In addition, truncating the CCHFV MLD is dispensable for the folding and trafficking of GPC [37]. In contrast, longer N-terminal truncation comprising the GP38 domain retains Gn in the ER, suggesting that the GP38 domain has a chaperone activity that assists PreGn folding or frees it from the ER [37].

3.3. L Segment: Ovarian Tumor Protease, Nuclease, and RdRp Activities

Nairoviruses have an unusually large L protein (~4000 amino acids) compared to those of other family *Bunyaviridae* members (~2500 amino acids). Additional sequences are present in the N-terminal region of nairovirus L proteins that are not found in other *Bunyaviridae* genomes. This finding suggests that amino acids 1–609 may contain domains with non-classical L function, such as the ovarian tumor (OTU) cysteine protease. Sequences located between the OTU domain and the RdRp conserved motifs contain a potential leucine zipper and a C2H2 zinc finger motif important for binding NP the N-terminal region of the L protein [64,85] (Figure 4).

The L protein regions involved in mRNA transcription and replication of the viral genome likely start with the internal endonuclease domain and include several conserved RdRp motifs [85,86]. The viral endonuclease cleaves host mRNAs and uses the resulting 5′ capped oligonucleotides as primers to initiate viral transcription. The 5′ termini of CCHFV vRNA are monophosphorylated [87,88], in contrast to many other RNA viruses that use the more common triphosphate group (5′-pppRNA). Monophosphorylated 5′ genome ends are likely created by a chain initiation mechanism called prime and realign, in which the viral endonuclease generates a 5′-pRNA by cleaving off the first nucleotide of the 5′ genomic end. This mechanism was previously suggested for the related Hantaan virus [89]. Processing of CCHFV genome 5′ termini to a monophosphate group (5′-p) is a possible strategy for evading the innate immune response by blunting the activation of retinoic acid-inducible gene I (RIG-I), which is preferentially activated by 5′-pppRNA [87,88]. Nevertheless, the type-I IFN response to CCHFV requires RIG-I [88]. RIG-I is believed to function not only as a sensor during viral infection, but also as an antiviral effector [90]. This effector function may partially explain why RNA viruses that induce poor IFN responses or do not have the preferred 5′-pp or 5′-pppRNA RIG-I ligands replicate more efficiently when RIG-I is knocked down [88,91].

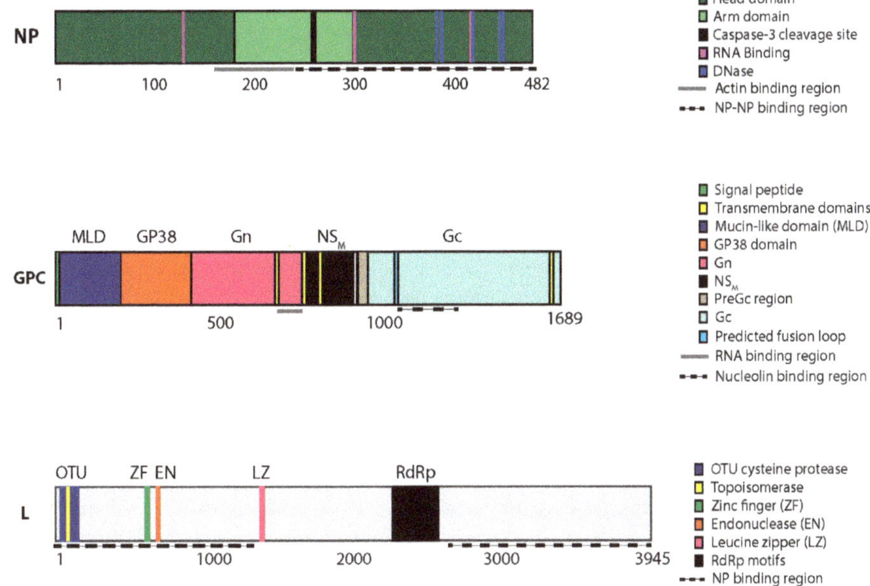

Figure 4. CCHFV protein domains, motifs and catalytic residues. The three CCHFV genomic segments (S, M and L) are translated into three proteins: NP, the GPC, and the L protein, respectively. The GPC is proteolytically processed to yield several additional proteins, including Gn and Gc. The approximate total size and location of motifs and catalytic residues of each protein is indicated below in amino acids.

The most extensively studied region of the CCHF L protein is the N-terminal OTU domain (residues 1–152). The OTU domain removes ubiquitin (Ub) and Ub-like protein IFN-stimulated gene-15 (ISG15) from their protein substrates [92–94]. Viral OTUs and papain-like proteases with similar activity have previously been found in both positive- and negative-stranded RNA viruses, including Dugbe virus, Nairobi sheep disease virus, rice stripe virus, porcine reproductive and respiratory syndrome virus, equine arteritis virus, murine hepatitis virus, severe acute respiratory syndrome coronavirus, human coronavirus NL63, and Middle East respiratory syndrome coronavirus [86,92,94–101]. In addition, deubiquitinases of positive-strand RNA viruses proteolytically process the viral polyproteins and are therefore necessary for replication [94]. In CCHFV, the L protein is not proteolytically processed by the OTU domain, and the OTU cysteine protease activity is dispensable for CCHFV transcription and replication of minigenomes [26].

Mammalian deubiquitinases are implicated as a negative feedback system of the IFN response [102], and viral OTUs appear to perform similar functions. The CCHFV OTU domain is thought to suppress innate immune signaling by deconjugating Ub or ISG15. Conjugation of Ub (ubiquitination) and ISG15 (ISGylation) to lysine residues regulates IFN signaling, and targets several key components of the innate immune response, including nuclear factor kappa-light-chain-enhancer of activated B cells (NFκB), RIG-I, MxA, interferon regulatory factor 3 (IRF3), signal transducer and activator of transcription 1 (STAT1), and protein kinase R (PKR) [103–108]. The crystal structure of CCHFV OTU domain provided insights into Ub and ISG15 binding specificity and allowed the design of CCHFV mutants specifically lacking activity against Ub or ISG15 [93,109,110]. Overexpression of the CCHFV OTU domain results in reduced general cellular ubiquitination and ISGylation levels. In addition, CCHFV OTU overexpression reduces the RIG-I/mitochondrial antiviral-signaling protein (MAVS)-mediated IFN-response, likely because this overexpression blocks ubiquitination of RIG-I [92,94]. The putative role of the OTU in infection remains paradoxical, as ubiquitination of RIG-I by tripartite motif-containing protein 25 (TRIM25) activates it, whereas ISGylation of RIG-I negatively regulates RIG-I activation by competing with ubiquitination sites [103,111]. More studies are required to understand how deubiquitination of RIG-I by a viral OTU may facilitate viral replication, while simultaneous removal of ISG15 moieties from RIG-I may also result in increased antiviral responses.

4. CCHFV Reverse Genetics

CCHFV reverse genetics systems are powerful tools for investigating underlying mechanisms of the viral replication cycle, host immune evasion, and numerous other pathogen-host interactions. These systems are particularly useful for dissecting and studying the basics of the viral replication cycle, including viral genome transcription, replication, particle assembly and egress. In addition, reverse genetics enables the generation of reporter viruses and allows mutational analyses, for example when studying viral drug resistance or pinpointing catalytic residues.

4.1. CCHFV Minigenome System

The minigenome system is useful for studying viral transcription, replication, and encapsidation using model RNAs. Instead of a full-length vRNA or cRNA, it uses genome or antigenome analogues called minigenomes. The minigenome contains the terminal NCRs of a genomic segment, but the CCHFV coding region is replaced with a reporter gene [25–27]. Since the minigenome system is non-infectious and does not use full CCHFV genomes, it allows the study of CCHFV outside of high biocontainment laboratories. In the minigenome system, DNA copies of the vRNA minigenomes are cloned into expression vectors under the control of a T7 promoter; co-transfection with a T7-encoding plasmid yields a minigenome RNA (Figure 5). As with other reverse genetics systems that rely on T7 promoter expression, a single terminal G is added to the 5′ RNA termini to enhance T7 activity, and a native 3′ terminus is generated using a hepatitis delta virus ribozyme [56]. The NP and L protein are provided either from expression plasmids or by superinfection with CCHFV. The minigenome RNA is encapsidated, and acts as a template for replication and for transcription, resulting in the production of mRNA and ultimately a reporter signal. The reporter signal provides a quantitative collective measure of genome replication, transcription and translation [112,113]. However, by using an L protein with a mutated catalytic D693 it is possible to monitor replication alone as this mutant is unable to transcribe mRNA due to a lack of cap snatching activity [25]. While highly useful, the CCHFV minigenome system is limited as it does not model all aspects of the viral replication and cannot be used to study processes requiring the glycoproteins, such as entry, virus assembly, and egress.

4.2. CCHFV Virus-like Particle System

In order to overcome some of the limitations inherent to the minigenome system, entry competent virus-like particle (VLP) systems have also been developed. A VLP contains all viral proteins and a minigenome. Therefore it is unable to express viral proteins upon entering a target cell. The particles in these systems can mimic a single-cycle infection; and because they do not encode CCHFV proteins, this system can be studied outside of high biocontainment laboratories. VLPs are generated by expressing three helper plasmids encoding the NP, GPC, and L protein together with a minigenome plasmid, resulting in the incorporation of encapsidated minigenome RNA into VLPs (Figure 6). The VLPs are able to enter target cells, and the RNPs can serve as templates for replication and transcription when the CCHFV NP and the L protein are supplied in trans. To date, two different CCHFV VLP systems have been developed [25]. In the first system, transcriptionally competent VLPs (tc-VLP) are incubated with target cells expressing the CCHFV NP and L protein in order to obtain robust *Renilla* reporter activity that can be used to study cell entry [25]. In the second system, transcriptionally and entry competent VLPs (tec-VLPs) are generated using a codon-optimized version of the L protein and GPC, and a NanoLuc reporter to produce VLPs. The resulting tec-VLPs can enter target cells to generate a robust NanoLuc signal without the need to express L protein and NP in the target cells, simplifying workflow for studying entry and primary transcription (Figure 6) [27].

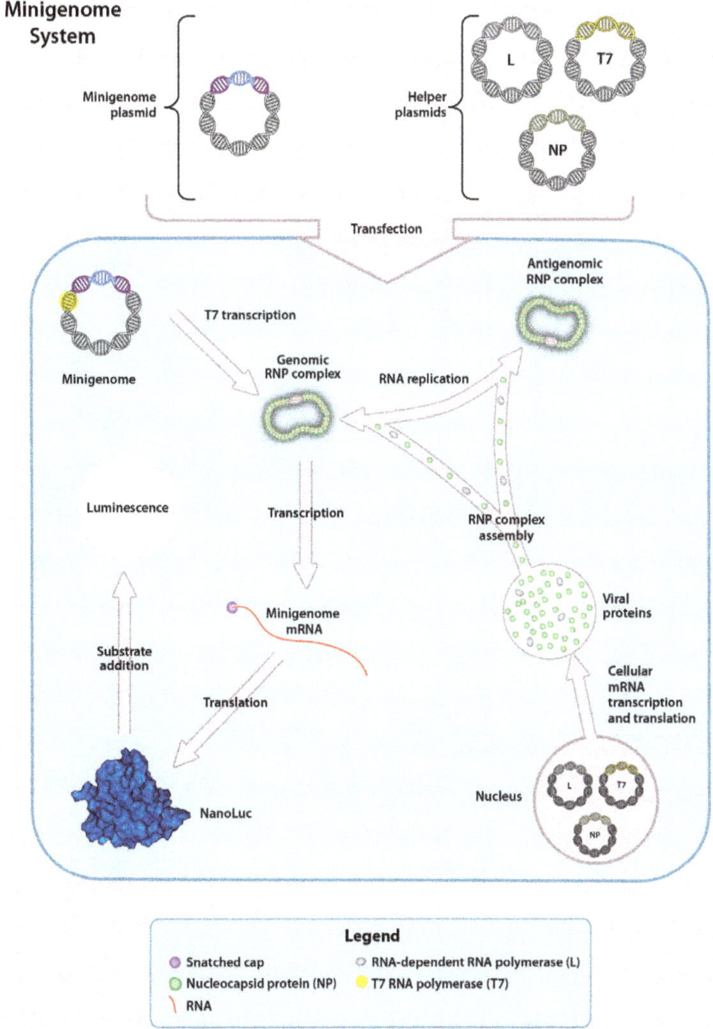

Figure 5. CCHFV minigenome system. The CCHFV minigenome system is composed of a plasmid encoding the minigenome and three helper plasmids encoding the CCHFV NP, L, and T7 RNA polymerase (T7) genes downstream of a RNA polymerase II promoter. Downstream of a T7 promoter, the minigenome plasmid contains the 5′ and 3′ non-coding regions (NCR) of a CCHFV genomic segment (S, M, or L) flanking a gene encoding a reporter protein (NanoLuc) in the negative orientation. Transfection of the helper plasmids yields the corresponding proteins to enable transcription of the minigenome plasmid and production of minigenome-derived vRNA. Following T7 transcription, the vRNA is encapsidated to form the genomic RNP. The RNP is subsequently transcribed into mRNA (secondary transcription) and translated to yield the reporter protein or replicated to produce additional vRNA. vRNA generated by both T7 transcription and RNA replication can be used as templates for transcription of reporter gene mRNA by NP and the L protein, resulting in enhanced reporter activity. A measurable luminescent signal is produced by hydrolysis of an externally provided reporter substrate.

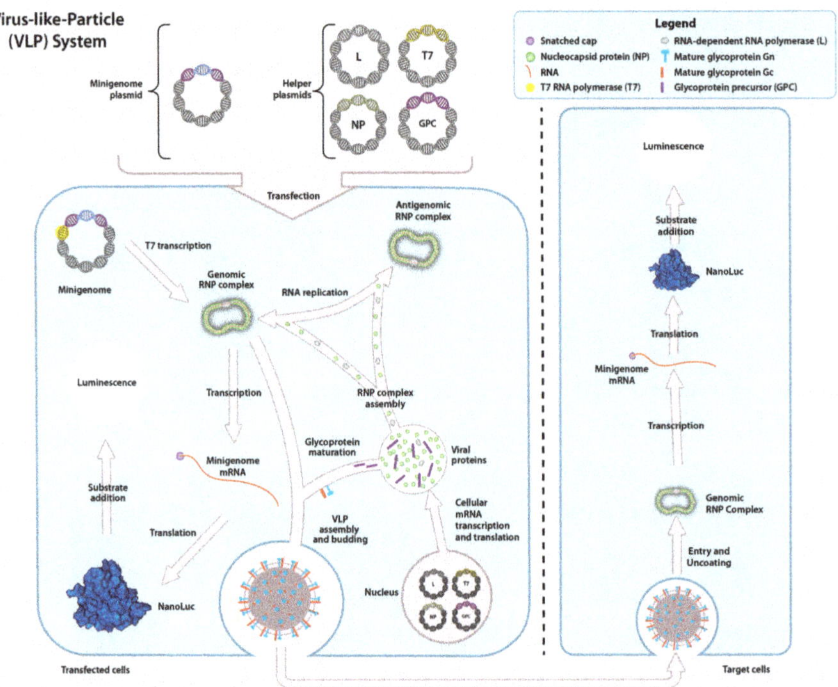

Figure 6. Transcriptionally and entry competent CCHF Virus-like-Particle (tec-VLP) system. The VLP system is composed of the minigenome plasmid and four helper plasmids that collectively encode the T7 polymerase, CCHFV NP, L, and GPC downstream of a cellular RNA polymerase II promoter. The corresponding proteins facilitate the generation of minigenome plasmid-derived vRNA. The minigenome plasmid is transcribed by T7 and the resulting vRNA is encapsidated to form the genomic RNP. The genomic vRNA is amplified from antigenomic RNPs, and subsequently transcribed and translated into the reporter protein. Reporter activity is monitored by adding a luciferase substrate and measuring the luminescent signal. VLPs assemble and bud from transfected cells and can subsequently enter other cells. Upon release of the genomic RNP from the VLP in the recipient cells, the minigenome can be amplified and transcribed into reporter mRNA, resulting in reporter protein expression.

4.3. CCHFV Infectious Clone System

The CCHFV infectious clone system allows the production of infectious recombinant CCHFV (rCCHFV) from DNA plasmids and contains the complete genome sequence under the control of a T7 promoter. Successful rescue of rCCHFV requires the addition of helper plasmids and transfection into cells. In the rCCHFV rescue process three plasmids, each containing the full sequence of one of the genome segments, are transfected into permissive cells (e.g., Huh7, BSR-T7/5). The genome is then transcribed by T7 into cRNA copies (Figure 7) that are used as a replication template for vRNA. Helper plasmids designed to produce NP and L proteins are provided in trans to synthesize and encapsidate vRNA, and finalize production of reconstituted genomic RNPs. After genomic RNPs are obtained, replication and transcription can be driven by NP and L protein produced from viral mRNA. The production of all viral proteins (NP, GPC and L protein) ultimately assembles into rCCHFV particles capable of infecting neighboring cells and performing multiple infection cycles. Since rCCHFV is infectious, all precautions and biosafety level restrictions associated with live CCHFV experimentation must be adhered to. A key to the success of this system is codon optimization of the DNA sequence of the L helper plasmid. Codon optimization of the L protein increases the activity and the amount of full-length L protein present in transfected cells [56].

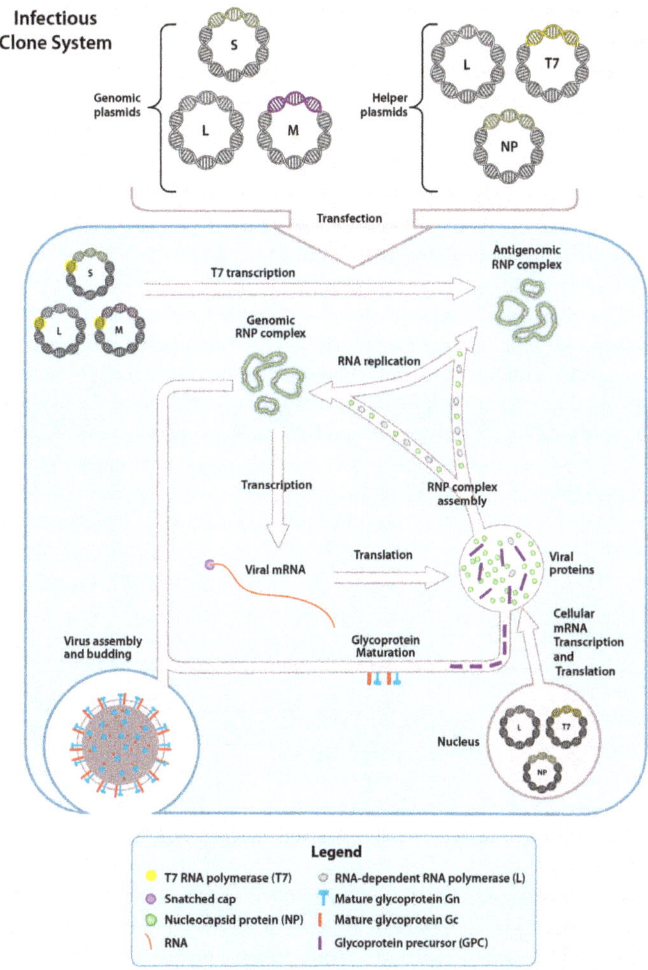

Figure 7. CCHFV infectious clone system. The CCHFV infectious clone system functions by co-transfection of genomic plasmids and helper plasmids into cells. The genomic plasmids each contain the DNA sequence of one of the CCHFV segments (S, M, or L) in the positive sense orientation downstream of a T7 promoter. The three helper plasmids encode the T7 RNA polymerase, the CCHFV NP and the L genes downstream of cellular RNA polymerase II promoters. Following T7 transcription of the genomic plasmid (enabled by the T7 helper plasmid), the cRNA is encapsidated by NP and L protein to form the antigenomic RNP, and is subsequently replicated by helper plasmid-derived NP and L protein. The vRNAs from the genomic RNPs are transcribed to mRNA and translated to yield CCHFV mRNA. The CCHFV mRNA are translated into additional NP, L, and GPC. The GPC undergoes processing and to yield the mature Gn and Gc. Following the accumulation of genomic RNPs and mature glycoproteins, infectious CCHFV particles assemble and are released.

While the infectious clone system is the most comprehensive model available of the virus replication cycle, the newly generated rCCHFV must functionally perform all the basic steps of the viral replication cycle for successful rescue of virus. Mutations that abolish a protein function that is critical for replication will prohibit virus rescue without insight into which specific aspects of the replication cycle have been disrupted. Minigenome and VLP systems can be used to more precisely dissect the effects of mutations or treatments on the individual steps in the viral replication cycle.

5. Conclusions and Future Directions

CCHF is a medically important tick-borne disease with a wide endemic distribution. At present, antiviral strategies to treat human CCHFV infection remain controversial or experimental. The most widely used antiviral, Ribavirin, has been shown to be effective against CCHFV *in vitro* and in animal models, but its clinical benefit remains unproven [114–116]. Although significant advances have been made in the field in recent years, many aspects of the CCHFV replicative cycle and pathogenesis still remain poorly defined. Additional studies to improve our knowledge of all viral proteins are warranted. Structural and/or sequence similarities of the CCHFV NP to the NP of related, better-studied arenaviruses may help to predict the putative functions of the CCHFV NP domains and guide future research. The advent of more advanced molecular tools to study NP in the context of a viral infection has the potential to greatly enhance our understanding of the roles of NP and the potential differences between various CCHFV genotypes.

In addition, studying CCHFV glycoprotein function is a critical area of research due to the significant gaps in our knowledge of the essential roles of the glycoproteins in the viral replication cycle. Improved knowledge of glycoprotein processing may provide targeted development of medical countermeasures. Since furin has a relatively minor role in the CCHFV replication cycle compared to SKI-1/S1P, pharmacological inhibitors of SKI-1/S1P may represent a promising drug target. Efficacy studies of proprotein convertase inhibitors *in vivo* and *in vitro* are critically needed to better assess the therapeutic potential of furin and SKI-1/S1P inhibition.

Further development of our understanding of the OTU domain will provide key information about CCHFV-mediated innate immune modulation, and OTU-defective or inactive mutants may be potential vaccine candidates. Reverse genetics can be used to investigate CCHFV OTU mutants that selectively cleave Ub or ISG15 conjugates, shedding light on the specific roles these conjugates play during CCHFV infection. Future experiments should focus on investigating OTU function in the context of virus infection. Previous experiments overexpressing the CCHFV OTU domain have provided valuable insights into its function, but may not accurately reflect events during a natural infection, where other factors, like spatiotemporal distribution, may affect the full-length L protein and OTU domain activity.

Advancement in all the aforementioned areas should be complemented and supported by additional investigation into viral interactions with the tick-vector and host. Studies on the molecular aspects of CCHFV replication are almost exclusively focused on cell culture systems of mammalian cells. Since CCHFV establishes a persistent infection in ticks, CCHFV replication and viral protein function should be studied in ticks and tick cell lines [117,118]. Additional research involving the tick vector and vector derived cells may aid in elucidating aspects of CCHFV that remain unclear in traditional mammalian cell based studies.

Existing interferon-α/β receptor (IFNAR)$^{-/-}$ and STAT1$^{-/-}$ mouse models of CCHF have been used to screen the efficacy of antivirals [119,120] and experimental vaccines [121–124]. The development of the existing disease models and studies of CCHFV-infected cells indicate the importance of the early innate immune response in disease [88,125]. In order to better model early events in human infection, continued efforts should be made in the development of immunocompetent CCHF animal models. An immunocompetent model may provide critical insight into correlates of protection and enhance our understanding of the molecular mechanisms underlying pathogenesis that are largely restricted to *in vitro* investigations at this time. Given the complex role of the immune response in CCHF pathogenesis, such models would be invaluable for screening new therapeutic and vaccine candidates.

CCHFV has been recognized as a significant emerging public health threat and is becoming a higher public health priority. Vaccine and therapeutic development for CCHF are rapidly expanding fields. However, a summary of these fields is beyond the scope of this review. Here we review recent *in vitro* experimental research efforts aimed at elucidating the mechanisms of CCHFV replication and pathogenesis. We provide a comprehensive summary of the known aspects of the CCHFV replication cycle, viral protein function, and cell-mediated host responses to infection. We also identify areas in CCHFV research that remain unknown or unclear, and merit additional investigation.

Acknowledgments: The authors would like to thank Tatyana Klimova for critical editing of the manuscript and Jim Walters for assistance with design of figures. The CDC and a CDC foundation project funded by NIAID grant R01AI109008 supported this work. This work was also supported in part by an appointment to the Research Participation Program at the Centers for Disease Control and Prevention administered by the Oak Ridge Institute for Science and Education through an interagency agreement between the U.S. Department of Energy and CDC (to J.R.S), and by the National Institutes of Health Loan Repayment Award (to J.R.S). The findings and conclusions in this report are those of the authors and do not necessarily represent the official position of the Centers for Disease Control and Prevention.

Conflicts of Interest: All authors report no conflicts of interest.

References

1. Ergönül, Ö. Crimean-Congo haemorrhagic fever. *Lancet Infect. Dis.* **2006**, *6*, 203–214.
2. Whitehouse, C. A Crimean-Congo hemorrhagic fever. *Antivir. Res.* **2004**, *64*, 145–160. [CrossRef]
3. Bente, D.A.; Forrester, N.L.; Watts, D.M.; McAuley, A.J.; Whitehouse, C.A.; Bray, M. Crimean-Congo hemorrhagic fever: History, epidemiology, pathogenesis, clinical syndrome and genetic diversity. *Antivir. Res.* **2013**, *100*, 159–189. [CrossRef] [PubMed]
4. Vorou, R.; Pierroutsakos, I.N.; Maltezou, H.C. Crimean-Congo hemorrhagic fever. *Curr. Opin. Infect. Dis.* **2007**, *20*, 495–500. [CrossRef] [PubMed]
5. Hoogstraal, H. The epidemiology of tick-borne Crimean-Congo hemorrhagic fever in Asia, Europe, and Africa. *J. Med. Entomol.* **1979**, *15*, 307–417. [CrossRef] [PubMed]
6. Wolfel, R.; Paweska, J.T.; Petersen, N.; Grobbelaar, A.A.; Leman, P.A.; Hewson, R.; Georges-Courbot, M.-C.; Papa, A.; Günther, S.; Drosten, C. Virus detection and monitoring of viral load in Crimean-Congo hemorrhagic fever virus patients. *Emerg. Infect. Dis.* **2007**, *13*, 1097–1100. [CrossRef] [PubMed]
7. Papa, A.; Drosten, C.; Bino, S.; Papadimitriou, E.; Panning, M.; Velo, E.; Kota, M.; Harxhi, A.; Antoniadis, A. Viral load and hemorrhagic fever. *Emerg. Infect. Dis.* **2007**, *13*, 805–806. [CrossRef] [PubMed]
8. Ergonul, O.; Tuncbilek, S.; Baykam, N.; Celikbas, A.; Dokuzoguz, B. Evaluation of serum levels of interleukin (IL)-6, IL-10, and tumor necrosis factor-alpha in patients with Crimean-Congo hemorrhagic fever. *J. Infect. Dis.* **2006**, *193*, 941–944. [CrossRef] [PubMed]
9. Ergonul, O.; Celikbas, A.; Baykam, N.; Eren, S.; Dokuzoguz, B. Analysis of risk-factors among patients with Crimean-Congo haemorrhagic fever virus infection: Severity criteria revisited. *Clin. Microbiol. Infect.* **2006**, *12*, 551–554. [CrossRef] [PubMed]
10. Duh, D.; Saksida, A.; Petrovec, M.; Ahmeti, S.; Dedushaj, I.; Panning, M.; Drosten, C.; Avsic-Zupanc, T. Viral load as predictor of Crimean-Congo hemorrhagic fever outcome. *Emerg. Infect. Dis.* **2007**, *13*, 1769–1772. [CrossRef] [PubMed]
11. Yilmaz, G.; Koksal, I.; Topbas, M.; Yilmaz, H.; Aksoy, F. The effectiveness of routine laboratory findings in determining disease severity in patients with Crimean-Congo hemorrhagic fever: Severity prediction criteria. *J. Clin. Virol.* **2010**, *47*, 361–365. [CrossRef] [PubMed]
12. Yesilyurt, M.; Gul, S.; Ozturk, B.; Kayhan, B.C.; Celik, M.; Uyar, C.; Erdogan, F. The early prediction of fatality in Crimean Congo hemorrhagic fever patients. *Saudi Med. J.* **2011**, *32*, 742–743. [PubMed]
13. Weber, F.; Mirazimi, A. Interferon and cytokine responses to Crimean Congo hemorrhagic fever virus; an emerging and neglected viral zonoosis. *Cytokine Growth Factor Rev.* **2008**, *19*, 395–404. [CrossRef] [PubMed]
14. Saksida, A.; Duh, D.; Wraber, B.; Dedushaj, I.; Ahmeti, S.; Avsic-Zupanc, T. Interacting roles of immune mechanisms and viral load in the pathogenesis of Crimean-Congo hemorrhagic fever. *Clin. Vaccine Immunol.* **2010**, *17*, 1086–1093. [CrossRef] [PubMed]

15. Papa, A.; Bino, S.; Velo, E.; Harxhi, A.; Kota, M.; Antoniadis, A. Cytokine levels in Crimean-Congo hemorrhagic fever. *J. Clin. Virol.* **2006**, *36*, 272–276. [CrossRef] [PubMed]
16. Onguru, P.; Dagdas, S.; Bodur, H.; Yilmaz, M.; Akinci, E.; Eren, S.; Ozet, G. Coagulopathy parameters in patients with Crimean-Congo hemorrhagic fever and its relation with mortality. *J. Clin. Lab. Anal.* **2010**, *24*, 163–166. [CrossRef] [PubMed]
17. Cevik, M.A.; Erbay, A.; Bodur, H.; Gülderen, E.; Baştuğ, A.; Kubar, A.; Akinci, E. Clinical and laboratory features of Crimean-Congo hemorrhagic fever: Predictors of fatality. *Int. J. Infect. Dis.* **2008**, *12*, 374–379. [CrossRef] [PubMed]
18. Cevik, M.A.; Erbay, A.; Bodur, H.; Eren, S.S.; Akinci, E.; Sener, K.; Ongürü, P.; Kubar, A. Viral load as a predictor of outcome in Crimean-Congo hemorrhagic fever. *Clin. Infect. Dis.* **2007**, *45*, e96–e100. [CrossRef] [PubMed]
19. Burt, F.J.; Swanepoel, R.; Shieh, W.J.; Smith, J.F.; Leman, P.A.; Greer, P.W.; Coffield, L.M.; Rollin, P.E.; Ksiazek, T.G.; Peters, C.J.; *et al.* Immunohistochemical and *in situ* localization of Crimean-Congo hemorrhagic fever (CCHF) virus in human tissues and implications for CCHF pathogenesis. *Arch. Pathol. Lab. Med.* **1997**, *121*, 839–846. [PubMed]
20. Peyrefitte, C.N.; Perret, M.; Garcia, S.; Rodrigues, R.; Bagnaud, A.; Lacote, S.; Crance, J.-M.; Vernet, G.; Garin, D.; Bouloy, M.; *et al.* Differential activation profiles of Crimean-Congo hemorrhagic fever virus- and Dugbe virus-infected antigen-presenting cells. *J. Gen. Virol.* **2010**, *91*, 189–198. [CrossRef] [PubMed]
21. Connolly-Andersen, A.-M.; Douagi, I.; Kraus, A.A.; Mirazimi, A. Crimean Congo hemorrhagic fever virus infects human monocyte-derived dendritic cells. *Virology* **2009**, *390*, 157–162. [CrossRef] [PubMed]
22. Connolly-Andersen, A.-M.; Moll, G.; Andersson, C.; Akerström, S.; Karlberg, H.; Douagi, I.; Mirazimi, A. Crimean-Congo hemorrhagic fever virus activates endothelial cells. *J. Virol.* **2011**, *85*, 7766–7774. [CrossRef] [PubMed]
23. ICTV. Virus Taxonomy 2014. Available online: http://www.ictvonline.org/virusTaxonomy.asp?bhcp=1 (accessed on 15 April 2016).
24. Hewlett, M.J.; Pettersson, R.F.; Baltimore, D. Circular forms of Uukuniemi virion RNA: An electron microscopic study. *J. Virol.* **1977**, *21*, 1085–1093. [PubMed]
25. Devignot, S.; Bergeron, E.; Nichol, S.; Mirazimi, A.; Weber, F. A Virus-like particle system identifies the endonuclease domain of Crimean-Congo hemorrhagic fever virus. *J. Virol.* **2015**, *89*, 5957–5967. [CrossRef] [PubMed]
26. Bergeron, E.; Albariño, C.G.; Khristova, M.L.; Nichol, S.T. Crimean-Congo hemorrhagic fever virus-encoded ovarian tumor protease activity is dispensable for virus RNA polymerase function. *J. Virol.* **2010**, *84*, 216–226. [CrossRef] [PubMed]
27. Zivcec, M.; Metcalfe, M.G.; Albariño, C.G.; Guerrero, L.W.; Pegan, S.D.; Spiropoulou, C.F.; Bergeron, É. Assessment of inhibitors of pathogenic Crimean-Congo hemorrhagic fever virus strains using virus-like particles. *PLoS Negl. Trop. Dis.* **2015**, *9*, e0004259. [CrossRef] [PubMed]
28. Barnwal, B.; Karlberg, H.; Mirazimi, A.; Tan, Y.-J. Non-structural protein of Crimean-Congo hemorrhagic fever virus disrupts mitochondrial membrane potential and induces apoptosis. *J. Biol. Chem.* **2015**, *291*, 582–592. [CrossRef] [PubMed]
29. Albariño, C.G.; Bird, B.H.; Nichol, S.T. A shared transcription termination signal on negative and ambisense RNA genome segments of Rift Valley fever, sandfly fever Sicilian, and Toscana viruses. *J. Virol.* **2007**, *81*, 5246–5256. [CrossRef] [PubMed]
30. Pinschewer, D.D.; Perez, M.; de La Torre, J.C. Dual role of the lymphocytic choriomeningitis virus intergenic region in transcription termination and virus propagation. *J. Virol.* **2005**, *79*, 4519–4526. [CrossRef] [PubMed]
31. Sanchez, A.J.; Vincent, M.J.; Nichol, S.T. Characterization of the glycoproteins of Crimean-Congo hemorrhagic fever virus. *J. Virol.* **2002**, *76*, 7263–7275. [CrossRef] [PubMed]
32. Zhou, Z.-R.; Wang, M.-L.; Deng, F.; Li, T.-X.; Hu, Z.-H.; Wang, H.-L. Production of CCHF virus-like particle by a baculovirus-insect cell expression system. *Virol. Sin.* **2011**, *26*, 338–346. [CrossRef] [PubMed]
33. Joubert, J.R.; King, J.B.; Rossouw, D.J.; Cooper, R. A nosocomial outbreak of Crimean-Congo haemorrhagic fever at Tygerberg Hospital Part III. Clinical pathology and pathogenesis. *S. Afr. Med. J.* **1985**, *68*, 722–728. [PubMed]

34. Hardestam, J.; Simon, M.; Hedlund, K.O.; Vaheri, A.; Klingström, J.; Lundkvist, A. *Ex vivo* stability of the rodent-borne Hantaan virus in comparison to that of arthropod-borne members of the Bunyaviridae family. *Appl. Environ. Microbiol.* **2007**, *73*, 2547–2551. [CrossRef] [PubMed]
35. Korolev, M.B.; Donets, M.A.; Rubin, S.G.; Chumakov, M.P. Morphology and morphogenesis of Crimean hemorrhagic fever virus. *Arch. Virol.* **1976**, *50*, 169–172. [CrossRef] [PubMed]
36. Donets, M.A.; Chumakov, M.P.; Korolev, M.B.; Rubin, S.G. Physicochemical characteristics, morphology and morphogenesis of virions of the causative agent of Crimean hemorrhagic fever. *Intervirology* **1977**, *8*, 294–308. [CrossRef] [PubMed]
37. Bertolotti-ciarlet, A.; Smith, J.; Strecker, K.; Paragas, J.; Altamura, L.A.; Mcfalls, J.M.; Frias-sta, N.; Schmaljohn, C.S.; Doms, R.W. Cellular localization and antigenic characterization of Crimean-Congo hemorrhagic fever virus glycoproteins. *J. Virol.* **2005**, *79*, 6152–6161. [CrossRef] [PubMed]
38. Garry, C.E.; Garry, R.F. Proteomics computational analyses suggest that the carboxyl terminal glycoproteins of Bunyaviruses are class II viral fusion protein (beta-penetrenes). *Theor. Biol. Med. Model.* **2004**, *1*, 10. [CrossRef] [PubMed]
39. Dessau, M.; Modis, Y. Crystal structure of glycoprotein C from Rift Valley fever virus. *Proc. Natl. Acad. Sci. USA* **2013**, *110*, 1696–1701. [CrossRef] [PubMed]
40. Xiao, X.; Feng, Y.; Zhu, Z.; Dimitrov, D.S. Identification of a putative Crimean-Congo hemorrhagic fever virus entry factor. *Biochem. Biophys. Res. Commun.* **2011**, *411*, 253–258. [CrossRef] [PubMed]
41. Tayyari, F.; Marchant, D.; Moraes, T.J.; Duan, W.; Mastrangelo, P.; Hegele, R.G. Identification of nucleolin as a cellular receptor for human respiratory syncytial virus. *Nat. Med.* **2011**, *17*, 1132–1135. [CrossRef] [PubMed]
42. Thongtan, T.; Wikan, N.; Wintachai, P.; Rattanarungsan, C.; Srisomsap, C.; Cheepsunthorn, P.; Smith, D.R. Characterization of putative Japanese encephalitis virus receptor molecules on microglial cells. *J. Med. Virol.* **2012**, *84*, 615–623. [CrossRef] [PubMed]
43. Tajrishi, M.M.; Tuteja, R.; Tuteja, N. Nucleolin: The most abundant multifunctional phosphoprotein of nucleolus. *Commun. Integr. Biol.* **2011**, *4*, 267–275. [CrossRef] [PubMed]
44. Ueno, T.; Tokunaga, K.; Sawa, H.; Maeda, M.; Chiba, J.; Kojima, A.; Hasegawa, H.; Shoya, Y.; Sata, T.; Kurata, T.; *et al.* Nucleolin and the packaging signal, psi, promote the budding of human immunodeficiency virus type-1 (HIV-1). *Microbiol. Immunol.* **2004**, *48*, 111–118. [CrossRef] [PubMed]
45. Simon, M.; Johansson, C.; Mirazimi, A. Crimean-Congo hemorrhagic fever virus entry and replication is clathrin-, pH- and cholesterol-dependent. *J. Gen. Virol.* **2009**, *90*, 210–215. [CrossRef] [PubMed]
46. Garrison, A.R.; Radoshitzky, S.R.; Kota, K.P.; Pegoraro, G.; Ruthel, G.; Kuhn, J.H.; Altamura, L.A.; Kwilas, S.A.; Bavari, S.; Haucke, V.; *et al.* Crimean-Congo hemorrhagic fever virus utilizes a clathrin- and early endosome-dependent entry pathway. *Virology* **2013**, *444*, 45–54. [CrossRef] [PubMed]
47. Shtanko, O.; Nikitina, R.A.; Altuntas, C.Z.; Chepurnov, A.A.; Davey, R.A. Crimean-Congo hemorrhagic fever virus entry into host cells occurs through the multivesicular body and requires ESCRT regulators. *PLoS Pathog.* **2014**, *10*, e1004390. [CrossRef] [PubMed]
48. Morin, B.; Coutard, B.; Lelke, M.; Ferron, F.; Kerber, R.; Jamal, S.; Frangeul, A.; Baronti, C.; Charrel, R.; de Lamballerie, X.; *et al.* The N-terminal domain of the arenavirus L protein is an RNA endonuclease essential in mRNA transcription. *PLoS Pathog.* **2010**, *6*, e1001038. [CrossRef] [PubMed]
49. Reguera, J.; Weber, F.; Cusack, S. Bunyaviridae RNA polymerases (L-protein) have an N-terminal, influenza-like endonuclease domain, essential for viral cap-dependent transcription. *PLoS Pathog.* **2010**, *6*, e10011101. [CrossRef] [PubMed]
50. Dias, A.; Bouvier, D.; Crépin, T.; McCarthy, A.A.; Hart, D.J.; Baudin, F.; Cusack, S.; Ruigrok, R.W.H. The cap-snatching endonuclease of influenza virus polymerase resides in the PA subunit. *Nature* **2009**, *458*, 914–918. [CrossRef] [PubMed]
51. Sanchez, A.J.; Vincent, M.J.; Erickson, B.R.; Nichol, S.T. Crimean-Congo hemorrhagic fever virus glycoprotein precursor is cleaved by Furin-like and SKI-1 proteases to generate a novel 38-kilodalton glycoprotein. *J. Virol.* **2006**, *80*, 514–525. [CrossRef] [PubMed]
52. Altamura, L.A.; Bertolotti-Ciarlet, A.; Teigler, J.; Paragas, J.; Schmaljohn, C.S.; Doms, R.W. Identification of a novel C-terminal cleavage of Crimean-Congo hemorrhagic fever virus PreGN that leads to generation of an NSM protein. *J. Virol.* **2007**, *81*, 6632–6642. [CrossRef] [PubMed]
53. Estrada, D.F.; De Guzman, R.N. Structural characterization of the Crimean-Congo hemorrhagic fever virus Gn tail provides insight into virus assembly. *J. Biol. Chem.* **2011**, *286*, 21678–21686. [CrossRef] [PubMed]

54. Bergeron, E.; Vincent, M.J.; Nichol, S.T. Crimean-Congo hemorrhagic fever virus glycoprotein processing by the endoprotease SKI-1/S1P is critical for virus infectivity. *J. Virol.* **2007**, *81*, 13271–13276. [CrossRef] [PubMed]
55. Erickson, B.R.; Deyde, V.; Sanchez, A.J.; Vincent, M.J.; Nichol, S.T. N-linked glycosylation of Gn (but not Gc) is important for Crimean Congo hemorrhagic fever virus glycoprotein localization and transport. *Virology* **2007**, *361*, 348–355. [CrossRef] [PubMed]
56. Bergeron, É.; Zivcec, M.; Chakrabarti, A.K.; Nichol, S.T.; Albariño, C.G.; Spiropoulou, C.F. Recovery of recombinant Crimean Congo hemorrhagic fever virus reveals a function for non-structural glycoproteins cleavage by furin. *PLoS Pathog.* **2015**, *11*, e1004879. [CrossRef] [PubMed]
57. Haferkamp, S.; Fernando, L.; Schwarz, T.F.; Feldmann, H.; Flick, R. Intracellular localization of Crimean-Congo hemorrhagic fever (CCHF) virus glycoproteins. *Virol. J.* **2005**, *2*, 42. [CrossRef] [PubMed]
58. Vincent, M.J.; Sanchez, A.J.; Erickson, B.R.; Basak, A.; Chretien, M.; Seidah, N.G.; Nichol, S.T. Crimean-Congo hemorrhagic fever virus glycoprotein proteolytic processing by subtilase SKI-1. *J. Virol.* **2003**, *77*, 8640–8649. [CrossRef] [PubMed]
59. Pasquato, A.; de Palma, J.R.; Galan, C.; Seidah, N.G.; Kunz, S. Viral envelope glycoprotein processing by proprotein convertases. *Antivir. Res.* **2013**, *99*, 49–60. [CrossRef] [PubMed]
60. Rojek, J.M.; Lee, A.M.; Nguyen, N.; Spiropoulou, C.F.; Kunz, S. Site 1 protease is required for proteolytic processing of the glycoproteins of the South American hemorrhagic fever viruses Junin, Machupo, and Guanarito. *J. Virol.* **2008**, *82*, 6045–6051. [CrossRef] [PubMed]
61. Andersson, I.; Simon, M.; Lundkvist, A.; Nilsson, M.; Holmström, A.; Elgh, F.; Mirazimi, A. Role of actin filaments in targeting of Crimean Congo hemorrhagic fever virus nucleocapsid protein to perinuclear regions of mammalian cells. *J. Med. Virol.* **2004**, *72*, 83–93. [CrossRef] [PubMed]
62. Andersson, I.; Bladh, L.; Mousavi-jazi, M.; Magnusson, K.; Lundkvist, A.; Haller, O.; Mirazimi, A. Human MxA protein inhibits the replication of Crimean-Congo hemorrhagic fever virus. *J. Virol.* **2004**, *78*, 4323–4329. [CrossRef] [PubMed]
63. Connolly-Andersen, A.M.; Magnusson, K.E.; Mirazimi, A. Basolateral entry and release of Crimean-Congo hemorrhagic fever virus in polarized MDCK-1 cells. *J. Virol.* **2007**, *81*, 2158–2164. [CrossRef] [PubMed]
64. Macleod, J.M.L.; Marmor, H.; García-Sastre, A.; Frias-Staheli, N. Mapping of the interaction domains of the Crimean-Congo hemorrhagic fever virus nucleocapsid protein. *J. Gen. Virol.* **2015**, *96*, 524–537. [CrossRef] [PubMed]
65. Carter, S.D.; Surtees, R.; Walter, C.T.; Ariza, A.; Bergeron, É.; Nichol, S.T.; Hiscox, J.A.; Edwards, T.A.; Barr, J.N. Structure, function, and evolution of the Crimean-Congo hemorrhagic fever virus nucleocapsid protein. *J. Virol.* **2012**, *86*, 10914–10923. [CrossRef] [PubMed]
66. Wang, W.; Liu, X.; Wang, X.; Dong, H.; Ma, C.; Wang, J.; Liu, B.; Mao, Y.; Wang, Y.; Li, T.; *et al.* Structural and functional diversity of nairovirus-encoded nucleoproteins. *J. Virol.* **2015**, *89*, 11740–11749. [CrossRef] [PubMed]
67. Guo, Y.; Wang, W.; Ji, W.; Deng, M.; Sun, Y.; Zhou, H.; Yang, C.; Deng, F.; Wang, H.; Hu, Z.; *et al.* Crimean-Congo hemorrhagic fever virus nucleoprotein reveals endonuclease activity in bunyaviruses. *Proc. Natl. Acad. Sci. USA* **2012**, *109*, 5046–5051. [CrossRef] [PubMed]
68. Hastie, K.M.; Liu, T.; Li, S.; King, L.B.; Ngo, N.; Zandonatti, M.A.; Woods, V.L.; de la Torre, J.C.; Saphire, E.O. Crystal structure of the Lassa virus nucleoprotein-RNA complex reveals a gating mechanism for RNA binding. *Proc. Natl. Acad. Sci. USA* **2011**, *108*, 19365–19370. [CrossRef] [PubMed]
69. Qi, X.; Lan, S.; Wang, W.; Schelde, L.M.; Dong, H.; Wallat, G.D.; Ly, H.; Liang, Y.; Dong, C. Cap binding and immune evasion revealed by Lassa nucleoprotein structure. *Nature* **2010**, *468*, 779–783. [CrossRef] [PubMed]
70. Raymond, D.D.; Piper, M.E.; Gerrard, S.R.; Smith, J.L. Structure of the Rift Valley fever virus nucleocapsid protein reveals another architecture for RNA encapsidation. *Proc. Natl. Acad. Sci. USA* **2010**, *107*, 11769–11774. [CrossRef] [PubMed]
71. Olal, D.; Dick, A.; Woods, V.L.; Liu, T.; Li, S.; Devignot, S.; Weber, F.; Saphire, E.O.; Daumke, O. Structural insights into RNA encapsidation and helical assembly of the Toscana virus nucleoprotein. *Nucleic Acids Res.* **2014**, *42*, 6025–6037. [CrossRef] [PubMed]
72. Reguera, J.; Malet, H.; Weber, F.; Cusack, S. Structural basis for encapsidation of genomic RNA by La Crosse Orthobunyavirus nucleoprotein. *Proc. Natl. Acad. Sci. USA* **2013**, *110*, 7246–7251. [CrossRef] [PubMed]

73. Wang, Y.; Dutta, S.; Karlberg, H.; Devignot, S.; Weber, F.; Hao, Q.; Tan, Y.J.; Mirazimi, A.; Kotaka, M. Structure of Crimean-Congo hemorrhagic fever virus nucleoprotein: Superhelical homo-oligomers and the role of caspase-3 cleavage. *J. Virol.* **2012**, *86*, 12294–12303. [CrossRef] [PubMed]
74. Burt, F.J.; Swanepoel, R.; Braack, L.E.O. Enzyme-linked immunosorbent assays for the detection of antibody to Crimean-Congo haemorrhagic fever virus in the sera of livestock and wild vertebrates. *Epidemiol. Infect.* **2009**, *111*, 547–558. [CrossRef]
75. Burt, F.J.; Samudzi, R.R.; Randall, C.; Pieters, D.; Vermeulen, J.; Knox, C.M. Human defined antigenic region on the nucleoprotein of Crimean-Congo hemorrhagic fever virus identified using truncated proteins and a bioinformatics approach. *J. Virol. Methods* **2013**, *193*, 706–712. [CrossRef] [PubMed]
76. Wei, P.; Luo, Y.; Li, T.; Wang, H.; Hu, Z.; Zhang, F.; Zhang, Y.; Deng, F.; Sun, S. Serial expression of the truncated fragments of the nucleocapsid protein of CCHFV and identification of the epitope region. *Virol. Sin.* **2010**, *25*, 45–51. [CrossRef] [PubMed]
77. Mousavi-Jazi, M.; Karlberg, H.; Papa, A.; Christova, I.; Mirazimi, A. Healthy individuals' immune response to the Bulgarian Crimean-Congo hemorrhagic fever virus vaccine. *Vaccine* **2012**, *30*, 6225–6229. [CrossRef] [PubMed]
78. Levine, J.R.; Prescott, J.; Brown, K.S.; Best, S.M.; Ebihara, H.; Feldmann, H. Antagonism of type I interferon responses by new world hantaviruses. *J. Virol.* **2010**, *84*, 11790–11801. [CrossRef] [PubMed]
79. Karlberg, H.; Tan, Y.J.; Mirazimi, A. Crimean-Congo haemorrhagic fever replication interplays with regulation mechanisms of apoptosis. *J. Gen. Virol.* **2015**, *96*, 538–546. [CrossRef] [PubMed]
80. Karlberg, H.; Tan, Y.-J.; Mirazimi, A. Induction of caspase activation and cleavage of the viral nucleocapsid protein in different cell types during Crimean-Congo hemorrhagic fever virus infection. *J. Biol. Chem.* **2011**, *286*, 3227–3234. [CrossRef] [PubMed]
81. Wolff, S.; Becker, S.; Groseth, A. Cleavage of the Junin virus nucleoprotein serves a decoy function to inhibit the induction of apoptosis during infection. *J. Virol.* **2013**, *87*, 224–233. [CrossRef] [PubMed]
82. Volchkov, V.E.; Feldmann, H.; Volchkova, V.A.; Klenk, H.D. Processing of the Ebola virus glycoprotein by the proprotein convertase furin. *Proc. Natl. Acad. Sci. USA* **1998**, *95*, 5762–5767. [CrossRef] [PubMed]
83. Volchkov, V.E.; Volchkova, V.A.; Ströher, U.; Becker, S.; Dolnik, O.; Cieplik, M.; Garten, W.; Klenk, H.D.; Feldmann, H. Proteolytic processing of Marburg virus glycoprotein. *Virology* **2000**, *268*, 1–6. [CrossRef] [PubMed]
84. Fusco, M.L.; Hashiguchi, T.; Cassan, R.; Biggins, J.E.; Murin, C.D.; Warfield, K.L.; Li, S.; Holtsberg, F.W.; Shulenin, S.; Vu, H.; *et al.* Protective mAbs and cross-reactive mAbs raised by immunization with engineered marburg virus GPs. *PLoS Pathog.* **2015**, *11*, e1005016.
85. Honig, J.E.; Osborne, J.C.; Nichol, S.T. Crimean-Congo hemorrhagic fever virus genome L RNA segment and encoded protein. *Virology* **2004**, *321*, 29–35. [CrossRef] [PubMed]
86. Kinsella, E.; Martin, S.G.; Grolla, A.; Czub, M.; Feldmann, H.; Flick, R. Sequence determination of the Crimean-Congo hemorrhagic fever virus L segment. *Virology* **2004**, *321*, 23–28. [CrossRef] [PubMed]
87. Habjan, M.; Andersson, I.; Klingström, J.; Schümann, M.; Martin, A.; Zimmermann, P.; Wagner, V.; Pichlmair, A.; Schneider, U.; Mühlberger, E.; *et al.* Processing of genome 5′ termini as a strategy of negative-strand RNA viruses to avoid RIG-I-dependent interferon induction. *PLoS ONE* **2008**, *3*, 1–8. [CrossRef] [PubMed]
88. Spengler, J.R.; Patel, J.R.; Chakrabarti, A.K.; Zivcec, M.; García-Sastre, A.; Spiropoulou, C.F.; Bergeron, E. RIG-I mediates an antiviral response to Crimean-Congo hemorrhagic fever virus. *J. Virol.* **2015**, *89*, 10119–10229. [CrossRef] [PubMed]
89. Garcin, D.; Lezzi, M.; Dobbs, M.; Elliott, R.M.; Schmaljohn, C.; Kang, C.Y.; Kolakofsky, D. The 5′ ends of Hantaan virus (Bunyaviridae) RNAs suggest a prime-and-realign mechanism for the initiation of RNA synthesis. *J. Virol.* **1995**, *69*, 5754–5762. [PubMed]
90. Yao, H.; Dittmann, M.; Peisley, A.; Hoffmann, H.H.; Gilmore, R.H.; Schmidt, T.; Schmid-Burgk, J.L.; Hornung, V.; Rice, C.M.; Hur, S. ATP-Dependent effector-like functions of RIG-I-like receptors. *Mol. Cell* **2015**, *58*, 541–548. [CrossRef] [PubMed]
91. Lee, M.-H.; Lalwani, P.; Raftery, M.J.; Matthaei, M.; Lutteke, N.; Kirsanovs, S.; Binder, M.; Ulrich, R.G.; Giese, T.; Wolff, T.; *et al.* RNA helicase retinoic acid-inducible gene I (RIG-I) as a sensor of Hantaan virus (HTNV) replication. *J. Gen. Virol.* **2011**, *92*, 2191–2200. [CrossRef] [PubMed]

92. Frias-Staheli, N.; Giannakopoulos, N.V.; Kikkert, M.; Taylor, S.L.; Bridgen, A.; Paragas, J.; Richt, J.A.; Rowland, R.R.; Schmaljohn, C.S.; Lenschow, D.J.; *et al.* Ovarian tumor domain-containing viral proteases evade ubiquitin- and ISG15-dependent innate immune responses. *Cell Host Microbe* **2007**, *2*, 404–416. [CrossRef] [PubMed]
93. James, T.W.; Frias-Staheli, N.; Bacik, J.-P.; Levingston Macleod, J.M.; Khajehpour, M.; García-Sastre, A.; Mark, B.L. Structural basis for the removal of ubiquitin and interferon-stimulated gene 15 by a viral ovarian tumor domain-containing protease. *Proc. Natl. Acad. Sci. USA* **2011**, *108*, 2222–2227. [CrossRef] [PubMed]
94. Van Kasteren, P.B.; Beugeling, C.; Ninaber, D.K.; Frias-Staheli, N.; van Boheemen, S.; García-Sastre, A.; Snijder, E.J.; Kikkert, M. Arterivirus and nairovirus ovarian tumor domain-containing Deubiquitinases target activated RIG-I to control innate immune signaling. *J. Virol.* **2012**, *86*, 773–785. [CrossRef] [PubMed]
95. Makarova, K. A novel superfamily of predicted cysteine proteases from eukaryotes, viruses and Chlamydia pneumoniae. *Trends Biochem. Sci.* **2000**, *25*, 50–52. [CrossRef]
96. Holzer, B.; Bakshi, S.; Bridgen, A.; Baron, M.D. Inhibition of interferon induction and action by the nairovirus Nairobi sheep disease virus/Ganjam virus. *PLoS ONE* **2011**, *6*, e28594. [CrossRef] [PubMed]
97. Mielech, A.M.; Kilianski, A.; Baez-Santos, Y.M.; Mesecar, A.D.; Baker, S.C. MERS-CoV papain-like protease has deISGylating and deubiquitinating activities. *Virology* **2014**, *450–451*, 64–70. [CrossRef] [PubMed]
98. Zheng, D.; Chen, G.; Guo, B.; Cheng, G.; Tang, H. PLP2, a potent deubiquitinase from murine hepatitis virus, strongly inhibits cellular type I interferon production. *Cell Res.* **2008**, *18*, 1105–1113. [PubMed]
99. Clementz, M.A.; Chen, Z.; Banach, B.S.; Wang, Y.; Sun, L.; Ratia, K.; Baez-Santos, Y.M.; Wang, J.; Takayama, J.; Ghosh, A.K.; *et al.* Deubiquitinating and interferon antagonism activities of coronavirus papain-like proteases. *J. Virol.* **2010**, *84*, 4619–4629. [CrossRef] [PubMed]
100. Zhang, H.-M.; Yang, J.; Sun, H.-R.; Xin, X.; Wang, H.-D.; Chen, J.-P.; Adams, M.J. Genomic analysis of rice stripe virus Zhejiang isolate shows the presence of an OTU-like domain in the RNA1 protein and a novel sequence motif conserved within the intergenic regions of ambisense segments of tenuiviruses. *Arch. Virol.* **2007**, *152*, 1917–1923. [CrossRef] [PubMed]
101. Lombardi, C.; Ayach, M.; Beaurepaire, L.; Chenon, M.; Andreani, J.; Guerois, R.; Jupin, I.; Bressanelli, S. A compact viral processing proteinase/ubiquitin hydrolase from the OTU family. *PLoS Pathog.* **2013**, *9*, e1003560. [CrossRef] [PubMed]
102. Kayagaki, N.; Phung, Q.; Chan, S.; Chaudhari, R.; Quan, C.; O'Rourke, K.M.; Eby, M.; Pietras, E.; Cheng, G.; Bazan, J.F.; *et al.* DUBA: A deubiquitinase that regulates type I interferon production. *Science* **2007**, *318*, 1628–1632. [CrossRef] [PubMed]
103. Jiang, X.; Kinch, L.N.; Brautigam, C.A.; Chen, X.; Du, F.; Grishin, N.V.; Chen, Z.J. Ubiquitin-induced oligomerization of the RNA sensors RIG-I and MDA5 activates antiviral innate immune response. *Immunity* **2012**, *36*, 973–959. [CrossRef] [PubMed]
104. Zhao, C.; Denison, C.; Huibregtse, J.M.; Gygi, S.; Krug, R.M. Human ISG15 conjugation targets both IFN-induced and constitutively expressed proteins functioning in diverse cellular pathways. *Proc. Natl. Acad. Sci. USA* **2005**, *102*, 10200–10205. [CrossRef] [PubMed]
105. Okumura, F.; Okumura, A.J.; Uematsu, K.; Hatakeyama, S.; Zhang, D.E.; Kamura, T. Activation of double-stranded RNA-activated protein kinase (PKR) by interferon-stimulated gene 15 (ISG15) modification down-regulates protein translation. *J. Biol. Chem.* **2013**, *288*, 2839–2847. [CrossRef] [PubMed]
106. Wang, P.; Zhao, W.; Zhao, K.; Zhang, L.; Gao, C. TRIM26 negatively regulates interferon-β production and antiviral response through polyubiquitination and degradation of nuclear IRF3. *PLoS Pathog.* **2015**, *11*, e1004726. [CrossRef] [PubMed]
107. Shi, H.-X.; Yang, K.; Liu, X.; Liu, X.-Y.; Wei, B.; Shan, Y.-F.; Zhu, L.-H.; Wang, C. Positive regulation of interferon regulatory factor 3 activation by Herc5 via ISG15 modification. *Mol. Cell. Biol.* **2010**, *30*, 2424–2436. [CrossRef] [PubMed]
108. Giannakopoulos, N.V.; Luo, J.-K.; Papov, V.; Zou, W.; Lenschow, D.J.; Jacobs, B.S.; Borden, E.C.; Li, J.; Virgin, H.W.; Zhang, D.-E. Proteomic identification of proteins conjugated to ISG15 in mouse and human cells. *Biochem. Biophys. Res. Commun.* **2005**, *336*, 496–506. [CrossRef] [PubMed]
109. Akutsu, M.; Ye, Y.; Virdee, S.; Chin, J.W.; Komander, D. Molecular basis for ubiquitin and ISG15 cross-reactivity in viral ovarian tumor domains. *Proc. Natl. Acad. Sci. USA* **2011**, *108*, 2228–2233. [CrossRef] [PubMed]

110. Capodagli, G.C.; McKercher, M.A.; Baker, E.A.; Masters, E.M.; Brunzelle, J.S.; Pegan, S.D. Structural analysis of a viral ovarian tumor domain protease from the Crimean-Congo hemorrhagic fever virus in complex with covalently bonded ubiquitin. *J. Virol.* **2011**, *85*, 3621–3630. [CrossRef] [PubMed]
111. Kim, M.-J.; Hwang, S.-Y.; Imaizumi, T.; Yoo, J.-Y. Negative feedback regulation of RIG-I-mediated antiviral signaling by interferon-induced ISG15 conjugation. *J. Virol.* **2008**, *82*, 1474–1483. [CrossRef] [PubMed]
112. Kohl, A.; Dunn, E.F.; Lowen, A.C.; Elliott, R.M. Complementarity, sequence and structural elements within the 3′ and 5′ non-coding regions of the Bunyamwera orthobunyavirus S segment determine promoter strength. *J. Gen. Virol.* **2004**, *85*, 3269–3278. [CrossRef] [PubMed]
113. Barr, J.N.; Wertz, G.W. Role of the conserved nucleotide mismatch within 3′- and 5′-terminal regions of Bunyamwera virus in signaling transcription. *J. Virol.* **2005**, *79*, 3586–3594. [CrossRef] [PubMed]
114. Duygu, F.; Kaya, T.; Baysan, P. Re-evaluation of 400 Crimean-Congo hemorrhagic fever cases in an endemic area: Is ribavirin treatment suitable? *Vector Borne Zoonotic Dis.* **2012**, *12*, 812–816. [CrossRef] [PubMed]
115. Koksal, I.; Yilmaz, G.; Aksoy, F.; Aydin, H.; Yavuz, I.; Iskender, S.; Akcay, K.; Erensoy, S.; Caylan, R.; Aydin, K. The efficacy of ribavirin in the treatment of Crimean-Congo hemorrhagic fever in Eastern Black Sea region in Turkey. *J. Clin. Virol.* **2010**, *47*, 65–68. [CrossRef] [PubMed]
116. Soares-Weiser, K.; Thomas, S.; Thomson, G.; Garner, P. Ribavirin for Crimean-Congo hemorrhagic fever: Systematic review and meta-analysis. *BMC Infect. Dis.* **2010**, *10*, 207. [CrossRef] [PubMed]
117. Bell-Sakyi, L.; Kohl, A.; Bente, D.A.; Fazakerley, J.K. Tick cell lines for study of Crimean-Congo hemorrhagic fever virus and other arboviruses. *Vector Borne Zoonotic Dis.* **2012**, *12*, 769–781. [CrossRef] [PubMed]
118. Gargili, A.; Thangamani, S.; Bente, D. Influence of laboratory animal hosts on the life cycle of Hyalomma marginatum and implications for an *in vivo* transmission model for Crimean-Congo hemorrhagic fever virus. *Front. Cell. Infect. Microbiol.* **2013**, *3*, 39. [CrossRef] [PubMed]
119. Bente, D.A.; Alimonti, J.B.; Shieh, W.-J.; Camus, G.; Ströher, U.; Zaki, S.; Jones, S.M. Pathogenesis and immune response of Crimean-Congo hemorrhagic fever virus in a STAT-1 knockout mouse model. *J. Virol.* **2010**, *84*, 11089–11100. [CrossRef] [PubMed]
120. Oestereich, L.; Rieger, T.; Neumann, M.; Bernreuther, C.; Lehmann, M.; Krasemann, S.; Wurr, S.; Emmerich, P.; de Lamballerie, X.; Ölschläger, S.; *et al.* Evaluation of antiviral efficacy of ribavirin, arbidol, and T-705 (favipiravir) in a mouse model for Crimean-Congo hemorrhagic fever. *PLoS Negl. Trop. Dis.* **2014**, *8*, e2804. [CrossRef] [PubMed]
121. Dowall, S.D.; Buttigieg, K.R.; Findlay-Wilson, S.J.D.; Rayner, E.; Pearson, G.; Miloszewska, A.; Graham, V.A.; Carroll, M.; Hewson, R. A Crimean-Congo Haemorrhagic Fever (CCHF) viral vaccine expressing nucleoprotein is immunogenic but fails to confer protection against lethal disease. *Hum. Vaccin. Immunother.* **2016**. [CrossRef] [PubMed]
122. Kortekaas, J.; Vloet, R.P.M.; McAuley, A.J.; Shen, X.; Bosch, B.J.; de Vries, L.; Moormann, R.J.M.; Bente, D.A. Crimean-Congo hemorrhagic fever virus subunit vaccines induce high levels of neutralizing Antibodies but no protection in STAT1 knockout mice. *Vector Borne Zoonotic Dis.* **2015**, *15*, 759–764. [CrossRef] [PubMed]
123. Canakoglu, N.; Berber, E.; Tonbak, S.; Ertek, M.; Sozdutmaz, I.; Aktas, M.; Kalkan, A.; Ozdarendeli, A. Immunization of knock-out α/β interferon receptor mice against high lethal dose of Crimean-Congo hemorrhagic fever virus with a cell culture based vaccine. *PLoS Negl. Trop. Dis.* **2015**, *11*, e0003579. [CrossRef] [PubMed]
124. Buttigieg, K.R.; Dowall, S.D.; Findlay-Wilson, S.; Miloszewska, A.; Rayner, E.; Hewson, R.; Carroll, M.W. A novel vaccine against Crimean-Congo haemorrhagic fever protects 100% of animals against lethal challenge in a mouse model. *PLoS ONE* **2014**, *9*, e91596. [CrossRef] [PubMed]
125. Andersson, I.; Lundkvist, Å.; Haller, O.; Mirazimi, A. Type I Interferon inhibits Crimean-Congo hemorrhagic fever virus in human target cells. *J. Med. Virol.* **2006**, *222*, 216–222. [CrossRef] [PubMed]

© 2016 by the authors. Licensee MDPI, Basel, Switzerland. This article is an open access article distributed under the terms and conditions of the Creative Commons Attribution (CC BY) license (http://creativecommons.org/licenses/by/4.0/).

Article

Experimental Infection of Calves by Two Genetically-Distinct Strains of Rift Valley Fever Virus

William C. Wilson [1,†,*], A. Sally Davis [2,†], Natasha N. Gaudreault [1,3], Bonto Faburay [2], Jessie D. Trujillo [2], Vinay Shivanna [2], Sun Young Sunwoo [2], Aaron Balogh [2], Abaineh Endalew [2], Wenjun Ma [2], Barbara S. Drolet [1], Mark G. Ruder [1,4], Igor Morozov [2], D. Scott McVey [1] and Juergen A. Richt [2,*]

1. United States Department of Agriculture, Agricultural Research Service, Arthropod Borne Animal Disease Research Unit, 1515 College Ave., Manhattan, KS 66502, USA; nng5757@vet.k-state.edu (N.N.G.); Barbara.Drolet@ars.usda.gov (B.S.D.); mgruder@uga.edu (M.G.R.); Scott.McVey@ars.usda.gov (D.S.M.)
2. Department of Diagnostic Medicine/Pathobiology, College of Veterinary Medicine, Kansas State University, Manhattan, KS 66502, USA; asally@vet.k-state.edu (A.S.D.); bfaburay@vet.k-state.edu (B.F.); jdtrujillo@vet.k-state.edu (J.D.T.); vinays@vet.k-state.edu (V.S.); sunwoo@ksu.edu (S.Y.S.); balogh@vet.k-state.edu (A.B.); adendale@vet.k-state.edu (A.E.); wma@vet.k-state.edu (W.M.); imorozov@vet.k-state.edu (I.M.)
3. Department of Diagnostic Medicine/Pathobiology, College of Veterinary Medicine, Kansas State University, Manhattan, KS 66502, USA
4. Southeastern Cooperative Wildlife Disease Study, College of Veterinary Medicine, University of Georgia, Athens, GA 30602, USA

* Correspondence: William.Wilson@ars.usda.gov (W.C.W.); jricht@ksu.edu (J.A.R.); Tel.: +1-785-537-5570 (W.C.W.); +1-785-532-4408 (J.A.R.); Fax: +1-785-537-5560 (W.C.W.)

† These authors contributed equally to this work.

Academic Editors: Jane Tao and Pierre-Yves Lozach
Received: 1 April 2016; Accepted: 12 May 2016; Published: 23 May 2016

Abstract: Recent outbreaks of Rift Valley fever in ruminant livestock, characterized by mass abortion and high mortality rates in neonates, have raised international interest in improving vaccine control strategies. Previously, we developed a reliable challenge model for sheep that improves the evaluation of existing and novel vaccines in sheep. This sheep model demonstrated differences in the pathogenesis of Rift Valley fever virus (RVFV) infection between two genetically-distinct wild-type strains of the virus, Saudi Arabia 2001 (SA01) and Kenya 2006 (Ken06). Here, we evaluated the pathogenicity of these two RVFV strains in mixed breed beef calves. There was a transient increase in rectal temperatures with both virus strains, but this clinical sign was less consistent than previously reported with sheep. Three of the five Ken06-infected animals had an early-onset viremia, one day post-infection (dpi), with viremia lasting at least three days. The same number of SA01-infected animals developed viremia at 2 dpi, but it only persisted through 3 dpi in one animal. The average virus titer for the SA01-infected calves was 1.6 logs less than for the Ken06-infected calves. Calves, inoculated with either strain, seroconverted by 5 dpi and showed time-dependent increases in their virus-neutralizing antibody titers. Consistent with the results obtained in the previous sheep study, elevated liver enzyme levels, more severe liver pathology and higher virus titers occurred with the Ken06 strain as compared to the SA01 strain. These results demonstrate the establishment of a virulent challenge model for vaccine evaluation in calves.

Keywords: Challenge model; Rift Valley fever; Rift Valley fever virus; Cattle; pathogenicity

1. Introduction

The mosquito-borne Rift Valley fever virus (RVFV) is a zoonotic pathogen within the genus *Phlebovirus* family *Bunyaviridae*. The virus has a tripartite, single-stranded negative-sense RNA genome. The large segment (L) encodes the RNA-dependent RNA polymerase. The medium segment (M) encodes two major envelope glycoproteins, Gn and Gc, and a non-structural protein, NSm. The S-segment utilizes an ambisense strategy to encode the nucleocapsid protein and the non-structural protein, NSs, which has been shown to be an interferon antagonist [1]. Although large outbreaks have predominantly occurred in Sub-Saharan Africa, outbreaks outside of the African continent, in the Arabian Peninsula, have raised concerns about the potential spread of the virus to Europe and the Americas [2,3]. These concerns are warranted given that North America has large, economically-important populations of susceptible animals and indigenous vector populations, experimentally shown to be competent for virus infection and transmission [4–6]. Typically, outbreaks of disease are characterized by abortion storms and high rates of mortality in young sheep, goats and cattle [7,8]. Livestock are considered to be the amplifying and/or reservoir hosts for human infections. Thus, outbreaks in livestock lead to human infections, resulting in acute febrile illness that in some cases can progress to more severe disease, including retinal vasculitis, resulting in blindness, encephalitis and fatal hepatitis with hemorrhagic fever [9]. Although reported human case fatality rates are generally low, higher fatality rates (20%–40%) have been noted [10,11].

RVFV is a Category A pathogen [12], and the potential for RVFV as a bioterrorism agent is widely recognized [13]. Therefore, the intentional or unintentional introduction of RVFV into the U.S. is both an agricultural and public health concern and requires a "One Health" approach. One primary control method is through livestock vaccination [14,15]. Killed or live attenuated vaccines are currently used in endemic countries. Historically, the commonly-used veterinary vaccine is an attenuated vaccine developed in 1949 known as the Smithburn strain [16]. A more recently introduced attenuated vaccine is Clone 13, which has a deletion in the NSs interferon antagonist gene and has been experimentally proven to be safe and efficacious [17,18]. Another attenuated vaccine that is considered to be safe is MP12 [19–21], which is conditionally licensed for use in the U.S. [22]. MP12 has also been evaluated for potential human use [23,24]. There are concerns with the use of attenuated vaccines due to potential teratogenicity [25] and potential reversion to virulence or reassortment with RVFV field strains. A number of novel vaccine approaches are in development [1], the most recent being an adenovirus-vectored vaccine [26]. With multiple inactivated, live attenuated and other new generation potential vaccines, there is a critical need for consistent, reliable vaccine efficacy animal models based on the target species (*i.e.*, sheep, goats or cattle). Animal models are available [27], and because abortion is a hallmark of RVFV of livestock, a common model is the pregnant ewe [28]. However, the pregnancy synchronization of ewes and the scheduling of limited high biosecurity animal space makes this model logistically difficult. Another model that has proven useful is vaccine-age (~4-month-old) sheep and goats [29]. This study involved different breeds of sheep (Suffolk cross, Rideau Arcott cross, Ile-de-France cross with Rideau Arcott) and a human-derived virus, the Egyptian 1977 virus strain ZH501. Although no severe pathology was reported in the ZH501-infected sheep, viremia and febrile response allowed for the evaluation of vaccine efficacy [30].

We recently adopted this model to evaluate the pathogenicity of two genetically-distinct RVFV strains in sheep [31]. The first strain, Saudi Arabia 2001 (SA01), was originally isolated from *Aedes vexans arabiensis* during the 2000–2001 outbreak in the Kingdom of Saudi Arabia [32]. This outbreak had higher human case fatality rates than previously noted (13.9%–33.9%) [33]. The second strain, Kenya 2006 (Ken06), was isolated from *Aedes ochraceus* from Kenya in 2006 [34]. While Ken06 was found to be closely related to viruses isolated in 1991 in Madagascar and in 1997 in Kenya [32], it is genetically distinct from these viruses [31]. Our sheep challenge model for SA01 and Ken06 RVFV demonstrated differences in viremia, microscopic and macroscopic pathological changes and aberrations in hematology and liver enzyme chemistry values. The aim of the present study was to further compare and confirm these virulence differences by evaluating SA01 and Ken06 infections in another important target species, cattle.

2. Materials and Methods

2.1. Virus Strains and Cell Culture

The RVFV Saudi Arabia 2000–2001 (SA01) [34] and Kenya 2006–2007 (Ken06) isolates [32] were provided by Barry Miller, Centers for Disease Control, Fort Collins, CO, through Richard Bowen, Colorado State University. The two virus strains were passaged once in a C6/36 *Aedes albopictus* cell line (ATCC, Manassas, VA, USA) at 30 °C with MEM culture medium (Life Technologies, Grand Island, NY, USA) supplemented with 10% fetal bovine serum (FBS; Sigma-Aldrich, St. Louis, MO, USA) and 1× antibiotic-antimycotics (Penicillin/Streptomycin/Amphotericin B (PSF); Gibco, Watham, MA, USA). Plaque assays were performed using the Vero MARU (Middle America Research Unit, Panama) cell line grown in M199E culture medium (Sigma-Aldrich, St. Louis, MO, USA) supplemented with 10% FBS and 1× PSF and maintained in a 37 °C, 5% CO_2 incubator.

2.2. Animals and Experimental Design

Twelve healthy Hereford cross or Angus cross cattle, 4–5 months old, were obtained from private breeders in Kansas, USA. The animals were acclimated for one week at the Large Animal Research Center (Kansas State University (KSU), Manhattan, KS, USA) prior to relocation to a Biosafety Level-3 Agriculture (BSL-3Ag) facility at the KSU Biosecurity Research Institute for the virus inoculation experiment. In the BSL-3Ag facility, the animals were divided into two experimental groups of five each. Two additional animals were included as mock-inoculated controls. Cattle in each experimental group were inoculated subcutaneously with 2 mL of 1×10^6 plaque-forming units (pfu)/mL of the Ken06 or SA01 strain. After RVFV exposure, all animals were monitored daily for temperature changes and clinical signs. Nasal swabs for virological analysis and blood samples for virological, immunological, and blood chemistry analyses were collected daily from Days 0–10 and additionally at 14 and 21 days post-infection (dpi). One animal per experimental infection group was euthanized and necropsied at 3, 4, 5, 10 and 21 dpi. Animals were randomly pre-selected for necropsy; however, severe clinical illness required one animal to be substituted earlier in the schedule. The two mock-inoculated control cattle were necropsied at 20 dpi in order to allow for complete and thorough necropsies on all animals under BSL-3Ag conditions. Tissues were collected for viral titer determinations and histopathology. The research was performed under an Institutional Animal Care and Use Committee-approved protocol of KSU in compliance with the Animal Welfare Act and other regulations relating to animals and experiments involving animals.

2.3. Virus Isolation and Plaque Titration

Tissue samples of brain, kidney, liver and spleen were collected at necropsy and frozen at −80 °C. Approximately 10 mg of tissue were added to 1 mL M199E supplemented with 10% FBS and 1× PSF and homogenized by high-speed shaking dissociation with steel beads using the TissueLyser instrument (QIAGEN Inc.; Valencia, CA, USA). Virus stocks, cattle sera, nasal swabs and homogenized tissue samples were titered by standard plaque assay on Vero MARU cells, as previously described [35]. Briefly, confluent cell monolayers were inoculated with ten-fold serial diluted samples in M199E and incubated for 1 h. Following adsorption, the inocula were replaced with a 1:1 mixture of 2% carboxymethyl cellulose (Sigma-Aldrich, St. Louis, MO, USA) in 2× M199E (2% FBS and 2× PSF) and returned to the incubator. After five days, cells were fixed and stained with crystal violet fixative (25% formaldehyde, 10% ethanol, 5% acetic acid, 1% crystal violet).

2.4. Viral RNA Extraction and Real-Time RT-PCR

Total RNA from serum, nasal swabs or homogenized tissue samples was extracted using TRIzol-LS reagent (Life Technologies, Grand Island, NY, USA) and the magnetic-bead capture MagMAX-96 total RNA Isolation kit (Life Technologies). Briefly, 100 μL of aqueous phase were added to 90 μL of isopropanol and 10 μL bead mix (Beckman Coulter, Danvers, MA, USA). Total sample RNA was washed four times with wash buffer (150 μL) and then eluted in 30 μL of elution buffer. A quadruplex real-time reverse transcriptase-polymerase chain reaction (RT-PCR) assay was used to detect each of the three RVFV RNA genome segments and an external RNA control [36].

2.4.1. RNA Copy Number Determination

In vitro transcribed RNA (IVT RNA) was generated using the T7 transcription kit (MEGAscript, ThermoFisher) from a PCR-generated amplicon derived from a DNA plasmid (pBluescript III) using cDNA SuperMix (Quanta Biosciences) and T7 promoter and terminator primers (Integrated Technologies). The RVFV L plasmid (provided by Hana Weingartl, National Centre for Foreign Animal Diseases, Canadian Food Inspection Agency, Manitoba, Canada) contains 3482 base pairs of the L segment of RVFV (nucleotides 1–3482 of the ZH501 RVFV strain). IVT RNA was DNAse treated 3×, column purified (MEGAclear, ThermoFisher) and quantitated with spectrophotometry. The copy number was calculated using an online calculator [37]. Ten-fold serial dilutions of IVT stock RNA (10^4–10^{-1} copies) were utilized to generate a six-point standard curve using six PCR well replicates per dilution using quantitative RVFV real-time RT-PCR [36]. Copy numbers for samples were mathematically determined using the PCR-determined mean Ct for the L segment (three PCR well replicates) and the slope and intercept of the L segment IVT RNA standard curve. Data are reported as PCR-determined copy number per reaction. Calculated copy numbers less than 15 (equivalent to Ct greater than 35) are considered past the limits of detection for this assay, are classified as equivocal and, thus, are not reported as true positives.

2.5. RVFV Serology

2.5.1. Anti-RVFV IgG Antibody Response

The serum was inactivated prior to serological testing by adjusting to 0.25% Tween and incubating at 60 °C for 2 h. Each sample was safety tested by 3 blind passes in cell culture. Only samples that demonstrated no cytopathic effect on the third passage were removed for serology. Anti-RVFV antibody response was measured by an anti-RVFV total IgG indirect enzyme linked immunosorbent assay (ELISA) using recombinant baculovirus-expressed RVFV Gn and N proteins as diagnostic antigens. Briefly, each plate was coated overnight at 4 °C with approximately 150 ng of each purified recombinant protein/antigen, and the ELISA was performed as previously described [38]. The cut-off point for the specific ELISAs was determined by the addition of three standard deviations to the corresponding mean OD value of the pre-vaccination for all of the animals and the control serum. Mean OD values equal to or greater than the cut-off value were considered positive.

2.5.2. Plaque Reduction Neutralization Test

To assess the anti-RVFV neutralizing antibody response to RVFV inoculation, a plaque reduction neutralization test was performed as previously described [39]. Briefly, the stock of MP12 RVFV was diluted to 50 pfu in 250 μL of 1× MEM containing 4% bovine serum albumin (BSA; Sigma-Aldrich). Separately, aliquots of serum from each animal were serially diluted from 1:10–1:1280 in 1× MEM containing 2% BSA and 1% penicillin-streptomycin (Gibco). Diluted serum (250 μL) was mixed with an equal volume of diluted MP12 virus and incubated at 37 °C for 1 h. Thereafter, each mixture of serum plus RVFV was used to infect confluent monolayers of Vero E6 cells (ATCC, Manassas, VA, USA) in 12-well plates. After 1 h of adsorption at 37 °C and 5% CO_2, the mixture was removed, and 1.5 mL of nutrient agarose overlay (1× MEM, 4% FBS and 0.9% SeaPlaque agar (Lonza Rockland

Inc., Rockland, ME, USA)) were added to the monolayers. After 4 days of incubation, the cells were fixed with 10% neutral buffered formalin for 3 h prior to removal of the agarose overlay. The monolayer was stained with 0.5% crystal violet in PBS, and plaques were enumerated. The calculated plaque reduction neutralization test ($PRNT_{80}$) corresponded to the reciprocal titer of the highest serum dilution, which reduced the number of plaques by 80% or more relative to the virus control. As the positive neutralizing serum control, a 1:40 dilution of Day 28 neutralizing serum (titer >1280) obtained from sheep previously immunized with the RVFV glycoprotein subunit vaccine was used [39].

2.6. Blood Chemistry Analyses

An aliquot of serum was frozen immediately upon processing for later comprehensive large animal diagnostic panel (ALB, ALP, BUN, CA, CK, GGT, GLOB, MG, PHOS and TP) analyses with a Vetscan VS2 instrument (Abaxis, Union City, CA, USA) according to the manufacturer's instructions. Vendor-provided normal ranges for cattle were used as reference values.

2.7. Liver Pathology

Cattle liver samples were collected at necropsy and placed in 10% neutral buffered formalin for at least 7 days. Liver tissue was trimmed, processed and paraffin-embedded. All histochemical stains and immunohistochemistry were done on 4-μm sections placed on positively-charged slides. Hematoxylin and eosin (H&E)-stained tissues were reviewed by a veterinary pathologist in a blinded fashion and the liver pathology scored per slide (2–3 slides, minimum 3 tissues sections per animal) for lesion severity on a semi-quantitative scale from 0–4, where 0 signified no lesions and 1–4 progressively more severe pathology with a greater degree of liver parenchyma involvement (Table 1). Immunohistochemistry (IHC) for RVFV antigen using the polyclonal rabbit anti-RVFV nucleocapsid protein antibody [40] and an avidin-biotinylated peroxidase complex (ABC) detection technique was conducted on all liver sections as described [31], with the addition of a chromogen-enhancing step. Briefly, slides were deparaffinized and rehydrated, the antigen retrieved using a vegetable steamer technique in pH 6.0 citrate buffer with detergents (DAKO, Carpinteria, CA, USA) for 20 min, blocked with 3% hydrogen peroxide for 10 min, serum blocked as per the kit, incubated overnight at 4 °C with a 1:1000 dilution of primary antibody, secondary antibody, and ABC reagent was applied as per kit, with 3,3'-Diaminobenzidine (DAB) followed by DAB enhancing solution applied as per the vendor instructions (Vector Labs), counterstained with hematoxylin and mounted in Permount (Electron Microscopy Systems, Hatfield, PA, USA). Throughout, the following controls were employed: reagent control slides, with and without equivalent concentrations of primary antibody matched animal serum, and uninfected control cattle liver. Additionally, a Hall's histochemical stain for bilirubin was run on select liver sections. All gross tissue images were captured with a Canon G12 camera (Cannon, USA Inc., Melville, NY, USA), and microscopic images were captured with a DP25 camera (Olympus, Tokyo, Japan) on a BX46 light microscope (Olympus) using CellSens Standard Version 1.12 (Olympus). All microscopic images were further color calibrated using ChromaCal software ver 2.5 (Datacolor Inc., Lawrenceville, NJ, USA) as per the manufacturer's instructions and published recommendations (Linden, Sedgewick and Ericson, 2015); the figure panels were composed in Adobe Photoshop and InDesign CC (Adobe, San Jose, CA, USA).

Table 1. Liver histopathology score descriptions.

Histopathology Score	Description
0	Multifocal, peri-portal, mild lymphoplasmacytic (lymphocytes and plasma cells) inflammation (background lesion)
1	Multifocal, mid-zonal to central foci of lymphohistiocytic (lymphocytes and macrophages) inflammation with lesser numbers of plasma cells and occasional single hepatocyte necrosis accompanied by low numbers of neutrophils
2	Multifocal, up to 1-mm areas of mid-zonal to central lymphohistiocytic inflammation involving up to 5% of the examined parenchyma; in the foci with central necrosis, the inflammation shifts to predominantly neutrophils; less than 5% of examined parenchyma involved
3	As prior, but including scattered necrotic foci that have >1 mm-diameter areas and involving up to 20% of the hepatic tissue reviewed; scattered hepatocyte apoptosis is additionally present
4	Greater than 20% of the hepatic parenchyma involved with lesions, as described previously; additionally, there is prominent multifocal hemorrhage

2.8. Statistical Analysis

Differences in values of key experimental parameters were analyzed statistically. The values were analyzed using the *t*-test for independent samples. Mean and standard deviations were calculated for animals available per days post-infection (dpi). The small number of animals used due to space constraints (BSL-3Ag) reduced the value of statistical analysis.

3. Results

3.1. Rectal Temperatures

Rectal temperatures of calves inoculated with SA01 or Ken06 were monitored from 0–10, 14 and 21 dpi (Table 2). Two out of the five animals inoculated with SA01 had a fever (>40 °C) at 2 dpi, while the others were normal throughout the study. At 2 dpi or later, four out of five animals inoculated with Ken06 had a fever for at least one day, and one had a consistent fever for three days (Table 2). Rectal temperatures were within normal limits for all of the animals remaining past 7 dpi.

Table 2. Kinetics of the rectal temperature (°C) of calves infected with Rift Valley fever virus (RVFV) strains, Saudi Arabia 2001 (SA01) and Kenya 2006 (Ken06) (red indicates above normal for cattle, >40 °C).

		Days Post-Infection							
Strain	**No.**	**0**	**1**	**2**	**3**	**4**	**5**	**6**	**7**
SA01	33	39.4	38.7	41.3	38.8	39.0	38.6	39.8	38.3
	34	39.2	38.8	39.6	38.9	38.7	38.5		
	37	38.9	38.9	38.8	39.0	38.7			
	39	39.1	38.4	40.9	39.8	39.2	38.8	38.9	38.6
	43	39.8	38.9	39.5					
Ken06	36	40.4	39.4	39.3	38.3	38.7	38.8		
	38	38.9	38.9	41.5	41.6	41.1			
	40	39.0	38.8	38.8	38.9	38.8	40.8	39.3	38.5
	41	39.2	39.1	40.8	39.8				
	44	39.4	39.7	40.7	39.5	39.1	38.3	38.3	38.5
Control	35	39.1	39.4	38.8	38.6	38.4	38.7	38.3	38.3
	42	39.0	40.3	38.5	38.9	39.3	38.6	38.8	38.5

3.2. *RT-RCR and Viremia*

Viremia was determined in the cattle sera by both real-time RT-PCR and virus titration up to 21 dpi. Virus was detectable in the serum starting at 1 dpi by both real-time RT-PCR (Figure 1A–C) and the plaque assay (Figure 1D) for the Ken06 group. Viral RNA and virus were detected from the sera of Animal #38 of the Ken06 group at 1 dpi. By 2 dpi, four of five in the SA01 group and three of five in the Ken06 group were positive by real-time RT-PCR, and also, three of five in both groups were viremic (Figure 1A–D). All infected animals were at least weakly positive by real-time RT-PCR (cycle threshold (Ct) ⩽37; the standard cut-off threshold is 35), but only three of five animals per group had detectable viremia by virus titration. There was individual animal variation in the detection of viral RNA and virus isolation (Figure 2). The detection of virus occurred just prior to the appearance of fever, at 1 dpi for the Ken06 group, but there was viral RNA and viremia in the SA01 group in the absence of fever. One calf in the Ken06 group had a late fever at 5 dpi and was never positive for viral RNA using the quantitative RT-PCR test criteria Ct ⩽ 35 for two of three RVFV gene segments or virus isolation. This animal did have some questionable positive quantitative RT-PCR results (Ct = 35–39) from 5–7 dpi and did have a viral RNA positive nasal swab at 5 dpi.

Peak viremia determined by the plaque assay occurred at 2 and 3 dpi (Figure 1D). Cattle infected with the SA01 strain were all negative by both real-time RT-PCR (Figure 1A–C) and the plaque assay by 4 dpi (Figure 1D). Cattle inoculated with the Ken06 strain were negative by real-time RT-PCR (Figure 1A–C) by 5 dpi, but some weakly positive Ct values above the standard threshold cut-off of 35 were detected out to 7 dpi. No virus plaques were detected by Day 5 dpi for Ken06 (Figure 1D). One animal in the Ken06 group was viremic from 1–4 dpi and died immediately prior to necropsy at 4 dpi. This animal had a peak viremia of 3.4×10^8 pfu/mL at 2 dpi with real-time RT-PCR Ct values of 14–19 depending on the viral segment used for detection. Control co-housed cattle sera remained negative by RT-PCR and virus titration throughout the study.

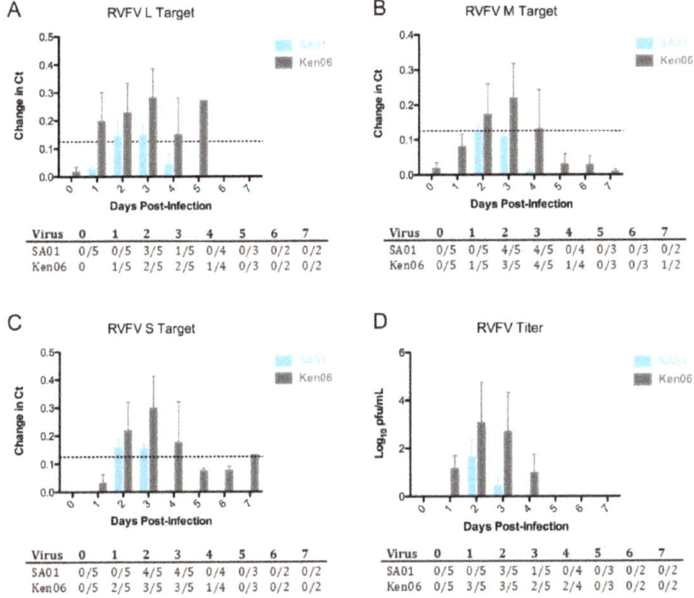

Virus	0	1	2	3	4	5	6	7
SA01	0/5	0/5	3/5	1/5	0/4	0/3	0/2	0/2
Ken06	0	1/5	2/5	2/5	1/4	0/3	0/2	0/2

Virus	0	1	2	3	4	5	6	7
SA01	0/5	0/5	4/5	4/5	0/4	0/3	0/3	0/2
Ken06	0/5	1/5	3/5	4/5	1/4	0/3	0/3	1/2

Virus	0	1	2	3	4	5	6	7
SA01	0/5	0/5	4/5	4/5	0/4	0/3	0/2	0/2
Ken06	0/5	2/5	3/5	3/5	1/4	0/3	0/2	0/2

Virus	0	1	2	3	4	5	6	7
SA01	0/5	0/5	3/5	1/5	0/4	0/3	0/2	0/2
Ken06	0/5	3/5	3/5	2/5	2/4	0/3	0/2	0/2

Figure 1. Kinetics of viral RNA and virus titers for calves infected with RVFV strains, SA01 (blue) and Ken06 (gray). The mean with the standard deviation of change in Ct values; (40-Ct)/40 (**A**) RVFV L RNA segment (dashed line indicates standard Ct = 35 cut-off or 0.13 change in Ct), (**B**) RVFV M RNA segment and (**C**) RVFV S RNA segment from calf serum are shown; (**D**) viral titers in pfu/mL. To be considered positive by the multiplex real-time RT-PCR, the Ct value for at least two of the three RVFV genome segments must be less than or equal to 35 [36].

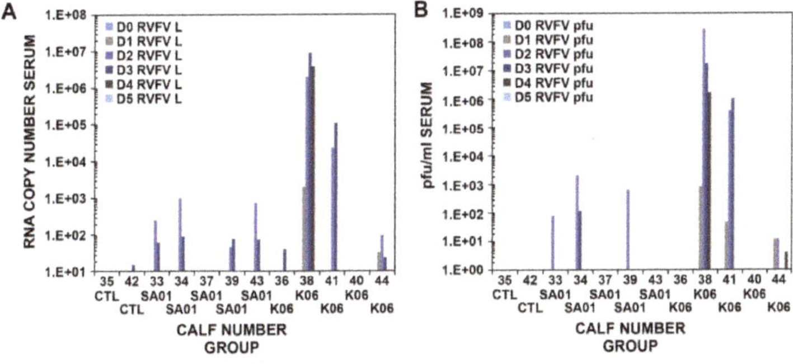

Figure 2. Viral replication dynamics in individual calf serum as determined by molecular and traditional virological methods. (**A**) Quantitative real-time RT-PCR-determined RVFV L segment copy number per reaction calculated from the mean Ct for the RVFV L segment from Day 0–Day 5 post-infection; (**B**) RVF virus titer determined from the plaque assay reported as pfu/mL of serum. PCR Ct equal to or less than 35 equates to greater than 27 copies, which is considered positive and within the quantitative range for quantitative RT-PCR. The X-axis includes calf numbers and group designations; CTL = mock controls, SA01 = infected with RVFV Saudi Arabia 2000–2001; K06 is Ken06 = infected with RVFV Kenya 2006–2007. Calves 43 and 41 were necropsied at dpi 3 and, thus, are not included in the datasets for dpi 4 or 5. Calves 37 and 38 were necropsied on dpi 4 and, thus, are not included in datasets for dpi 5.

3.3. *Viral Load or Titers in Tissues*

Brain, kidney, liver and spleen samples collected at necropsy were also tested for virus presence (Table 3). The Ken06 group Animals #41 at 3 dpi and #38 at 4 dpi had viral RNA-positive brain tissue (average of L, M and S Ct = 34 and 31, respectively). Viral RNA was also found in the kidneys of the Ken06 group cattle, #41 at 3 dpi (average of L, M and S Ct = 28) and #38 at 4 dpi (average of L, M and S Ct = 23), as well as the SA01 group Animal #34 at 5 dpi (average of L, M and S Ct = 26). Virus was isolated from the kidneys of #38 (1.4×10^5 pfu/mL) and #34 ($1.5 \text{x} 10^2$ pfu/mL). Liver samples from both groups were positive for viral RNA from 3–4 dpi and at 5 dpi for the SA01 group (Ct = 18–30). At 5 dpi, the SA01 animal's liver tissue was positive for viral RNA (Ct = 29), but the Ken06 was inconclusive (Ct = 38). Virus was isolated from the liver of both groups at 3 dpi and for a Ken06 animal at 4 dpi (Table 3). The spleen results reflected the liver results for viral RNA detection with Ct values from 21–33 from 3–5 dpi. Virus was isolated only from the spleens of Ken06 animals at 3 and 4 dpi with titers of 1.2×10^3 pfu/mL and 2.9×10^5 pfu/mL, respectively. The spleens were positive for viral RNA at 10 dpi from both groups (SA01, Ct = 33; Ken06, Ct = 35). The spleens from both groups were weakly positive at 21 dpi (Ken06, Ct = 36) and inconclusive (SA01, Ct = 38). Nasal swabs from both groups were sporadically positive for viral RNA from 3–6 dpi (total of nine positive samples; Ct = 26–34; data not shown). One swab from Animal #38 was found virus positive (32 pfu/mL). The greatest number of positive nasal swabs was at 3 dpi with three out of five for the SA01 group (average of L, M and S Ct = 32) and two out of five for the Ken06 group (average of L, M and S Ct = 34). Virus was isolated from the nasal swab of only the Ken06 group Calf #38 that also had the highest serum virus titer of 3.5×10^8 pfu/mL, whereas the highest titer of the SA01 group found in Calf #34 was 2.2×10^3 pfu/mL.

Table 3. Presence of RVFV RNA and virus in tissues at days post-infection.

		Days Post-Infection									
		3		**4**		**5**		**10**		**20**	
		Ct	*Titer*	*Ct*	*Titer*	*Ct*	*Titer*	*Ct*	*Titer*	*Ct*	*Titer*
SA01	Brain	ND	-	ND	-	ND	-	ND	-	ND	-
	Kidney	ND	-	37	-	26	1.5×10^2	ND	-	ND	-
	Liver	30	$4.0 \times 10^{\circ}$	30	-	29	-	37	-	ND	-
	Spleen	31	-	33	-	32	-	33	-	38	-
Ken06	Brain	34	-	31	-	38	-	ND	-	ND	-
	Kidney	28	-	23	1.4×10^5	ND	-	ND	-	ND	-
	Liver	20	2.1×10^5	18	3.9×10^6	38	-	ND	-	ND	-
	Spleen	24	1.2×10^3	22	2.9×10^5	ND	-	35	-	36	-

Titer: pfu/mL from 10 mg homogenate in 1 mL of media; Ct = cycle threshold mean of S, L, M, real-time RT-PCR; ND = not detected or Ct of 40; - = no plaque formation.

3.4. *Serological Responses*

Calves showed the first indication of seroconversion at 6 dpi with OD values above the cut-off points for reactivity to N and 10–14 dpi for reactivity to Gn (Figure 3A, B). Serum samples from 1–3 dpi were not included in the analysis because our standard inactivation procedure failed to fully inactivate the virus due to the presence of high virus titers in some of the sera. Neutralizing antibody responses were detected in remaining animals at 5 dpi and increased until peaking at 10–14 dpi (Table 4) consistent with the ELISA data and indicating serological conversion in response to the experimental RVFV inoculations. None of the co-housed control cattle seroconverted.

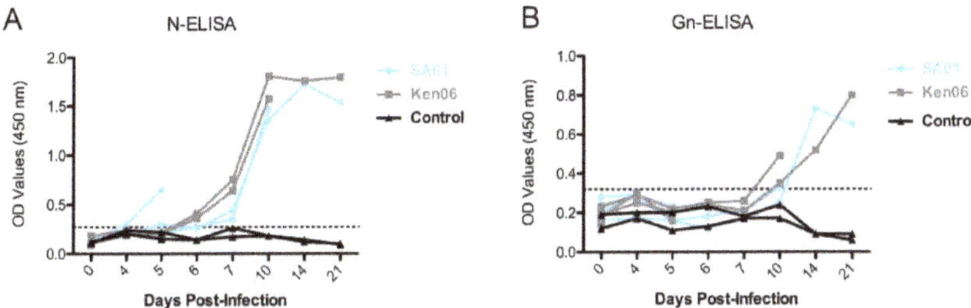

Figure 3. Development of the antibody response of individual calves infected with RVFV strains, SA01 (blue) and Ken06 (gray). Specific indirect ELISA shows the kinetics of total IgG antibody responses in sheep inoculated with wild-type RVFV strains, SA01 and Ken06: (**A**) RVFV N-ELISA; (**B**) RVFV Gn-ELISA. The dashed line indicates the calculated cut-off values (N-ELISA = 0.27; Gn-ELISA = 0.33).

Table 4. Reciprocal plaque reduction neutralization test ($PRNT_{80}$) titers in calves infected with RVFV strains, SA01 and Ken06.

		Days Post Infection							
Strain	**No.**	**0**	**4**	**5**	**6**	**7**	**10**	**14**	**21**
SA01	33	-	-	10	40	160	640	1280	1280
	39	-	-	10	160	320	>1280		
	34	-	-	-					
	37	-	-						
	43	-							
Mean				10	100	240	960		
Ken06	40	-	-	40	40	80	1280	640	>1280
	44	-	-	40	80	320	1280		
	36	-	-	-					
	38	-	ND						
	41	-							
Mean				40	60	200	1280		

- = negative; ND = not determined.

3.5. Pathology

In general, gross pathology observations at 3–5 dpi revealed no consistent pattern of difference between the SA01 and Ken06 virus strains groups; all had 1–3-mm multifocal tan foci (necrotic foci) disseminated throughout their hepatic parenchyma (Figure 4A). However, gross findings for Animal #38 (4 dpi Ken06 inoculated) were more severe. This animal had multifocal to coalescing hepatic necrosis accompanied by marked, multifocal hemorrhage disseminated throughout its hepatic parenchyma (Figure 4B). Additionally, #38 had clinical signs and lesions suggestive of disseminated intravascular coagulation, including hemorrhage and edema in multiple viscera, many edematous and congested lymph nodes throughout its body, renal pelvic hemorrhage, splenic ecchymoses, marked and diffuse pulmonary congestion, multifocal endocardial hemorrhage and red urine observed ante-mortem. In all animals necropsied at 10 dpi and later, gross changes in hepatic parenchyma were no longer evident.

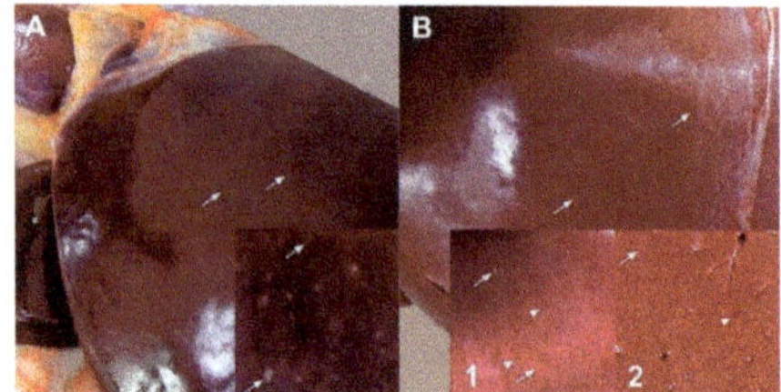

Figure 4. Gross pathology of acute post-infection, time-point livers from virus-inoculated cattle. (**A**) This liver from SA01-inoculated, 3 dpi, Animal #43 shows the typical acute time-point hepatic pathology seen in the 3–5 dpi animals regardless of virus inoculum. Disseminated throughout the parenchyma are myriads of 1–2-mm tan foci (white arrows), necrosis. (**B**) This liver from severely affected Ken06-inoculated, 4 dpi, Animal #38 is diffusely pale. Disseminated throughout the parenchyma are multifocal to coalescing foci of necrosis (white arrows) and hemorrhage (white arrowheads). Inset 1 is a capsular liver view similar to the inset in (A), while Inset 2 shows a cross-section of hepatic parenchyma.

Similar to the gross lesion results, we did not find RVFV strain-dependent patterns between the two groups in terms of hepatic histopathology changes. RVFV-infected cattle had severe hepatic lesions at 3–5 dpi, averaging no less than 2.5 (scale 0–4) for their hepatic histopathology score (Table 5 and Figure 5). All of these animals had multifocal mid-zonal to central hepatocellular necrosis disseminated throughout their hepatic parenchyma accompanied by predominantly lymphohistiocytic, inflammation and increased numbers of degenerate and viable neutrophils in larger necrotic foci. The lesions tapered off by 10 dpi (score $\leqslant$2) and by 21 dpi were characterized by occasional foci of inflammation, predominantly aggregates of lymphocytes and macrophages.

In contrast, while the sample size per time-point prevented us from appreciating histopathologic differences attributable to virus strain, we saw a consistent difference in immunohistochemical (IHC) labeling for RVFV nucleoprotein antigen in the hepatic lesions at 3 and 4 dpi SA01 cattle. As seen in our prior challenge model work in sheep [31], SA01 hepatic lesions contained notably less viral antigen signal than hepatic lesions in time-point matched Ken06 cattle. Figure 5F shows a typical Ken06 hepatic lesion IHC signal (Figure 5F), whereas Figure 5I shows a stronger than average labeling of a hepatic lesion in an SA01 animal (Figure 5I). Finally, all examined cattle livers, regardless of inoculum type, had mild, peri-portal lymphoplasmacytic inflammation, interpreted as background inflammation unrelated to the study. Additionally, mock-inoculated, co-housed Animal #42, unlike its mock-inoculated peer, Animal #35, had low numbers of scattered foci of mixed inflammation in central and mid-zonal areas. These foci were negative for RVFV antigen and may be attributable to an unrelated etiology.

Table 5. Liver histopathology and immunohistochemistry for RVFV antigen.

Strain	Calf No.	Days PI	H Score	IHC	PCR	Titer
SA01	43	3	3	+	+	4.0×10^{0}
	37	4	2.5	+	+	-
	34	5	3	+	+	-
	39	10	2	-	-	-
	33	21	1	-	-	-
Ken06	41	3	3	+	+	2.1×10^{5}
	38	4	4	+	+	3.9×10^{6}
	36	5	1	+ *	-	-
	44	10	2	-	-	-
	40	21	1	-	-	-
Mock	35	20	0	-	-	-
	42	20	0	-	-	-

H Score is the average hepatic histopathology score on a scale from 0, no lesions, to 4, severe lesions (Table 1), for 3–5 reviewed sections of liver per animal. IHC is the anti-RVFV immunohistochemistry (IHC) result on liver sections; a positive "+" is given when at least one section of liver was positive for viral antigen. Key: + = positive for viral antigen by IHC; - = negative for viral antigen by IHC. PCR is the quantitative RT-PCR result for liver tissue. * No viral antigen-positive lesions; a positive cytoplasmic signal is present in low numbers of circulating Kupffer cells (macrophages). The titer is pfu/mL of tissue homogenate from 10 mg of tissue in 1 mL of media (- = no plaques).

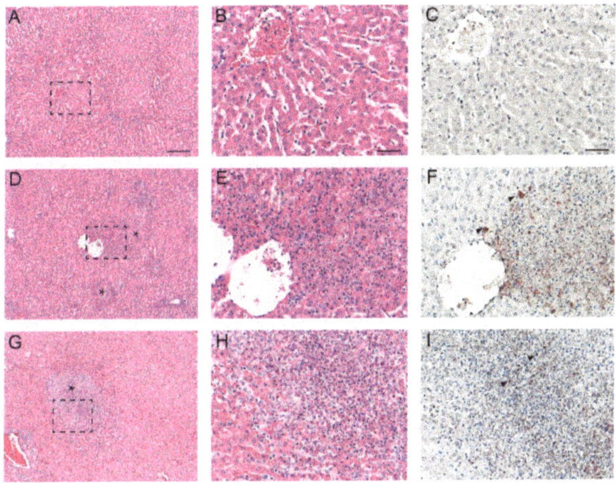

Figure 5. Acute time-point liver histopathology and viral antigen immunohistochemistry. (**A**) Shown is a low magnification hematoxylin and eosin (H&E) stain of the normal liver histology seen in mock-inoculated Animal #42; the black broken line box outlines the area magnified in (**B**); (**C**) viral antigen immunohistochemistry is negative for RVFV antigen; (**D**) the depicted Ken06, 3 dpi, Animal #41 liver H&E is representative of the acute time-point histopathology seen in animals inoculated by this virus; the black stars mark necrotic foci; a portion of one focus is magnified in (**E**); (**F**) shown is the same necrotic focus, which is strongly positive for RVFV antigen in hepatocytes, macrophages and cellular debris; black arrowheads indicate cytoplasmic positive labeling; (**G**) the depicted SA01, 3 dpi, Animal #43 liver H&E is typical of acute time-point liver histopathology for this virus strain; the black star denotes a single large necrotic focus further magnified in (**H**); acute time-point SA01 liver foci labeled sporadically and less strongly for RVFV antigen when compared to Ken06 foci; (**I**) a stronger than average viral antigen labeling for this larger than average necrotic focus in Animal #43; black arrowheads denote RVFV antigen labeling. Column 1 images are 100× magnification. The bar is 20 μm. Columns 2 and 3 images are 400× magnification. Bars are 50 μm.

3.6. Blood Chemistry

Blood chemistries were run on all sera available from 0–7 dpi. Of the 14 parameters assessed, significant elevations were seen in multiple hepatic enzymes, BUN, an indicator of renal damage, as well as creatinine kinase (CK), an indicator of skeletal muscle damage (data not shown). These changes were all found in RVFV-infected animals with the exception of CK. Aspartate amino transferase (AST), the hepatic leakage enzyme consistent with acute liver damage, was elevated above normal range (66–211 U/L) in two of the Ken06 group animals, #38 (2–4 dpi; 279, 1959, 1742 U/L, respectively) and #41 (3 dpi; 452 U/L). Using ALP values normalized to 0 dpi values for individual animals, the only animal with a significant elevation in its ALP was Ken06 Animal #38 starting at 3 dpi (0–4 dpi; 149, 151, 155, 451, 476 U/L, respectively). Animal #38 was also the only one with an elevation in gamma-glutamyl transferase (GGT) at 3–4 dpi (90 and 100 U/L, respectively). This animal had the most severe hepatic histopathology as documented above, but no evidence of cholestasis (associated with high ALP and GGT values) was seen during the histopathology analysis. An additional check with a Hall's bilirubin stain on multiple liver sections confirmed an absence of bilirubin in bile canaliculi, a hallmark of cholestasis. Animal #38 also had high BUN levels at 4 dpi (32 mg/dL), immediately prior to death. Finally, Animal #38 and SA01 group Animal #37 both had a single elevated CK value accompanied by a high normal value on a consecutive day indicating muscle damage, #37 (3–4 dpi; 797 and 504 U/L, respectively) and #38 (3–4 dpi; 585 and 834, respectively). Interestingly, while Animal #38 had severe clinical signs, Animal #37 did not show obvious clinical signs; it did have significant hepatic pathology (see the pathology section). One of the mock-inoculated, co-housed cattle had a single elevated CK value (1175 U/L), most likely due to difficult restraining procedures during sampling.

4. Discussion

The development of reliable challenge models for arboviral disease is often challenging because needle inoculation does not mimic natural infection via insect vectors. There has been significant effort made in the development of vaccines for RVF [23,26,41–43]; thus, it is important that reliable animal models using a variety of challenge viruses are available for vaccine evaluation in all target species. For example, previous RVFV challenge models have used the 1977 strain from Egypt, ZH501, in goats and sheep [29]. While this model does provide a viremia and fever response in sheep sufficient to evaluate vaccines [30], it does not produce clinical disease. In a previous study using two more recently-isolated, genetically-distinct strains of RVFV (SA01 and Ken06) and sheep (Dorper/Katahdin cross and Polypay), we observed clinical disease [31]. In this sheep study, we also found that the Ken06 produced greater liver pathology based on blood chemistry markers and histopathology. This finding was surprising because preliminary studies that compared ZH501 with another RVFV isolate from Kenya did not demonstrate any differences in calf responses to infection (Weingartl *et al.*, unpublished data). Therefore, the current study was performed to confirm and determine if the differences in clinical signs after infection with genetically-distinct RVFV strains noted in sheep would also be found in cattle.

In the previous sheep studies, the febrile response was found to be consistent across all RVFV-infected animals. In this calf study, the febrile response was less consistent with both isolates. Only two out of five SA01-inoculated calves developed a mild fever and only for one day. In the Ken06 group, four out of five had a fever for one day, and one animal was febrile for three days (Table 2). In most cases, the fever occurred at 2 dpi when viremia was beginning to peak. Viral RNA was detectable in the liver and spleen for both virus groups at 3 dpi and in the brain and kidneys of Ken06 group animals at 3 and 4 dpi. The spleens of both virus groups had detectable viral RNA starting at 3 dpi up to 21 dpi. Virus was detected at low titer in SA01 animals in the liver and kidney at 3 and 5 dpi, respectively. Virus was isolated with titers from 10^3–10^6 pfu/mL from the liver and spleen at 3 and 4 dpi and from the kidney at 4 dpi for the Ken06 animals. Although viral RNA was detected in all four tissues examined (brain, kidney, liver and spleen), the liver and spleen were the most consistently positive by both RT-PCR and virus isolation. The presence of viral RNA in the spleen post-viremia potentially out to 20 dpi suggests this may be a good target tissue for diagnosis. There was considerably more variation in the calf responses (Figure 2) than seen in the sheep study [31]. This may have been in part because of the use of mixed breed calves due to their availability at the time of the study. No breed susceptibility conclusions can be drawn due to the small number of animals used.

Nasal swabs from a few animals of both groups were sporadically positive for viral RNA. Ken06 Animal #38 with the highest viremia was the only animal that also had a virus-positive nasal swab (32 pfu/mL). Mock-inoculated control calves co-housed with the challenged animals remained negative for all RVFV-specific parameters throughout the study, suggesting that RVFV was not shed at levels necessary for transmission among cattle. This is consistent with the view that RVFV is most often transmitted through an infected mosquito bite. However, although the potential for transmission through nasal discharge is low, it should not be entirely ruled out.

The blood chemistry analysis indicated primarily liver involvement in most affected animals. As with the previous study in sheep [31], AST appears to be the most consistent marker of clinical disease. The small number of animals in the present study and the removal of animals for necropsy at 3, 4 and 5 dpi during the peak of viremia could have affected these results; however, the liver function results are consistent with a previous experimental infection of calves using seven- and 21-day-old animals [44]. Because the renal function marker BUN was also elevated in the most severely affected animal, kidney function may also be worth monitoring in future studies.

Regardless of virus strain, at the acute time-point (3–5 dpi) post-infection, calf livers had grossly visible multifocal necrosis that in the most severe case, Animal #38, was accompanied by marked hemorrhage (Figures 4 and 5). These lesions contained RVFV antigen-positive cells. Differences in hepatic histopathology due to virus strain were not noted. This could be due to the study design of one animal per virus, per time-point. However, similar to what was seen in our sheep study [31], the comparison of hepatic lesions from acute time-point animals for both viruses revealed that Ken06 lesions on average labeled more strongly positive for RVFV antigen (Figure 5). Lesions found in this study were consistent with those reported in previous cattle studies [28,44,45].

Most recent studies of RVFV pathogenesis and vaccine evaluation focus on more susceptible target species, such as sheep. This is the first study to examine virus replication and pathological development of two distinct RVFV strains during the clinical stage of infection in cattle. Although the fever and viremia were less consistent in cattle than sheep, the vaccine-age calf model did display sufficient clinical responses to be useful for efficacy evaluation of cattle vaccines against RVF.

Acknowledgments: The authors wish to thank the staff of KSU Biosecurity Research Institute, the Comparative Medicine Group and the staff of the Histopathology Unit at the Veterinary Diagnostic Laboratory. We wish to thank Dane Jasperson, Lindsey Reister-Hendricks (Arthropod-Borne Animal Diseases Research Unit, USDA/Agricultural Research Service) and Chester McDowell, Tammy Koopman, Haixia Liu, Maira Cotton-Caballero, Elizabeth Stietzle, Monica Gamez and Mal Hoover (KSU) for technical assistance. We also thank Mal Rooks Hoover for assistance in medical illustration. This work was funded by the grants of the Department of Homeland Security Center of Excellence for Emerging and Zoonotic Animal Diseases (CEEZAD), Grant No. 2010-ST061-AG0001, the Kansas Bioscience Authority, the National Bio and Agro-Defense Facility (NBAF) Transition Funds and USDA Agricultural Services Project 3020-32000-005-00D. Mention of trade names or commercial products in this publication is solely for the purpose of providing specific information and does not imply recommendation or endorsement by the U.S. Department of Agriculture. USDA is an equal opportunity provider and employer.

Author Contributions: This study was conceived by JAR and WCW. ASD, WCW, BF, VS, SYS, AB, AE, WM MR, IM, DSM and JAR performed the animal sample collection and necropsies. The laboratory analysis was performed by ASD, WCW NNG, BF, BSD, and JDT. The manuscript was written by WCW, ASD, and JAR with assistance and review of the coauthors.

Conflicts of Interest: The authors declare no conflict of interest.

References

1. Ikegami, T.; Makino, S. Rift valley fever vaccines. *Vaccine* **2009**, *27*, D69–D72. [CrossRef] [PubMed]
2. Chevalier, V. Relevance of Rift Valley fever to public health in the European Union. *Clin. Microbiol. Infect.* **2013**, *19*, 705–708. [CrossRef] [PubMed]

3. Rolin, A.I.; Berrang-Ford, L.; K, M.A. The risk of Rift Valley fever virus introduction and establishment in the United States and European Union. *Emerg Microb Infect* **2013**, *2*, e81. [CrossRef] [PubMed]
4. Iranpour, M.; Turell, M.J.; Lindsay, L.R. Potential for Canadian mosquitoes to transmit Rift Valley fever virus. *J. Am. Mosq. Control Assoc.* **2011**, *27*, 363–369. [CrossRef] [PubMed]
5. Turell, M.J.; Wilson, W.C.; Bennett, K.E. Potential for North American mosquitoes (Diptera: Culicidae) to transmit rift valley fever virus. *J. Med. Entomol.* **2010**, *47*, 884–889. [CrossRef] [PubMed]
6. Turell, M.J.; Dohm, D.J.; Mores, C.N.; Terracina, L.; Wallette, D.L.; Hribar, L.J.; Pecor, J.E.; Blow, J.A. Potential for North American mosquitoes to transmit Rift Valley fever virus. *J. Am. Mosq. Control Assoc.* **2008**, *24*, 502–507. [CrossRef] [PubMed]
7. Gerdes, G.H. Rift Valley fever. *Rev. Sci. Technol.* **2004**, *23*, 613–623. [CrossRef]
8. Pepin, M.; Bouloy, M.; Bird, B.H.; Kemp, A.; Paweska, J. Rift Valley fever virus (Bunyaviridae: Phlebovirus): An update on pathogenesis, molecular epidemiology, vectors, diagnostics and prevention. *Vet. Res.* **2010**, *41*, 61. [CrossRef] [PubMed]
9. Ikegami, T.; Makino, S. The pathogenesis of Rift Valley fever. *Viruses* **2011**, *3*, 493–519. [CrossRef] [PubMed]
10. Nguku, P.M.; Sharif, S.K.; Mutonga, D.; Amwayi, S.; Omolo, J.; Mohammed, O.; Farnon, E.C.; Gould, L.H.; Lederman, E.; Rao, C.; *et al.* An investigation of a major outbreak of Rift Valley fever in Kenya: 2006–2007. *Am. J. Trop. Med. Hyg.* **2010**, *83*, 5–13. [CrossRef] [PubMed]
11. Nanyingi, M.O.; Munyua, P.; Kiama, S.G.; Muchemi, G.M.; Thumbi, S.M.; Bitek, A.O.; Bett, B.; Muriithi, R.M.; Njenga, M.K. A systematic review of Rift Valley Fever epidemiology 1931–2014. *Infect. Ecol. Epidemiol.* **2015**, *5*, 1–12. [CrossRef] [PubMed]
12. NIH, NIAID Emerging Infectious Diseases/Pathogens. Available online: http://www.niaid.nih.gov/topics/biodefenserelated/biodefense/pages/cata.aspx (accessed on 16 March 2016).
13. Mandell, R.B.; Flick, R. Rift Valley fever virus: A real bioterror threat. *J. Bioterror. Biodef.* **2011**, *2*, 108. [CrossRef]
14. Kortekaas, J. One Health approach to Rift Valley fever vaccine development. *Antivir. Res.* **2014**, *106*, 24–32. [CrossRef] [PubMed]
15. Faburay, B. The case for a "one health" approach to combatting vector-borne diseases. *Infect. Ecol. Epidemiol.* **2015**, *5*, 1–4. [CrossRef] [PubMed]
16. Smithburn, K.C. Rift Valley fever: The neurotropic adaptation of the virus and experimental use of the modified virus as a vaccine. *Br. J. Exp. Pathol.* **1949**, *30*, 1–16. [PubMed]
17. Dungu, B.; Louw, I.; Lubisi, A.; Hunter, P.; von Teichman, B.F.; Bouloy, M. Evaluation of the efficacy and safety of the Rift Valley Fever Clone 13 vaccine in sheep. *Vaccine* **2010**, *28*, 4581–4587. [CrossRef] [PubMed]
18. Von Teichman, B.; Engelbrecht, A.; Zulu, G.; Dungu, B.; Pardini, A.; Bouloy, M. Safety and efficacy of Rift Valley fever Smithburn and Clone 13 vaccines in calves. *Vaccine* **2011**, *29*, 5771–5777. [CrossRef] [PubMed]
19. Morrill, J.C.; Jennings, G.B.; Caplen, H.; Turell, M.J.; Johnson, A.J.; Peters, C.J. Pathogenicity and immunogenicity of a mutagen-attenuated Rift Valley fever virus immunogen in pregnant ewes. *Am. J. Vet. Res.* **1987**, *48*, 1042–1047. [PubMed]
20. Morrill, J.C.; Peters, C.J. Protection of MP-12-vaccinated rhesus macaques against parenteral and aerosol challenge with virulent Rift Valley fever virus. *J. Infect. Dis.* **2011**, *204*, 229–236. [CrossRef] [PubMed]
21. Caplen, H.; Peters, C.J.; Bishop, D.H. Mutagen-directed attenuation of Rift Valley fever virus as a method for vaccine development. *J Gen. Virol.* **1985**, *66*, 2271–2277. [CrossRef] [PubMed]
22. Hill, R.E. *Issuance of a Conditional License for Rift Valley Fever Vaccine, Modified Live Virus*; CVB Notices 13–12; United States Department of Agriculture: Washington, DC, USA, 2013.
23. Pittman, P.R.; McClain, D.; Quinn, X.; Coonan, K.M.; Mangiafico, J.; Makuch, R.S.; Morrill, J.; Peters, C.J. Safety and immunogenicity of a mutagenized, live attenuated Rift Valley fever vaccine, MP-12, in a Phase 1 dose escalation and route comparison study in humans. *Vaccine* **2016**, *34*, 424–429. [CrossRef] [PubMed]
24. Pittman, P.R.; Norris, S.L.; Brown, E.S.; Ranadive, M.V.; Schibly, B.A.; Bettinger, G.E.; Lokugamage, N.; Korman, L.; Morrill, J.C.; Peters, C.J. Rift Valley fever MP-12 vaccine Phase 2 clinical trial: Safety, immunogenicity, and genetic characterization of virus isolates. *Vaccine* **2016**, *34*, 523–530. [CrossRef] [PubMed]
25. Hunter, P.; Erasmus, B.J.; Vorster, J.H. Teratogenicity of a mutagenised Rift Valley fever virus (MVP 12) in sheep. *Onderstepoort J. Vet. Res.* **2002**, *69*, 95–98. [PubMed]

26. Warimwe, G.M.; Gesharisha, J.; Carr, B.V.; Otieno, S.; Otingah, K.; Wright, D.; Charleston, B.; Okoth, E.; Elena, L.-G.; Lorenzo, G.; *et al.* Chimpanzee Adenovirus Vaccine Provides Multispecies Protection against Rift Valley Fever. *Sci. Rep.* **2016**, *6*, 1–7. [CrossRef] [PubMed]
27. Ross, T.M.; Bhardwaj, N.; Bissel, S.J.; Hartman, A.L.; Smith, D.R. Animal models of Rift Valley fever virus infection. *Virus Res.* **2012**, *163*, 417–423. [CrossRef] [PubMed]
28. Bird, B.H.; Maartens, L.H.; Campbell, S.; Erasmus, B.J.; Erickson, B.R.; Dodd, K.A.; Spiropoulou, C.F.; Cannon, D.; Drew, C.P.; Knust, B.; *et al.* Rift Valley fever virus vaccine lacking the NSs and NSm genes is safe, nonteratogenic, and confers protection from viremia, pyrexia, and abortion following challenge in adult and pregnant sheep. *J. Virol.* **2011**, *85*, 12901–12909. [CrossRef] [PubMed]
29. Weingartl, H.M.; Miller, M.; Nfon, C.; Wilson, W.C. Development of a Rift Valley fever virus viremia challenge model in sheep and goats. *Vaccine* **2014**, *32*, 2337–2344. [CrossRef] [PubMed]
30. Weingartl, H.M.; Nfon, C.K.; Zhang, S.; Marszal, P.; Wilson, W.C.; Morrill, J.C.; Bettinger, G.E.; Peters, C.J. Efficacy of a recombinant Rift Valley fever virus MP-12 with NSm deletion as a vaccine candidate in sheep. *Vaccine* **2014**, *32*, 2345–2349. [CrossRef] [PubMed]
31. Faburay, B.; Gaudreault, N.N.; Liu, Q.; Davis, A.S.; Shivanna, V.; Sunwoo, S.Y.; Lang, Y.; Morozov, I.; Ruder, M.; Drolet, B.; *et al.* Development of a sheep challenge model for Rift Valley fever. *Virology* **2016**, *489*, 128–140. [CrossRef] [PubMed]
32. Miller, B.R.; Godsey, M.S.; Crabtree, M.B.; Savage, H.M.; Al-Mazrao, Y.; Al-Jeffri, M.H.; Abdoon, A.-M.M.; Al-Seghayer, S.M.; Al-Shahrani, A.M.; Ksiazek, T.G. Isolation and genetic characterization of Rift Valley fever virus from Aedes vexans arabiensis, Kingdom of Saudi Arabia. *Emerg. Infect. Dis.* **2002**, *8*, 1492–1494. [CrossRef] [PubMed]
33. Madani, T.A.; Al-Mazrou, Y.Y.; Al-Jeffri, M.H.; Mishkhas, A.A.; Al-Rabeah, A.M.; Turkistani, A.M.; Al-Sayed, M.O.; Abodahish, A.A.; Khan, A.S.; Ksiazek, T.G.; *et al.* Rift Valley fever epidemic in Saudi Arabia: Epidemiological, clinical, and laboratory characteristics. *Clin. Infect. Dis.* **2003**, *37*, 1084–1092. [CrossRef] [PubMed]
34. Sang, R.; Kioko, E.; Lutomiah, J.; Warigia, M.; Ochieng, C.; O'Guinn, M.; Lee, J.S.; Koka, H.; Godsey, M.; Hoel, D.; *et al.* Rift Valley fever virus epidemic in Kenya, 2006/2007: The entomologic investigations. *Am. J. Trop. Med. Hyg.* **2010**, *83*, 28–37. [CrossRef] [PubMed]
35. Gaudreault, N.N.; Indran, S.V.; Bryant, P.K.; Richt, J.A.; Wilson, W.C. Comparison of Rift Valley fever virus replication in North American livestock and wildlife cell lines. *Front. Microbiol.* **2015**, *6*, 664. [CrossRef] [PubMed]
36. Wilson, W.C.; Romito, M.; Jasperson, D.C.; Weingartl, H.; Binepal, Y.S.; Maluleke, M.R.; Wallace, D.B.; van Vuren, P.J.; Paweska, J.T. Development of a Rift Valley fever real-time RT-PCR assay that can detect all three genome segments. *J. Virol. Methods* **2013**, *193*, 426–431. [CrossRef] [PubMed]
37. Copy number calculator for realtime PCR. Available online: http://scienceprimer.com/copy-number-calculator-for-realtime-pcr (accessed on 7 May 2016).
38. Faburay, B.; Wilson, W.; McVey, D.S.; Drolet, B.S.; Weingartl, H.; Madden, D.; Young, A.; Ma, W.; Richt, J.A. Rift Valley fever virus structural and nonstructural proteins: Recombinant protein expression and immunoreactivity against antisera from sheep. *Vector Borne Zoonotic Dis.* **2013**, *13*, 619–629. [CrossRef] [PubMed]
39. Faburay, B.; Lebedev, M.; McVey, D.S.; Wilson, W.; Morozov, I.; Young, A.; Richt, J.A. A glycoprotein subunit vaccine elicits a strong rift valley Fever virus neutralizing antibody response in sheep. *Vector Borne Zoonotic Dis.* **2014**, *14*, 746–756. [CrossRef] [PubMed]
40. Drolet, B.S.; Weingartl, H.M.; Jiang, J.; Neufeld, J.; Marszal, P.; Lindsay, R.; Miller, M.M.; Czub, M.; Wilson, W.C. Development and evaluation of one-step rRT-PCR and immunohistochemical methods for detection of Rift Valley fever virus in biosafety level 2 diagnostic laboratories. *J. Virol. Methods* **2012**, *179*, 373–382. [CrossRef] [PubMed]
41. Indran, S.V.; Ikegami, T. Novel approaches to develop Rift Valley fever vaccines. *Front. Cell. Infect. Microbiol.* **2012**, *2*, 131. [CrossRef] [PubMed]
42. Morrill, J.C.; Laughlin, R.C.; Lokugamage, N.; Wu, J.; Pugh, R.; Kanani, P.; Adams, L.G.; Makino, S.; Peters, C.J. Immunogenicity of a recombinant Rift Valley fever MP-12-NSm deletion vaccine candidate in calves. *Vaccine* **2013**, *31*, 4988–4994. [CrossRef] [PubMed]

43. Mansfield, K.L.; Banyard, A.C.; McElhinney, L.; Johnson, N.; Horton, D.L.; Hernández-Triana, L.M.; Fooks, A.R. Rift Valley fever virus: A review of diagnosis and vaccination, and implications for emergence in Europe. *Vaccine* **2015**, *33*, 5520–5531. [CrossRef] [PubMed]
44. Rippy, M.K.; Topper, M.J.; Mebus, C.A.; Morrill, J.C. Rift Valley fever virus-induced encephalomyelitis and hepatitis in calves. *Vet. Pathol.* **1992**, *29*, 495–502. [CrossRef] [PubMed]
45. Smith, D.R.; Bird, B.H.; Lewis, B.; Johnston, S.C.; McCarthy, S.; Keeney, A.; Botto, M.; Donnelly, G.; Shamblin, J.; Albarino, C.G.; *et al.* Development of a novel nonhuman primate model for Rift Valley fever. *J. Virol.* **2012**, *86*, 2109–2120. [CrossRef] [PubMed]

© 2016 by the authors. Licensee MDPI, Basel, Switzerland. This article is an open access article distributed under the terms and conditions of the Creative Commons Attribution (CC BY) license (http://creativecommons.org/licenses/by/4.0/).

Review

Phleboviruses and the Type I Interferon Response

Jennifer Deborah Wuerth and Friedemann Weber *

Institute for Virology, FB10-Veterinary Medicine, Justus-Liebig University, Giessen 35392, Germany; wuerthje@staff.uni-marburg.de

* Correspondence: friedemann.weber@vetmed.uni-giessen.de; Tel.: +49-641-99-383-50

Academic Editors: Jane Tao and Pierre-Yves Lozach
Received: 8 May 2016; Accepted: 20 June 2016; Published: 22 June 2016

Abstract: The genus *Phlebovirus* of the family *Bunyaviridae* contains a number of emerging virus species which pose a threat to both human and animal health. Most prominent members include Rift Valley fever virus (RVFV), sandfly fever Naples virus (SFNV), sandfly fever Sicilian virus (SFSV), Toscana virus (TOSV), Punta Toro virus (PTV), and the two new members severe fever with thrombocytopenia syndrome virus (SFTSV) and Heartland virus (HRTV). The nonstructural protein NSs is well established as the main phleboviral virulence factor in the mammalian host. NSs acts as antagonist of the antiviral type I interferon (IFN) system. Recent progress in the elucidation of the molecular functions of a growing list of NSs proteins highlights the astonishing variety of strategies employed by phleboviruses to evade the IFN system.

Keywords: phlebovirus; NSs protein; interferon; RIG-I; PKR

---

1. Introduction

The family *Bunyaviridae* contains five genera, among which the *Orthobunyavirus*, *Phlebovirus*, *Nairovirus*, and *Hantavirus* all contain species that are pathogenic to humans and animals, while the genus *Tospovirus* contains plant-infecting viruses [1]. According to the International Committee on Taxonomy of Viruses (ICTV), the genus *Phlebovirus* comprises more than 70 accepted members that are grouped into ten species complexes, namely Bujaru virus (BUJV), Candiru virus (CDUV), Chilibre virus (CHIV), Frijoles virus (FRIV), Punta Toro virus (PTV), Rift Valley fever virus (RVFV), Salehabad virus (SALV), sandfly fever Naples virus (SFNV), severe fever with thrombocytopenia syndrome virus (SFTSV), and Uukuniemi virus (UUKV), as well as unassigned viruses [2]. Phleboviruses were traditionally classified by serologic methods, but recently extensive efforts were undertaken to refine phlebovirus taxonomy by genome sequencing [3–21].

Phleboviruses can cause a wide spectrum of symptoms, ranging from mild febrile disease up to hemorrhagic fever and death [22,23]. RVFV, for example, causes disease in cattle, sheep, and other ruminants, with symptoms including hepatitis, hemorrhage, and abortion [24]. Humans exposed to RVFV can present febrile illness, but in 1% to 2% of the cases it can progress to retinitis with persisting visual impairment, meningoencephalitis or hemorrhagic fever, resulting in mortality of up to 20% in hospitalized patients [25]. Since its original description during an outbreak of RVFV in ruminants in Kenya in 1931 [26], periodic outbreaks have been observed throughout the African continent, accompanied by so-called 'abortion storms' in livestock populations and simultaneously occurring illness in humans. Notably, RVFV has spread to the Arabian Peninsula in 2000 [27].

Sandfly fever Sicilian virus (SFSV) and SFNV were isolated from foreign soldiers stationed in Italy during 1943 and 1944. In spite of a full recovery after the so-called 'three-day' or 'Pappataci fever', the febrile illness provoked by SFSV and SFNV can be incapacitating due to headaches, myalgia, and general malaise [22]. The strongly neurotropic Toscana virus (TOSV) was also isolated in Italy first. It is the predominant cause of meningitis or meningoencephalitis during the summer season in countries bordering the Mediterranean Sea [28]. Similarly to SFSV and SFNV, several phleboviruses in Central America have been isolated from febrile soldiers or patients, such as PTV, Chagres virus (CHGV), and Alenquer virus (ALEV) [17,29,30].

Despite their obvious capacity for causing human and veterinary disease, as well as potential associated economic losses, only few phleboviruses are adequately characterized in terms of their interaction with the mammalian host organism. In this review, we will attempt to provide an overview spanning both the current knowledge about the activation of the type I interferon (IFN) system by phleboviruses, as well as the broadening spectrum of their IFN-antagonistic strategies.

2. Phleboviruses—An Emerging Group of Arthropod-Transmitted Pathogens

Phleboviruses are arboviruses that are taxonomically divided into dipteran- and tick-borne viruses. Dipteran-borne phleboviruses are generally found in eponymous *Phlebotomus* sandflies [22,31], with RVFV representing an outlier that is associated with *Aedes* and *Culex* mosquitoes, and more promiscuous in its vector range. The specific vector species are thought to be predominantly responsible for the maintenance of the viruses by vertical (transovarial) transmission, the geographic distribution of the virus and the spatial and temporal occurrence of the specific disease. Given the increasing spread of competent vector species, concerns have been raised about the potential introduction of RVFV into new areas with both susceptible vectors and hosts, and potential consequences for the human population and massive economic loss caused among affected livestock [32,33].

The epidemiological potential of the phleboviruses has been underscored by the recent identification of two new members as the causative agents of severe human disease [23]. In rural regions of China, cumulative cases of a febrile illness accompanied by thrombocytopenia, leukocytopenia, multiple organ dysfunction, and a high case-fatality rate led to the discovery of a novel phlebovirus, SFTSV, transmitted by *Haemaphysalis longicornis* ticks [34–38]. Since its discovery, SFTSV and associated cases have also been reported from Japan and Korea [39–41]. In North America, nearly simultaneously-occurring cases of a similar set of symptoms were shown to be caused by a related, tick-transmitted phlebovirus termed Heartland virus (HRTV) [42,43]. Thus, while tick-borne phleboviruses were long thought to be negligible with respect to public health, the emergence of SFTSV and HRTV suggested that this perception needed reevaluation. As one result, the genome sequences of members of the Bhanja virus (BHAV) serogroup, which has been associated with febrile illness, were determined and re-classified into the tick-borne phlebovirus group [10,44].

Accumulating reports indicate novel associations of diseases with phleboviruses in the Mediterranean area, such as sandfly fever Turkey virus (SFTV) [45–47] and Adria virus (ADRV) [48], or describe still more novel phleboviruses, such as Granada virus (GRV) [49], Adana virus (ADAV) [3] and Medjerda Valley virus (MVV) [7], to name only a few examples.

3. Viral Replication in the Mammalian Host

Phleboviruses have spherical particles of approximately 100 nm diameter [23,50,51]. They are enveloped by a host-derived lipid membrane with the two viral glycoproteins Gn and Gc decorating the surface of the virus particle, and contain three distinct single-stranded RNA genome segments which are packaged into ribonucleoprotein particles (RNPs) by the nucleocapsid protein N and associated with the RNA-dependent RNA polymerase (RdRp) L (Figure 1A). The tripartite genome consists of the large (L), medium (M), and small (S) segments. The L and M segments are of negative polarity and code for the polymerase L and a polyprotein precursor spanning the two glycoproteins and the nonstructural protein NSm, respectively (Figure 1B). The S segment uses an ambisense coding strategy, *i.e.*, it contains two genes with opposite polarities. The nucleocapsid protein N is thereby translated from a mRNA that is directly transcribed from the genomic S segment, whereas the nonstructural protein NSs mRNA is transcribed from the antigenomic S segment. Gene expression from the ambisense segments is regulated by an intergenic region (IGR), a sequence stretch that is proposed to form an irregular double-stranded RNA (dsRNA) structure [24], and by pentanucleotide transcription termination motifs [52–54]. The genome segments further contain conserved complementary oligonucleotide sequences at their 5′- and 3′-ends, allowing the formation of "panhandle" structures and the pseudocircularization of the RNPs [55].

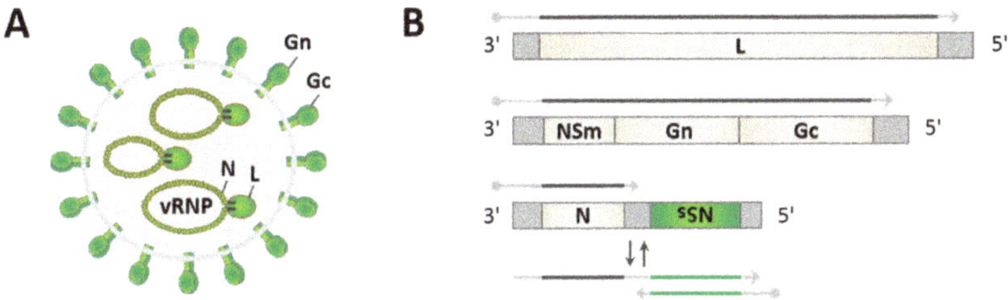

Figure 1. Prototypic phlebovirus virion and genome organization. (**A**) Virus particles contain the pseudocircularized tripartite single-stranded RNA genome, packaged into virus-sense RNPs (vRNPs) by nucleocapsid protein N and associated with the viral RNA-dependent RNA polymerase (RdRp) L, within a lipid envelope covered by heterodimers of glycoproteins Gn and Gc; and (**B**) the three viral genome segments large (L), medium (M) (both being purely negative-sense), and small (S) (ambisense) code for the structural proteins L, the Gn and Gc, and N, respectively. Viral mRNAs contain a 5′-cap (dot) and short heterogenous host-derived sequences. mRNAs transcribed from genomic RNAs are shown as grey arrows. The nonstructural protein NSs mRNA (green arrow) is synthesized from antigenomic RNA (two-colored arrow). Dipteran-borne phleboviruses also encode a nonstructural protein on the M segment (NSm).

Viral replication occurs entirely in the cytoplasm of infected mammalian host cells. Central features of the transmission from vector to host and the entry of phlebo- and other bunyaviruses has recently been reviewed elsewhere [56]. In short, after attachment of virus particles, uncoating is mediated by the fusion of the viral envelope with host membranes in the acidified compartments of the endocytic system [57,58]. Incoming RNPs then first serve as templates for primary transcription. To this end, the endonuclease domain [59,60] within the L protein cleaves host mRNAs 10–20 nucleotides downstream of the 5′-cap to use the resulting short fragments as primers for the synthesis of viral transcripts (cap snatching). Primary transcription is terminated prior to the segment termini via a specific sequence motif [52–54]. Phleboviral transcripts thus contain a 5′-cap and a short stretch of a heterogenous, host-derived sequence, but no poly(A) tail. Translation of viral proteins in the cytoplasm and at the endoplasmic reticulum (ER) is accompanied by cleavage of the polyprotein encoded by the M segment into Gn and Gc (and depending on the virus species, some other proteins e.g., NSm), heterodimerization of Gn and Gc, and their transport to the Golgi apparatus. For replication of the viral genome, the viral polymerase switches to primer-independent synthesis of full-length antigenomic RNA, which then, in turn, serves as a template for the synthesis of progeny genomic RNA. In a process called secondary transcription, these newly-generated genomes then produce even more viral mRNAs. Both the genomic and antigenomic RNA segments carry a 5′-triphosphate moiety and are packaged into RNPs. Assembly and budding finally take places at membranes of the Golgi apparatus, followed by release of virions via the secretory pathway.

The nonstructural proteins NSm and NSs are dispensable for viral replication [61–64]. sandfly-borne phleboviruses encode an NSm protein which may have a role in the regulation of apoptosis (as shown for RVFV [65]). The NSs protein is remarkable in its low conservation across the *Phlebovirus* genus compared to other viral proteins, with sequence similarities ranging only from approximately 10% to 30% [7,37]. As will be outlined below, the NSs protein is an important virulence determinant, acting as an inhibitor of the antiviral type I IFN system of the mammalian host [24,66,67].

4. The Type I Interferon System in RNA Virus Infection

Type I IFNs are cytokines that are produced by virus-infected cells [68]. In humans, there are thirteen IFN-α subtypes, a single IFN-β, and the less well-characterized IFN-ε, -τ, -κ, -ω, -δ, which activate the transcription of hundreds of IFN-stimulated genes (ISGs) [68,69]. Characterization of an ever-increasing number of ISGs shows that many of their products not only exert antiviral activity at every step of the viral replication cycle, but also possess antiproliferative and immunomodulatory functions.

The production of type I IFN is induced in response to conserved pathogen-associated molecular patterns (PAMPs), which are sensed by germline-encoded, so-called pattern-recognition receptors (PRRs). As PRRs of the cytoplasm, the RNA helicases retinoic acid-inducible gene 1 (RIG-I) and melanoma differentiation-associated protein 5 (MDA5) react to infection by distinct sets of RNA viruses [70]. RIG-I and MDA5 primarily recognize short 5′-triphosphate dsRNA, or long (preferentially of higher-order structure) dsRNA and its analogue polyinosinic:polycytidylic acid (poly(I:C)), respectively [71–73]. The prototypical RIG-I possesses two N-terminal caspase recruitment domains (CARDs), a central helicase domain and a C-terminal domain, and is kept in an auto-inhibited conformation by intramolecular interactions involving the CARDs and the helicase domains. Ligand binding by the helicase and C-terminal domains induces both ATP-dependent RIG-I oligomerization and a conformational switch, resulting in the exposure of the CARDs [73,74]. The latter then engage in K63-polyubiquitin-mediated homotypic CARD-CARD interaction with the adaptor mitochondrial antiviral signaling (MAVS) which in turn assembles prion-like fibrillary aggregates that are sufficient and necessary for the recruitment of tumor necrosis factor (TNF) receptor associated factor (TRAF) 2, 5, and 6 for downstream signaling [75,76]. The kinases TRAF family member-associated nuclear factor kappa-light-chain-enhancer of activated B cells (NF-κB) activator (TANK)-binding kinase 1 (TBK1) and inhibitor of kappa B kinase epsilon (IKKε) subsequently activate the transcription factor IFN regulatory factor 3 (IRF3) by phosphorylation, followed by its dimerization and nuclear accumulation, where it activates the production of type I IFN expression together with the transcription factors nuclear factor kappa-light-chain-enhancer of activated B cells (NF-κB) and activator protein (AP-1) [73].

Within the endosomal compartments, Toll-like receptor 3 (TLR3) recognizes viral dsRNA and poly(I:C), and signals via the adaptor Toll-interleukin 1 receptor (TIR) domain-containing adapter-inducing IFN-β (TRIF) to activate IRF3, NF-κB, and AP-1, and consequently induce the production of type I IFNs as well as inflammatory cytokines [77]. Further, recognition of single-stranded RNA by TLR7/8 and subsequent signaling via the adaptor myeloid differentiation primary response gene 88 (MyD88) results in the secretion of IFN-α, particularly by specialized plasmacytoid dendritic cells [78].

IFN-α/β bind to a common heterodimeric receptor, consisting of the subunits interferon-α/β receptor IFNAR1 and IFNAR2, on both infected and uninfected bystander cells. Signaling via the receptor-associated tyrosine kinases Janus kinase 1 (JAK1) and tyrosine kinase 2 (TYK2) leads to phosphorylation of Signal Transducer and Activator of Transcription 1 (STAT1) and STAT2, which then undergo heterodimerization and translocation to the nucleus. There, in a complex with IRF9, they bind to IFN-stimulated response elements (ISRE) within ISG promoters, finally resulting in the transcription of ISGs [68,69].

As the functions of the well characterized ISGs have been reviewed extensively elsewhere [68,79], only a few examples of antiviral ISGs will be described here. IFN-inducible transmembrane (IFITM) proteins interfere with fusion of the viral envelope at the plasma membrane (IFITM1) or in the endosomal pathway (IFITM2, 3) and, thus, the release of viral RNPs into the cytoplasm of infected cells [80]. The family of dynamin-like Mx GTPases are capable of restricting a wide range of viruses, presumably via trapping and missorting of incoming viral RNPs [81]. In contrast to other ISGs, Mx proteins are not expressed at low constitutive levels or in response to virus infection, but depend entirely on IFN signaling, rendering the abrogation of IFN induction and signaling an effective means of evading Mx activity. Protein kinase R (PKR) is expressed at low levels in an inactive form [82]. Binding of dsRNA results in PKR activation, leading to phosphorylation of its target eukaryotic initiation factor 2α (eIF2α) and, in consequence, the inhibition of the translation of both viral and cellular mRNA. PKR has also been implicated in NF-κB activation and the induction of apoptosis [83]. Interferon-induced protein with tetratricopeptide repeats (IFIT) proteins IFIT1, 2 and 3 are involved in translation inhibition and innate recognition of RNAs that lack proper 2′-*O* methylation or contain a 5′ ppp end [80,84].

Expression of the transcription factor IRF7 is also enhanced by IFN signaling. While the aforementioned activation of IRF3 leads to an initial wave of type I IFN secretion, including IFN-β and (in mice) IFN-α4, enhanced IRF7 expression and activation generates a second wave of type I IFN production which involves additional IFN-α subtypes [85].

In addition to direct antiviral effects of ISGs and the positive feedback loop via IRF7, type I IFN signaling also induces the production of a range of cytokines and chemokines, pro- and antiapoptotic factors, and affects multiple other signaling pathways. Through modulation of the differentiation and function of dendritic cells, T cells, natural killer (NK) cells, and B cells, type I IFNs shape the antiviral immune response beyond the initial innate immune response [68,79,86].

5. Activation of the Interferon System by Phleboviruses

Like other negative-strand RNA viruses, phleboviruses do not produce substantial amounts of dsRNA during infection [87,88]. As shown for RVFV, their naked virion RNA is, nonetheless, a strong activator of RIG-I due to the presence of the 5′-triphosphorylated dsRNA panhandle formed by the genome ends [89]. Moreover, also when packaged into RNPs, the RNA of RVFV particles can activate the RIG-I signaling pathway [90]. In fact, incoming RNPs already trigger RIG-I conformational switching and oligomerization, as well as IRF3 activation. Additionally, in vivo, the cytoplasmic RNA helicase/MAVS axis was demonstrated to be the primary IFN induction pathway for RVFV [91]. The *in vivo* role of TLRs, by contrast, is less clear. While Ermler et al. found for RVFV that neither the TLR7/8-MyD88 nor the TLR3-TRIF pathway play a significant role in IFN induction [91], Gowen *et al.* showed for PTV that TLR3 was activated and contributed to increased liver damage and mortality [92]. It remains to future studies to reveal whether these discrepancies are due to different experimental conditions or a differential ability of distinct phleboviruses to activate or inhibit TLR3.

Studies in a range of animal models suggested a protective effect of type I IFN in phleboviral infection. Treatment with synthetic type I IFN inducers, such as poly(I:C) or polyinosinic-polycytidylic acid, poly-L-lysine and carboxymethylcellulose (poly(ICLC)) in a prophylactic or therapeutic regimen was reported to protect mice and hamsters from lethal RVFV infection [93–95]. Similarly, administration of poly(I:C), poly(ICLC), or of IFN itself protect against PTV-induced liver damage and mortality in a mouse model [96–98], whereas treatment with IFN-neutralizing antibodies rendered otherwise resistant mice susceptible to PTV-associated death [99]. Several in vivo studies correlated the onset of type I IFN synthesis with increased survival after lethal RVFV challenge [100,101]. Lastly, mice deficient in IFN signaling are more prone to infections with RVFV and PTV [102,103]. Thus, induction of sufficient amounts of type I IFNs at an early point during infection is crucial for protective effects.

It is known that different viruses are targeted by distinct sets of ISGs [104,105]. Additionally, for phleboviruses, a number of inhibitory ISG products have been described (Table 1). Mx proteins drastically inhibit the replication of several phleboviruses, including RVFV, TOSV, and SFSV [106,107]. For human MxA it was shown that it sequesters RVFV N into large perinuclear complexes, thereby inhibiting primary and secondary transcription [108,109]. Replication of RVFV is also affected by IFITM2 and IFITM3, but not IFITM1, in accordance with their localization in the endocytic pathway and at the plasma membrane, respectively [110]. PKR is activated during phleboviral infection and can act as potent restriction factor [93,111]. Therefore, it is not surprising that PKR is targeted by different phleboviruses, as discussed below. Furthermore, IFIT proteins (mostly IFIT1 and IFIT2), long isoform of poly(ADP-ribose) polymerase 12 (PARP12L), 2'-5' oligoadenylate synthetase-like 2 (OASL2), and ISG15 influence the replication of RVFV [100,112,113]. The latter two ISGs are not upregulated in embryonic fibroblasts derived from a mouse strain with increased susceptibility to RVFV (MBT/Pas) and a generally decreased and delayed ISG response, compared to BALB/cByJ, C57BL/6J and 129/Sv/Pas mice. Small interfering RNA (siRNA)-mediated reduction of Oasl2 and ISG15, however, resulted in only slightly increased titers of recombinant NSs-deficient RVFV [100].

Table 1. Interferon (IFN)-induced proteins acting as restriction factors for phleboviruses.

ISG	Affected Step in Replication	Affected Phleboviruses (Strains)	References
IFITM2, 3	uncoating	RVFV (ZH501, MP12)	[110]
Mx	primary and secondary transcription	RVFV (MP12, Clone 13), TOSV, SFSV	[106–108]
OASL2	?	RVFV (rZH548ΔNSs)	[100]
PKR	viral protein translation	NSs-deficient RVFV mutants (e.g., Clone 13)	[93,111]
IFIT1-3	viral protein translation	RVFV (Clone 13)	[113]
mPARP12	?	RVFV (MP12)	[112]
ISG15	?	RVFV (rZH548ΔNSs)	[100]

ISG: IFN-stimulated genes; IFITM: IFN-inducible transmembrane; OASL2: 2′-5′ oligoadenylate synthetase-like 2; PKR: protein kinase R; IFIT: interferon-induced protein with tetratricopeptide repeats; mPARP12: murine poly(ADP-ribose) polymerase 12; RVFV: Rift Valley fever virus; TOSV: Toscana virus; SFSV: sandfly fever Sicilian virus.

6. Viral Countermeasures

As described above, phleboviruses are sensitive to IFN and an early induction of type I IFN seems to be a determinant of disease outcome in animal models. Furthermore, given the segmented nature of their genome, phleboviruses carry at least three RIG-I-activating moieties (5′ppp-dsRNA panhandle) per virus particle. Thus, in order to compensate for their stimulatory potential and to prevent or sufficiently delay a type I IFN response, they require highly efficient counterstrategies (Figure 2, Tables 2 and 3).

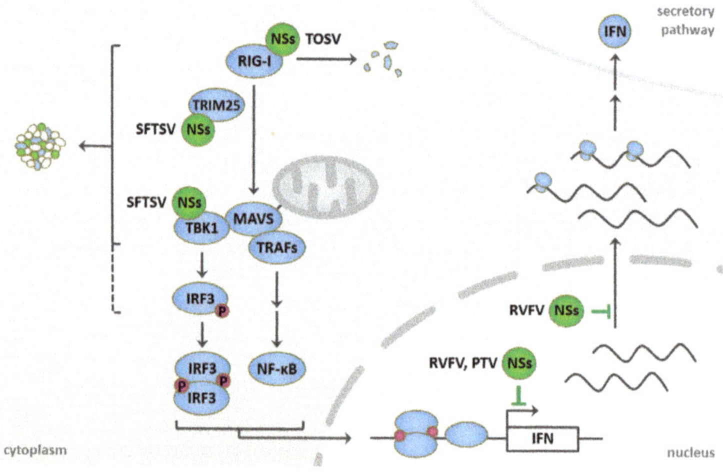

Figure 2. Known host targets of phleboviral NSs proteins in retinoic acid-inducible gene 1 (RIG-I) signaling and type I IFN induction. Incoming phleboviral RNPs are sensed by RIG-I, potentially leading to interferon induction via mitochondrial antiviral signaling (MAVS)-mediated activation of the transcription factors interferon regulatory factor 3 (IRF3) and nuclear factor kappa-light-chain-enhancer of activated B cells (NF-κB). NSs proteins, however, mediate the escape from the induction and the antiviral effects of the IFN system. Rift Valley fever virus (RVFV) NSs acts in the nucleus, where it blocks both the transcription and the export of host mRNAs. Toscana virus (TOSV) NSs localizes to the cytoplasm, where it interacts with and induces proteasomal degradation of RIG-I. Characteristic cytoplasmic structures are formed by severe fever with thrombocytopenia syndrome virus (SFTSV) NSs and serve as site of sequestration for several signaling factors of the RIG-I pathway. Punta Toro virus (PTV) NSs also inhibits host transcription.

The NSs protein of RVFV was the first to be identified as an IFN antagonist and still remains the most extensively studied phleboviral virulence factor. Comparative studies using the naturally-attenuated strain Clone 13 and the virulent RVFV isolate ZH548, as well as reassortants between these two viruses, showed that the S segment carries the determinant for attenuation and interference with IFN-α/β production in a murine model [102,114]. Since the S segment encodes the NSs and Clone 13 is a natural NSs deletion mutant, it was concluded that NSs confers an anti-IFN activity.

Although phleboviruses replicate exclusively in the cytoplasm, RVFV NSs is localized in the nucleus, forming characteristic filaments [115,116]. In contrast, Clone 13 contains a large in-frame deletion within the NSs open reading frame, resulting in a loss of 69% of the ORF [117]. Hence, Clone 13 NSs does not form nuclear filaments but instead is rapidly degraded [114]. The NSs of the RVFV wild-type strain ZH548 alone was then shown to almost completely block IFN-β promoter activation in response to poly(I:C), while Clone 13 NSs had no inhibitory effect [118]. Further, ZH548 did not affect IRF3 dimerization or nuclear accumulation, yet impaired IFN-β, NF-κB-driven, AP-1-driven, and even SV40 promoter activity, suggesting that RVFV NSs broadly inhibits both inducible and constitutive host cell transcription. Indeed, RVFV NSs targets the host mRNA synthesis machinery to induce a general cellular shutoff, including sequestration of general transcription factor II H (TFIIH) subunit p44 and, thus, prevention of TFIIH assembly [119]. In addition, NSs triggers the rapid proteasomal degradation of the TFIIH subunit p62 early in infection [120]. Proteomic analyses led to the identification of the F-box protein FBXO3 as host cell interactor of RVFV NSs [121]. F-box proteins are the substrate recognition component of modular E3 ubiquitin ligases of the Skp1, Cullin1, F-box (SCF) protein type [122], and FBXO3 was shown to be recruited by NSs to achieve TFIIH-p62 degradation [123]. The interaction with TFIIH-p62 thereby depends on a ΩXaV motif (where Ω: aromatic, X: any, a: acidic, V: valine) located in the C-terminal region of RVFV NSs [124]. Moreover, a nuclear mRNA export block was observed in RVFV NSs-expressing cells [125]. In contrast to these broadly-acting host cell shutoff mechanisms, RVFV NSs was also reported to specifically inhibit IFN induction by recruiting a transcriptional suppressor complex containing Sin3A associated protein 30 (SAP30) to the IFN-β promoter [126].

If not counteracted by viral measures, PKR has a strong restrictive effect on the replication of phleboviruses [93,111]. RVFV solves this problem by proteasomal degradation of PKR, thereby avoiding eIF2α phosphorylation and inhibition of translation [93,111]. Recent studies revealed that RVFV NSs recruits the F-box proteins FBXW11 and FBXW1 (also called beta-transducin repeat containing protein 1 (β-TRCP1)) as specific adaptors to mediate PKR degradation [127,128]. NSs thereby directly interacts with FBXW11/β-TRCP1 through a "degron" sequence [128]. Remarkably, this degron motif (DDGFVE) overlaps with the aforementioned ΩXaV motif (FVEV) necessary for TFIIH-p62 degradation, suggesting that RVFV NSs utilizes the very C-terminal part of the protein for the degradation of multiple host target factors, each time recruiting specific F-box proteins.

Infection of hamsters and mice with the PTV strain Adames is lethal, whereas PTV strain Balliet produces beneficial outcomes [103,129]. Reassortants between these two strains again identified the S segment genotype and NSs expression as correlates for lethality and suppression of type I IFN production [130]. Similar to RVFV, the NSs of PTV Adames has also been found to inhibit host transcription [131]. In contrast to RVFV NSs, however, PTV NSs does not form nuclear filaments or share the C-terminal ΩXaV motif of RVFV NSs [124]. A further difference between RVFV and PTV NSs is that the latter does not affect the levels of PKR [131,132].

In contrast to RVFV NSs, TOSV NSs localizes exclusively to the cytoplasm and does not affect cellular transcription [132,133], but inhibits IFN induction [134]. Instead, it has been shown to interact with RIG-I and trigger its proteasomal degradation [135]. Interestingly, binding of RIG-I and proteasomal degradation appear to be mediated by different regions of the NSs protein [136]. Also contrary to RVFV NSs, levels of TOSV NSs were found to be increased under MG132 treatment [132,135]. In line with this, C-terminally-truncated TOSV NSs mutants that were incapable of degrading RIG-I, but still able to bind RIG-I, were also detected at higher levels than the full-length protein [136], allowing speculations that TOSV NSs might be degraded along with its host target.

Peculiarly, while TOSV NSs efficiently inhibited IRF3 activation and IFN induction when expressed via transfection or from a recombinant RVFV, infection with the parental Italian TOSV isolate resulted in IRF3 activation, IFN-β induction and Mx expression [134]. A Spanish isolate, by contrast, was a potent IFN suppressor as expected from NSs action [137]. This discrepancy might be attributable to strain-specific differences in the kinetics of NSs accumulation.

Like RVFV NSs, TOSV NSs has also been observed to induce degradation of PKR in a proteasome-dependent manner [131,132].

Additionally, the NSs protein of the intermediately-pathogenic sandfly-borne SFSV possesses the capacity for inhibiting type I IFN induction [93,131]. The levels of PKR, however, are not affected [93,131,132].

The recent identification of tick-borne SFTSV as human pathogenic phlebovirus was quickly followed by a number of reports concerning the anti-IFN mechanism employed by its NSs protein. Type I IFN and ISGs were only moderately induced in SFTSV-infected cells, as observed by microarray analysis [138]. Indeed, SFTSV NSs was identified by several groups as inhibitor of IFN-β promoter activity, presumably acting at the level of TBK1 and IKKε [138–141].

SFTSV NSs neither forms nuclear filaments, nor is it diffusely distributed in the cytoplasm as described for the NSs proteins of sandfly-borne phleboviruses. Instead, it is concentrated in unprecedented cytoplasmic structures of granular appearance after both infection and transfection of a wide range of cell lines. Although these 'viral inclusion bodies' or 'viroplasm-like structures' seem to be subject to dynamic fission and fusion [139] and were found to be positive for the autophagosome marker microtubule-associated protein 1A/1B-light chain 3 (LC3), their formation was independent of autophagy-related protein 7 (Atg7), suggesting that they are not classical autophagosomes [140]. The early endosome RAS-associated protein Rab5 showed co-localization, but neither its presence nor canonical function were required for the formation of NSs inclusion bodies. Further analysis ruled out an association with ER, Golgi, mitochondria, peroxisomes, EDEMosomes, lysosomes and late endosomes, as well as aggresomes [139,140]. Furthermore, the inclusion bodies appeared to co-localize with lipid droplets and their formation associated with fatty acid synthesis [142].

Despite the open questions concerning biochemical composition and compartmental identity of the SFTSV NSs inclusions, it has become clear that they represent a site of sequestration and spatial isolation of multiple components of the RIG-I signaling pathway [139–141]. While all studies agree on TBK1 (and IKKε, where tested) as host interactors of SFTSV NSs, individual studies reported additional interactions with tripartite motif-containing protein 25 (TRIM25) (an E3 ubiquitin ligase involved in RIG-I signaling [143]), RIG-I [140], and IRF3 [139,141]. Remarkably, SFTSV NSs also sequesters transcription factors STAT1 and STAT2 into the inclusion bodies and inhibits STAT2 phosphorylation, thus interfering with their nuclear translocation, the stimulation of the interferon-stimulated response element (ISRE) promoter and, consequently, the induction of ISGs [144].

The non-pathogenic UUKV is the prototype of tick-borne phleboviruses. Its NSs is distributed throughout the cytoplasm [145] and has only a weak IFN-antagonistic effect [146]. Currently, there are no reports concerning the IFN-inhibitory capacity or action of the NSs proteins of HRTV and BHAV, despite their association with human illness.

Given the ambisense coding strategy of the S segment, the NSs would be expected to be expressed only late, after production of viral antigenomic RNA. This would represent a considerable disadvantage for the virus and is contradictory to the NSs-mediated effects that occur early after infection. This contradiction is resolved by the observation that antigenomic RNA segments are packaged into virions in both dipteran-borne RVFV and tick-borne UUKV [145,147,148]. Thus, the respective NSs proteins are directly produced during primary transcription, despite being encoded on the antigenomic RNA.

Table 2. IFN-related host pathways targeted by diptera-borne phleboviruses.

Phlebovirus	Host Target	Mechanism	References
RVFV	TFIIH p44, XPB	sequestration	[119]
	TFIIH p62	proteasomal degradation by recruitment of a SKP1-FBXO3 E3-ubiquitin ligase complex	[120,123,124]
	SAP30	recruitment of suppressors to the IFN promoter	[126]
	mRNA export	unknown	[125]
	PKR	proteasomal degradation by recruitment of a SKP1-CUL1-FBXW11 E3 ligase complex	[93,111,127, 128]
TOSV	RIG-I	proteasomal degradation	[134–136]
	PKR	proteasomal degradation	[132]
PTV	IFN induction	unknown	[131]
SFSV	IFN induction	unknown	[93,131]

PTV: Punta Toro virus; TFIIH: transcription factor II H; XPB: *xeroderma pigmentosum* type B; SAP30: Sin3A associated protein 30; RIG-I: retinoic acid-inducible gene 1.

Table 3. IFN-related host pathways targeted by tick-borne phleboviruses.

Phlebovirus	Host Target	Mechanism	References
SFTSV	RIG-I, TRIM25, TBK1/IKKε, IRF3	sequestration into cytoplasmic inclusion bodies	[138–141]
	STAT1, STAT2	sequestration into cytoplasmic inclusion bodies	[144]
UUKV	unknown	unknown	[146]

SFTSV: severe fever with thrombocytopenia syndrome virus; UUKV: Uukuniemi virus; TRIM: tripartite motif-containing protein; TBK1: (TANK)-binding kinase 1; IKKε: inhibitor of kappa B kinase epsilon; STAT: Signal Transducer and Activator of Transcription.

7. Conclusions and Future Directions

Phleboviruses are emerging arboviruses, causing human diseases ranging from mild febrile illness to severe cases of hemorrhagic fever or multiple organ dysfunction and death. Further, RVFV is associated with livestock epidemics and substantial economic losses. Within the genus *Phlebovirus*, the NSs protein is only weakly conserved in terms of its amino acid sequences or subcellular localization. Nevertheless, NSs proteins are highly conserved in their function as IFN antagonist, with their variety in sequence and localization being mirrored by a plethora of different molecular strategies. The diversity of IFN-antagonistic mechanisms of distinct phleboviruses tempts one to speculate whether a correlation between the NSs action and the degree of virulence exists.

Among sandfly-borne viruses, a common strategy of the more pathogenic members, such as RVFV and TOSV, seems the proteasomal degradation of host target factors that are involved in IFN induction or antiviral effector functions. Interestingly, while the NSs of the highly-virulent RVFV is not negatively affected, the NSs of intermediately pathogenic TOSV NSs seems to be susceptible to the proteasomal degradation machinery as well. For the highly pathogenic tick-borne virus SFTSV, the NSs protein sequesters a major fraction of the host factors involved in the RIG-I signaling pathway, as well as IFN signaling factors, into characteristic granular structures in the cytoplasm. This might suggest that broader action on multiple host cell functions, such as the general transcription block caused by RVFV or the deactivation of entire signaling chains, as seen for SFTSV, might be a correlate of increased phleboviral virulence.

Much of our current understanding of the functioning of the NSs protein has been achieved employing reverse genetics, such as the rMP12 and rZH548 rescue systems for the dipteran-borne RVFV [63,149–151], allowing the study of NSs-deficient mutants or chimeric viruses. Recently, reverse genetic systems were also established for the tick-borne phleboviruses UUKV and SFTSV [146,147].

It remains to future studies to further expand and specify the molecular characterization of the NSs proteins of both familiar and newly-emerging phleboviruses.

Acknowledgments: Our work is kindly supported by the SFB 1021 of the Deutsche Forschungsgemeinschaft (DFG) and by the Infect-ERA grant "ESCential" by the Bundesministerium für Bildung und Forschung (BMBF).

Conflicts of Interest: The authors declare no conflict of interest.

References

1. Walter, C.T.; Barr, J.N. Recent advances in the molecular and cellular biology of bunyaviruses. *J. Gen. Virol.* **2011**, *92*, 2467–2484. [CrossRef] [PubMed]
2. International Committee on Taxonomy of Viruses. Phleboviruses. Available online: http://ictvonline.org/virusTaxonomy.asp?version=2012 (accessed on 15 July 2012).
3. Alkan, C.; Alwassouf, S.; Piorkowski, G.; Bichaud, L.; Tezcan, S.; Dincer, E.; Ergunay, K.; Ozbel, Y.; Alten, B.; de Lamballerie, X.; *et al.* Isolation, genetic characterization, and seroprevalence of Adana virus, a novel phlebovirus belonging to the Salehabad virus complex, in Turkey. *J. Virol.* **2015**, *89*, 4080–4091. [CrossRef] [PubMed]
4. Alkan, C.; Kasap, O.E.; Alten, B.; de Lamballerie, X.; Charrel, R.N. Sandfly-borne phlebovirus isolations from Turkey: New insight into the sandfly fever Sicilian and sandfly fever Naples species. *PLoS Negl. Trop. Dis.* **2016**, *10*, e0004519. [CrossRef] [PubMed]
5. Amaro, F.; Hanke, D.; Ze-Ze, L.; Alves, M.J.; Becker, S.C.; Hoper, D. Genetic characterization of Arrabida virus, a novel phlebovirus isolated in South Portugal. *Virus Res.* **2016**, *214*, 19–25. [CrossRef] [PubMed]
6. Amaro, F.; Ze-Ze, L.; Alves, M.J.; Borstler, J.; Clos, J.; Lorenzen, S.; Becker, S.C.; Schmidt-Chanasit, J.; Cadar, D. Co-circulation of a novel phlebovirus and Massilia virus in sandflies, Portugal. *Virol. J.* **2015**, *12*, 174. [CrossRef] [PubMed]
7. Bichaud, L.; Dachraoui, K.; Alwassouf, S.; Alkan, C.; Mensi, M.; Piorkowski, G.; Sakhria, S.; Seston, M.; Fares, W.; de Lamballerie, X.; *et al.* Isolation, full genomic characterization and neutralization-based human seroprevalence of Medjerda Valley virus, a novel sandfly-borne phlebovirus belonging to the Salehabad virus complex in northern Tunisia. *J. Gen. Virol.* **2016**, *97*, 602–610. [CrossRef] [PubMed]
8. Charrel, R.N.; Moureau, G.; Temmam, S.; Izri, A.; Marty, P.; Parola, P.; da Rosa, A.T.; Tesh, R.B.; de Lamballerie, X. Massilia virus, a novel *Phlebovirus* (Bunyaviridae) isolated from sandflies in the Mediterranean. *Vector Borne Zoonotic Dis.* **2009**, *9*, 519–530. [CrossRef] [PubMed]
9. Dachraoui, K.; Fares, W.; Bichaud, L.; Barhoumi, W.; Beier, J.C.; Derbali, M.; Cherni, S.; Lamballerie, X.; Chelbi, I.; Charrel, R.N.; *et al.* Phleboviruses associated with sand flies in arid bio-geographical areas of Central Tunisia. *Acta Trop.* **2016**, *158*, 13–19. [CrossRef] [PubMed]
10. Matsuno, K.; Weisend, C.; Travassos da Rosa, A.P.; Anzick, S.L.; Dahlstrom, E.; Porcella, S.F.; Dorward, D.W.; Yu, X.J.; Tesh, R.B.; Ebihara, H. Characterization of the Bhanja serogroup viruses (Bunyaviridae): A novel species of the genus *Phlebovirus* and its relationship with other emerging tick-borne phleboviruses. *J. Virol.* **2013**, *87*, 3719–3728. [CrossRef] [PubMed]
11. Matsuno, K.; Weisend, C.; Kajihara, M.; Matysiak, C.; Williamson, B.N.; Simuunza, M.; Mweene, A.S.; Takada, A.; Tesh, R.B.; Ebihara, H. Comprehensive molecular detection of tick-borne phleboviruses leads to the retrospective identification of taxonomically unassigned bunyaviruses and the discovery of a novel member of the genus phlebovirus. *J. Virol.* **2015**, *89*, 594–604. [CrossRef] [PubMed]
12. Palacios, G.; da Rosa, A.T.; Savji, N.; Sze, W.; Wick, I.; Guzman, H.; Hutchison, S.; Tesh, R.; Lipkin, W.I. Aguacate virus, a new antigenic complex of the genus *Phlebovirus* (family Bunyaviridae). *J. Gen. Virol.* **2011**, *92*, 1445–1453. [CrossRef] [PubMed]
13. Palacios, G.; Tesh, R.; Travassos da Rosa, A.; Savji, N.; Sze, W.; Jain, K.; Serge, R.; Guzman, H.; Guevara, C.; Nunes, M.R.; *et al.* Characterization of the Candiru antigenic complex (Bunyaviridae: *Phlebovirus*), a highly diverse and reassorting group of viruses affecting humans in tropical America. *J. Virol.* **2011**, *85*, 3811–3820. [CrossRef] [PubMed]
14. Palacios, G.; Savji, N.; Travassos da Rosa, A.; Desai, A.; Sanchez-Seco, M.P.; Guzman, H.; Lipkin, W.I.; Tesh, R. Characterization of the Salehabad virus species complex of the genus *Phlebovirus* (Bunyaviridae). *J. Gen. Virol.* **2013**, *94*, 837–842. [CrossRef] [PubMed]

15. Palacios, G.; Savji, N.; Travassos da Rosa, A.; Guzman, H.; Yu, X.; Desai, A.; Rosen, G.E.; Hutchison, S.; Lipkin, W.I.; *et al.* Characterization of the Uukuniemi virus group (*Phlebovirus*: Bunyaviridae): Evidence for seven distinct species. *J. Virol.* **2013**, *87*, 3187–3195. [CrossRef] [PubMed]
16. Palacios, G.; Tesh, R.B.; Savji, N.; Travassos da Rosa, A.P.; Guzman, H.; Bussetti, A.V.; Desai, A.; Ladner, J.; Sanchez-Seco, M.; Lipkin, W.I. Characterization of the sandfly fever Naples species complex and description of a new Karimabad species complex (genus *Phlebovirus*, family Bunyaviridae). *J. Gen. Virol.* **2014**, *95*, 292–300. [CrossRef] [PubMed]
17. Palacios, G.; Wiley, M.R.; Travassos da Rosa, A.P.; Guzman, H.; Quiroz, E.; Savji, N.; Carrera, J.P.; Bussetti, A.V.; Ladner, J.T.; Lipkin, W.I.; *et al.* Characterization of the Punta Toro species complex (genus *Phlebovirus*, family Bunyaviridae). *J. Gen. Virol.* **2015**, *96*, 2079–2085. [CrossRef] [PubMed]
18. Remoli, M.E.; Fortuna, C.; Marchi, A.; Bucci, P.; Argentini, C.; Bongiorno, G.; Maroli, M.; Gradoni, L.; Gramiccia, M.; Ciufolini, M.G. Viral isolates of a novel putative phlebovirus in the Marche Region of Italy. *Am. J. Trop. Med. Hyg.* **2014**, *90*, 760–763. [CrossRef] [PubMed]
19. Xu, F.; Chen, H.; Travassos, A.P.; Tesh, R.B.; Xiao, S.Y. Phylogenetic relationships among sandfly fever group viruses (*Phlebovirus*: Bunyaviridae) based on the small genome segment. *J. Gen. Virol.* **2007**, *88*, 2312–2319. [CrossRef] [PubMed]
20. Zhao, G.Y.; Krishnamurthy, S.; Cai, Z.Q.; Popov, V.L.; da Rosa, A.P.T.; Guzman, H.; Cao, S.; Virgin, H.W.; Tesh, R.B.; Wang, D. Identification of novel viruses using VirusHunter—An automated data analysis pipeline. *PLoS ONE* **2013**, *8*, e78470. [CrossRef] [PubMed]
21. Zhioua, E.; Moureau, G.; Chelbi, I.; Ninove, L.; Bichaud, L.; Derbali, M.; Champs, M.; Cherni, S.; Salez, N.; Cook, S.; *et al.* Punique virus, a novel phlebovirus, related to sandfly fever Naples virus, isolated from sandflies collected in Tunisia. *J. Gen. Virol.* **2010**, *91*, 1275–1283. [CrossRef] [PubMed]
22. Alkan, C.; Bichaud, L.; de Lamballerie, X.; Alten, B.; Gould, E.A.; Charrel, R.N. Sandfly-borne phleboviruses of Eurasia and Africa: Epidemiology, genetic diversity, geographic range, control measures. *Antivir. Res.* **2013**, *100*, 54–74. [CrossRef] [PubMed]
23. Elliott, R.M.; Brennan, B. Emerging phleboviruses. *Curr. Opin. Virol.* **2014**, *5*, 50–57. [CrossRef] [PubMed]
24. Boshra, H.; Lorenzo, G.; Busquets, N.; Brun, A. Rift Valley fever: Recent insights into pathogenesis and prevention. *J. Virol.* **2011**, *85*, 6098–6105. [CrossRef] [PubMed]
25. Bird, B.H.; Ksiazek, T.G.; Nichol, S.T.; Maclachlan, N.J. Rift Valley fever virus. *J. Am. Vet. Med. Assoc.* **2009**, *234*, 883–893. [CrossRef] [PubMed]
26. Daubney, R.; Hudson, J.R.; Garnham, P.C. Enzootic hepatitis or Rift Valley fever. An undescribed virus disease of sheep cattle and man from East Africa. *J. Pathol. Bacteriol.* **1931**, *34*, 545–579. [CrossRef]
27. Arishi, H.; Ageel, A.; Rahman, M.A.; Hazmi, A.A.; Arishi, A.R.; Ayoola, B.; Menon, C.; Ashraf, J.; Frogusin, O.; Sawwan, F.; *et al.* Outbreak of Rift Valley fever—Saudi Arabia, August-October, 2000. *MMMW Rep.* **2000**, *49*, 905–908.
28. Charrel, R.N.; Bichaud, L.; de Lamballerie, X. Emergence of Toscana virus in the mediterranean area. *World J. Virol.* **2012**, *1*, 135–141. [CrossRef] [PubMed]
29. Srihongse, S.; Johnson, C.M. Human infections with Chagres virus in Panama. *Am. J. Trop. Med. Hyg.* **1974**, *23*, 690–693. [PubMed]
30. Travassos da Rosa, A.P.; Tesh, R.B.; Pinheiro, F.P.; Travassos da Rosa, J.F.; Peterson, N.E. Characterization of eight new phlebotomus fever serogroup arboviruses (Bunyaviridae: *Phlebovirus*) from the Amazon region of Brazil. *Am. J. Trop. Med. Hyg.* **1983**, *32*, 1164–1171. [PubMed]
31. Tesh, R.B. The genus Phlebovirus and its vectors. *Annu. Rev. Entomol.* **1988**, *33*, 169–181. [CrossRef] [PubMed]
32. Rolin, A.I.; Berrang-Ford, L.; Kulkarni, M.A. The risk of Rift Valley fever virus introduction and establishment in the United States and European Union. *Emerg. Microbes Infect.* **2013**, *2*, e81. [CrossRef] [PubMed]
33. Salman, M. Is the United States really at risk for introduction of Rift Valley fever virus? *J. Am. Vet. Med. Assoc.* **2013**, *242*, 606–608. [CrossRef] [PubMed]
34. Huang, X.; Liu, L.; Du, Y.; Ma, H.; Mu, Y.; Tang, X.; Wang, H.; Kang, K.; Zhang, S.; Wu, W.; *et al.* Detection of a novel bunyavirus associated with fever, thrombocytopenia and leukopenia syndrome in Henan Province, China, using real-time reverse transcription PCR. *J. Med. Microbiol.* **2013**, *62*, 1060–1064. [CrossRef] [PubMed]

35. Luo, L.M.; Zhao, L.; Wen, H.L.; Zhang, Z.T.; Liu, J.W.; Fang, L.Z.; Xue, Z.F.; Ma, D.Q.; Zhang, X.S.; Ding, S.J.; *et al.* Haemaphysalis longicornis ticks as reservoir and vector of severe fever with thrombocytopenia syndrome virus in China. *Emerg. Infect. Dis.* **2015**, *21*, 1770–1776. [CrossRef] [PubMed]
36. Xu, B.; Liu, L.; Huang, X.; Ma, H.; Zhang, Y.; Du, Y.; Wang, P.; Tang, X.; Wang, H.; Kang, K.; *et al.* Metagenomic analysis of fever, thrombocytopenia and leukopenia syndrome (FTLS) in Henan Province, China: Discovery of a new bunyavirus. *PLoS Pathog.* **2011**, *7*, e1002369. [CrossRef] [PubMed]
37. Yu, X.J.; Liang, M.F.; Zhang, S.Y.; Liu, Y.; Li, J.D.; Sun, Y.L.; Zhang, L.; Zhang, Q.F.; Popov, V.L.; Li, C.; *et al.* Fever with thrombocytopenia associated with a novel bunyavirus in China. *N. Engl. J. Med.* **2011**, *364*, 1523–1532. [CrossRef] [PubMed]
38. Zhang, Y.Z.; Zhou, D.J.; Qin, X.C.; Tian, J.H.; Xiong, Y.; Wang, J.B.; Chen, X.P.; Gao, D.Y.; He, Y.W.; Jin, D.; *et al.* The ecology, genetic diversity, and phylogeny of Huaiyangshan virus in China. *J. Virol.* **2012**, *86*, 2864–2868. [CrossRef] [PubMed]
39. Kim, K.H.; Yi, J.; Kim, G.; Choi, S.J.; Jun, K.I.; Kim, N.H.; Choe, P.G.; Kim, N.J.; Lee, J.K.; Oh, M.D. Severe fever with thrombocytopenia syndrome, South Korea, 2012. *Emerg. Infect. Dis.* **2013**, *19*, 1892–1894. [CrossRef] [PubMed]
40. Park, S.W.; Han, M.G.; Yun, S.M.; Park, C.; Lee, W.J.; Ryou, J. Severe Fever with thrombocytopenia syndrome virus, South Korea, 2013. *Emerg. Infect. Dis.* **2014**, *20*, 1880–1882. [CrossRef] [PubMed]
41. Takahashi, T.; Maeda, K.; Suzuki, T.; Ishido, A.; Shigeoka, T.; Tominaga, T.; Kamei, T.; Honda, M.; Ninomiya, D.; Sakai, T.; *et al.* The First Identification and retrospective study of severe fever with thrombocytopenia syndrome in Japan. *J. Infect. Dis.* **2014**, *209*, 816–827. [CrossRef] [PubMed]
42. McMullan, L.K.; Folk, S.M.; Kelly, A.J.; MacNeil, A.; Goldsmith, C.S.; Metcalfe, M.G.; Batten, B.C.; Albarino, C.G.; Zaki, S.R.; Rollin, P.E.; *et al.* A New phlebovirus associated with severe febrile illness in Missouri. *N. Engl. J. Med.* **2012**, *367*, 834–841. [CrossRef] [PubMed]
43. Savage, H.M.; Godsey, M.S.; Lambert, A.; Panella, N.A.; Burkhalter, K.L.; Harmon, J.R.; Lash, R.R.; Ashley, D.C.; Nicholson, W.L. First detection of Heartland virus (Bunyaviridae: *Phlebovirus*) from ffield collected arthropods. *Am. J. Trop. Med. Hyg.* **2013**, *89*, 445–452. [CrossRef] [PubMed]
44. Dilcher, M.; Alves, M.J.; Finkeisen, D.; Hufert, F.; Weidmann, M. Genetic characterization of Bhanja virus and Palma virus, two tick-borne phleboviruses. *Virus Genes* **2012**, *45*, 311–315. [CrossRef] [PubMed]
45. Carhan, A.; Uyar, Y.; Ozkaya, E.; Ertek, M.; Dobler, G.; Dilcher, M.; Wang, Y.; Spiegel, M.; Hufert, F.; Weidmann, M. Characterization of a sandfly fever Sicilian virus isolated during a sandfly fever epidemic in Turkey. *J. Clin. Virol. Off. Publ. Pan Am. Soc. Clin. Virol.* **2010**, *48*, 264–269. [CrossRef] [PubMed]
46. Ergunay, K.; Ismayilova, V.; Colpak, I.A.; Kansu, T.; Us, D. A case of central nervous system infection due to a novel sandfly fever virus (SFV) variant: Sandfly fever Turkey virus (SFTV). *J. Clin. Virol.* **2012**, *54*, 79–82. [CrossRef] [PubMed]
47. Kocak Tufan, Z.; Weidmann, M.; Bulut, C.; Kinikli, S.; Hufert, F.T.; Dobler, G.; Demiroz, A.P. Clinical and laboratory findings of a sandfly fever Turkey virus outbreak in Ankara. *J. Infect.* **2011**, *63*, 375–381. [CrossRef] [PubMed]
48. Anagnostou, V.; Pardalos, G.; Athanasiou-Metaxa, M.; Papa, A. Novel phlebovirus in febrile child, Greece. *Emerg. Infect. Dis.* **2011**, *17*, 940–941. [CrossRef] [PubMed]
49. Collao, X.; Palacios, G.; de Ory, F.; Sanbonmatsu, S.; Perez-Ruiz, M.; Navarro, J.M.; Molina, R.; Hutchison, S.K.; Lipkin, W.I.; Tenorio, A.; *et al.* Granada Virus: A natural phlebovirus reassortant of the sandfly fever Naples serocomplex with low seroprevalence in humans. *Am. J. Trop. Med. Hyg.* **2010**, *83*, 760–765. [CrossRef] [PubMed]
50. Huiskonen, J.T.; Overby, A.K.; Weber, F.; Grunewald, K. Electron cryo-microscopy and single-particle averaging of Rift Valley fever virus: Evidence for Gn-Gc glycoprotein heterodimers. *J. Virol.* **2009**, *83*, 3762–3769. [CrossRef] [PubMed]
51. Overby, A.K.; Pettersson, R.F.; Grunewald, K.; Huiskonen, J.T. Insights into bunyavirus architecture from electron cryotomography of Uukuniemi virus. *Proc. Natl. Acad. Sci. USA* **2008**, *105*, 2375–2379. [CrossRef] [PubMed]
52. Albarino, C.G.; Bird, B.H.; Nichol, S.T. A shared transcription termination signal on negative and ambisense RNA genome segments of Rift Valley fever, sandfly fever Sicilian, and Toscana viruses. *J. Virol.* **2007**, *81*, 5246–5256. [CrossRef] [PubMed]

53. Ikegami, T.; Won, S.; Peters, C.J.; Makino, S. Characterization of Rift Valley fever virus transcriptional terminations. *J. Virol.* **2007**, *81*, 8421–8438. [CrossRef] [PubMed]
54. Lara, E.; Billecocq, A.; Leger, P.; Bouloy, M. Characterization of wild-type and alternate transcription termination signals in the Rift Valley fever virus genome. *J. Virol.* **2011**, *85*, 12134–12145. [CrossRef] [PubMed]
55. Hewlett, M.J.; Pettersson, R.F.; Baltimore, D. Circular forms of Uukuniemi virion RNA: an electron microscopic study. *J. Virol.* **1977**, *21*, 1085–1093. [PubMed]
56. Leger, P.; Lozach, P.Y. Bunyaviruses: From transmission by arthropods to virus entry into the mammalian host first-target cells. *Future Virol.* **2015**, *10*, 859–881. [CrossRef]
57. Lozach, P.Y.; Mancini, R.; Bitto, D.; Meier, R.; Oestereich, L.; Overby, A.K.; Pettersson, R.F.; Helenius, A. Entry of bunyaviruses into mammalian cells. *Cell Host Microbe* **2010**, *7*, 488–499. [CrossRef] [PubMed]
58. Lozach, P.Y.; Kuhbacher, A.; Meier, R.; Mancini, R.; Bitto, D.; Bouloy, M.; Helenius, A. DC-SIGN as a receptor for phleboviruses. *Cell Host Microbe* **2011**, *10*, 75–88. [CrossRef] [PubMed]
59. Klemm, C.; Reguera, J.; Cusack, S.; Zielecki, F.; Kochs, G.; Weber, F. Systems To establish bunyavirus genome replication in the absence of transcription. *J. Virol.* **2013**, *87*, 8205–8212. [CrossRef] [PubMed]
60. Morin, B.; Coutard, B.; Lelke, M.; Ferron, F.; Kerber, R.; Jamal, S.; Frangeul, A.; Baronti, C.; Charrel, R.; de Lamballerie, X.; *et al.* The N-terminal domain of the Arenavirus L protein is an RNA endonuclease essential in mRNA transcription. *PLoS Pathog.* **2010**, *6*, 8990–8995. [CrossRef] [PubMed]
61. Bird, B.H.; Albarino, C.G.; Hartman, A.L.; Erickson, B.R.; Ksiazek, T.G.; Nichol, S.T. Rift Valley fever virus lacking the NSs and NSm genes is highly attenuated, confers protective immunity from virulent virus challenge, and allows for differential identification of infected and vaccinated animals. *J. Virol.* **2008**, *82*, 2681–2691. [CrossRef] [PubMed]
62. Bird, B.H.; Maartens, L.H.; Campbell, S.; Erasmus, B.J.; Erickson, B.R.; Dodd, K.A.; Spiropoulou, C.F.; Cannon, D.; Drew, C.P.; Knust, B.; *et al.* Rift Valley fever virus vaccine lacking the NSs and NSm genes is safe, nonteratogenic, and confers protection from viremia, pyrexia, and abortion following challenge in adult and pregnant sheep. *J. Virol.* **2011**, *85*, 12901–12909. [CrossRef] [PubMed]
63. Gerrard, S.R.; Bird, B.H.; Albarino, C.G.; Nichol, S.T. The NSm proteins of Rift Valley fever virus are dispensable for maturation, replication and infection. *Virology* **2007**, *359*, 459–465. [CrossRef] [PubMed]
64. Won, S.; Ikegami, T.; Peters, C.J.; Makino, S. NSm and 78-kilodalton proteins of Rift Valley fever virus are nonessential for viral replication in cell culture. *J. Virol.* **2006**, *80*, 8274–8278. [CrossRef] [PubMed]
65. Won, S.Y.; Ikegami, T.; Peters, C.J.; Makino, S. NSm protein of Rift Valley fever virus suppresses virus-induced apoptosis. *J. Virol.* **2007**, *81*, 13335–13345. [CrossRef] [PubMed]
66. Weber, M.; Weber, F. Segmented negative-strand RNA viruses and RIG-I: Divide (your genome) and rule. *Curr. Opin. Microbiol.* **2014**, *20*, 96–102. [CrossRef] [PubMed]
67. Weber, M.; Weber, F. RIG-I-like receptors and negative-strand RNA viruses: RLRly bird catches some worms. *Cytokine Growth Factor Rev.* **2014**, *25*, 621–628. [CrossRef] [PubMed]
68. Schneider, W.M.; Chevillotte, M.D.; Rice, C.M. Interferon-stimulated genes: A complex web of host defenses. *Annu. Rev. Immunol.* **2014**, *32*, 513–545. [CrossRef] [PubMed]
69. Ng, C.T.; Mendoza, J.L.; Garcia, K.C.; Oldstone, M.B. Alpha and beta type 1 interferon signaling: Passage for diverse biologic outcomes. *Cell* **2016**, *164*, 349–352. [CrossRef] [PubMed]
70. Kato, H.; Takeuchi, O.; Sato, S.; Yoneyama, M.; Yamamoto, M.; Matsui, K.; Uematsu, S.; Jung, A.; Kawai, T.; Ishii, K.J.; *et al.* Differential roles of MDA5 and RIG-I helicases in the recognition of RNA viruses. *Nature* **2006**, *441*, 101–105. [CrossRef] [PubMed]
71. Pichlmair, A.; Schulz, O.; Tan, C.P.; Rehwinkel, J.; Kato, H.; Takeuchi, O.; Akira, S.; Way, M.; Schiavo, G.; Sousa, C.R.E. Activation of MDA5 requires higher-order RNA structures generated during virus infection. *J. Virol.* **2009**, *83*, 10761–10769. [CrossRef] [PubMed]
72. Schlee, M. Master sensors of pathogenic RNA—RIG-I like receptors. *Immunobiology* **2013**, *218*, 1322–1335. [CrossRef] [PubMed]
73. Yoneyama, M.; Onomoto, K.; Jogi, M.; Akaboshi, T.; Fujita, T. Viral RNA detection by RIG-I-like receptors. *Curr. Opin. Immunol.* **2015**, *32*, 48–53. [CrossRef] [PubMed]
74. Sparrer, K.M.; Gack, M.U. Intracellular detection of viral nucleic acids. *Curr. Opin. Microbiol.* **2015**, *26*, 1–9. [CrossRef] [PubMed]

75. Davis, M.E.; Gack, M.U. Ubiquitination in the antiviral immune response. *Virology* **2015**, *479–480*, 52–65. [CrossRef] [PubMed]
76. Liu, S.; Chen, J.; Cai, X.; Wu, J.; Chen, X.; Wu, Y.T.; Sun, L.; Chen, Z.J. MAVS recruits multiple ubiquitin E3 ligases to activate antiviral signaling cascades. *eLife* **2013**, *2*, e00785. [CrossRef] [PubMed]
77. Mogensen, T.H. Pathogen recognition and inflammatory signaling in innate immune defenses. *Clin. Microbiol. Rev.* **2009**, *22*, 240–273. [CrossRef] [PubMed]
78. Goubau, D.; Deddouche, S.; Sousa, C.R.E. Cytosolic Sensing of Viruses. *Immunity* **2013**, *38*, 855–869. [CrossRef] [PubMed]
79. MacMicking, J.D. Interferon-inducible effector mechanisms in cell-autonomous immunity. *Nat. Rev. Immunol.* **2012**, *12*, 367–82. [CrossRef] [PubMed]
80. Diamond, M.S.; Farzan, M. The broad-spectrum antiviral functions of IFIT and IFITM proteins. *Nat. Rev. Immunol.* **2013**, *13*, 46–57. [CrossRef] [PubMed]
81. Haller, O.; Staeheli, P.; Schwemmle, M.; Kochs, G. Mx GTPases: Dynamin-like antiviral machines of innate immunity. *Trends Microbiol.* **2015**, *23*, 154–163. [CrossRef] [PubMed]
82. Garcia, M.A.; Gil, J.; Ventoso, I.; Guerra, S.; Domingo, E.; Rivas, C.; Esteban, M. Impact of protein kinase PKR in cell biology: From antiviral to antiproliferative action. *Microbiol. Mol. Biol. Rev. MMBR* **2006**, *70*, 1032–1060. [CrossRef] [PubMed]
83. Gil, J.; Garcia, M.A.; Gomez-Puertas, P.; Guerra, S.; Rullas, J.; Nakano, H.; Alcami, J.; Esteban, M. TRAF family proteins link PKR with NF-kappa B activation. *Mol. Cell. Biol.* **2004**, *24*, 4502–4512. [CrossRef] [PubMed]
84. Fensterl, V.; Sen, G.C. Interferon-induced IFIT Proteins: Their role in viral pathogenesis. *J. Virol.* **2015**, *89*, 2462–2468. [CrossRef] [PubMed]
85. Levy, D.E.; Marie, I.; Smith, E.; Prakash, A. Enhancement and diversification of IFN induction by IRF-7-mediated positive feedback. *J. Interferon Cytokine Res.* **2002**, *22*, 87–93. [CrossRef] [PubMed]
86. McNab, F.; Mayer-Barber, K.; Sher, A.; Wack, A.; O'Garra, A. Type I interferons in infectious disease. *Nat. Rev. Immunol.* **2015**, *15*, 87–103. [CrossRef] [PubMed]
87. Weber, F.; Wagner, V.; Rasmussen, S.B.; Hartmann, R.; Paludan, S.R. Double-stranded RNA is produced by positive-strand RNA viruses and DNA viruses but not in detectable amounts by negative-strand RNA viruses. *J. Virol.* **2006**, *80*, 5059–5064. [CrossRef] [PubMed]
88. Zielecki, F.; Weber, M.; Eickmann, M.; Spiegelberg, L.; Zaki, A.M.; Matrosovich, M.; Becker, S.; Weber, F. Human cell tropism and innate immune system interactions of human respiratory coronavirus EMC compared to those of severe acute respiratory syndrome coronavirus. *J. Virol.* **2013**, *87*, 5300–5304. [CrossRef] [PubMed]
89. Habjan, M.; Andersson, I.; Klingstrom, J.; Schumann, M.; Martin, A.; Zimmermann, P.; Wagner, V.; Pichlmair, A.; Schneider, U.; Muhlberger, E.; *et al.* Processing of genome 5′ termini as a strategy of negative-strand RNA viruses to avoid RIG-I-dependent interferon induction. *PLoS ONE* **2008**, *3*, e2032. [CrossRef] [PubMed]
90. Weber, M.; Gawanbacht, A.; Habjan, M.; Rang, A.; Bomer, C.; Schmidt, A.M.; Veitinger, S.; Jacob, R.; Devignot, S.; Kochs, G.; *et al.* Incoming RNA virus nucleocapsids containing a 5′-triphosphorylated genome activate RIG-I and antiviral signaling. *Cell Host Microbe* **2013**, *13*, 336–346. [CrossRef] [PubMed]
91. Ermler, M.E.; Yerukhim, E.; Schriewer, J.; Schattgen, S.; Traylor, Z.; Wespiser, A.R.; Caffrey, D.R.; Chen, Z.J.J.; King, C.H.; Gale, M.; *et al.* RNA helicase signaling is critical for type I interferon production and protection against Rift Valley fever virus during mucosal challenge. *J. Virol.* **2013**, *87*, 4846–4860. [CrossRef] [PubMed]
92. Gowen, B.B.; Hoopes, J.D.; Wong, M.H.; Jung, K.H.; Isakson, K.C.; Alexopoulou, L.; Flavell, R.A.; Sidwell, R.W. TLR3 deletion limits mortality and disease severity due to phlebovirus infection. *J. Immunol.* **2006**, *177*, 6301–6307. [CrossRef] [PubMed]
93. Habjan, M.; Pichlmair, A.; Elliott, R.M.; Overby, A.K.; Glatter, T.; Gstaiger, M.; Superti-Furga, G.; Unger, H.; Weber, F. NSs protein of Rift Valley fever virus induces the specific degradation of the double-stranded RNA-dependent protein kinase. *J. Virol.* **2009**, *83*, 4365–4375. [CrossRef] [PubMed]
94. Kende, M. Prophylactic and therapeutic efficacy of poly(I,C)-LC against Rift-Valley fever virus-infection in Mice. *J. Biol. Response Modif.* **1985**, *4*, 503–511.

95. Peters, C.J.; Reynolds, J.A.; Slone, T.W.; Jones, D.E.; Stephen, E.L. Prophylaxis of Rift Valley fever with antiviral drugs, immune serum, an interferon inducer, and a macrophage activator. *Antivir. Res.* **1986**, *6*, 285–297. [CrossRef]
96. Gowen, B.B.; Wong, M.H.; Jung, K.H.; Sanders, A.B.; Mitchell, W.M.; Alexopoulou, L.; Flavell, R.A.; Sidwell, R.W. TLR3 is essential for the induction of protective immunity against Punta Toro virus infection by the double-stranded RNA (dsRNA), poly(I:C12U), but not poly(I:C): Differential, recognition of synthetic dsRNA molecules. *J. Immunol.* **2007**, *178*, 5200–5208. [CrossRef] [PubMed]
97. Sidwell, R.W.; Huffman, J.H.; Smee, D.F.; Gilbert, J.; Gessaman, A.; Pease, A.; Warren, R.P.; Huggins, J.; Kende, M. Potential role of immunomodulators for treatment of phlebovirus infections of animals. *Ann. N. Y. Acad. Sci.* **1992**, *653*, 344–355. [CrossRef] [PubMed]
98. Sidwell, R.W.; Huffman, J.H.; Barnard, D.L.; Smee, D.F.; Warren, R.P.; Chirigos, M.A.; Kende, M.; Huggins, J. Antiviral and immunomodulating inhibitors of experimentally-induced Punta Toro virus infections. *Antivir. Res.* **1994**, *25*, 105–122. [CrossRef]
99. Pifat, D.Y.; Smith, J.F. Punta Toro virus infection of C57BL/6J mice—A model for phlebovirus-induced disease. *Microb Pathog.* **1987**, *3*, 409–422. [CrossRef]
100. Do Valle, T.Z.; Billecocq, A.; Guillemot, L.; Alberts, R.; Gommet, C.; Geffers, R.; Calabrese, K.; Schughart, K.; Bouloy, M.; Montagutelli, X.; *et al.* A new mouse model reveals a critical role for host innate immunity in resistance to Rift Valley fever. *J. Immunol.* **2010**, *185*, 6146–6156. [CrossRef] [PubMed]
101. Morrill, J.C.; Jennings, G.B.; Johnson, A.J.; Cosgriff, T.M.; Gibbs, P.H.; Peters, C.J. Pathogenesis of Rift Valley fever in Rhesus monkeys—Role of interferon response. *Arch. Virol.* **1990**, *110*, 195–212. [CrossRef] [PubMed]
102. Bouloy, M.; Janzen, C.; Vialat, P.; Khun, H.; Pavlovic, J.; Huerre, M.; Haller, O. Genetic evidence for an interferon-antagonistic function of Rift Valley fever virus nonstructural protein NSs. *J. Virol.* **2001**, *75*, 1371–1377. [CrossRef] [PubMed]
103. Mendenhall, M.; Wong, M.H.; Skirpstunas, R.; Morrey, J.D.; Gowen, B.B. Punta Toro virus (Bunyaviridae, *Phlebovirus*) infection in mice: Strain differences in pathogenesis and host interferon response. *Virology* **2009**, *395*, 143–151. [CrossRef] [PubMed]
104. Schoggins, J.W.; Wilson, S.J.; Panis, M.; Murphy, M.Y.; Jones, C.T.; Bieniasz, P.; Rice, C.M. A diverse range of gene products are effectors of the type I interferon antiviral response. *Nature* **2011**, *472*, 481–485. [CrossRef] [PubMed]
105. Schoggins, J.W.; MacDuff, D.A.; Imanaka, N.; Gainey, M.D.; Shrestha, B.; Eitson, J.L.; Mar, K.B.; Richardson, R.B.; Ratushny, A.V.; Litvak, V.; *et al.* Pan-viral specificity of IFN-induced genes reveals new roles for cGAS in innate immunity. *Nature* **2014**, *505*, 691–695. [CrossRef] [PubMed]
106. Frese, M.; Kochs, G.; Feldmann, H.; Hertkorn, C.; Haller, O. Inhibition of bunyaviruses, phleboviruses, and hantaviruses by human MxA protein. *J. Virol.* **1996**, *70*, 915–923. [PubMed]
107. Sandrock, M.; Frese, M.; Haller, O.; Kochs, G. Interferon-induced rat Mx proteins confer resistance to Rift Valley fever virus and other arthropod-borne viruses. *J. Interferon Cytokine Res.* **2001**, *21*, 663–668. [CrossRef] [PubMed]
108. Habjan, M.; Penski, N.; Wagner, V.; Spiegel, M.; Overby, A.K.; Kochs, G.; Huiskonen, J.T.; Weber, F. Efficient production of Rift Valley fever virus-like particles: The antiviral protein MxA can inhibit primary transcription of bunyaviruses. *Virology* **2009**, *385*, 400–408. [CrossRef] [PubMed]
109. Kochs, G.; Janzen, C.; Hohenberg, H.; Haller, O. Antivirally active MxA protein sequesters La Crosse virus nucleocapsid protein into perinuclear complexes. *Proc. Natl. Acad. Sci. USA* **2002**, *99*, 3153–3158. [CrossRef] [PubMed]
110. Mudhasani, R.; Tran, J.P.; Retterer, C.; Radoshitzky, S.R.; Kota, K.P.; Altamura, L.A.; Smith, J.M.; Packard, B.Z.; Kuhn, J.H.; Costantino, J.; *et al.* IFITM-2 and IFITM-3 but not IFITM-1 restrict Rift Valley fever virus. *J. Virol.* **2013**, *87*, 8451–8464. [CrossRef] [PubMed]
111. Ikegami, T.; Narayanan, K.; Won, S.; Kamitani, W.; Peters, C.J.; Makino, S. Rift Valley fever virus NSs protein promotes post-transcriptional downregulation of protein kinase PKR and inhibits eIF2alpha phosphorylation. *PLoS Pathog.* **2009**, *5*, e1000287. [CrossRef] [PubMed]
112. Atasheva, S.; Akhrymuk, M.; Frolova, E.I.; Frolov, I. New PARP gene with an anti-alphavirus function. *J. Virol.* **2012**, *86*, 8147–8160. [CrossRef] [PubMed]

113. Pichlmair, A.; Lassnig, C.; Eberle, C.A.; Gorna, M.W.; Baumann, C.L.; Burkard, T.R.; Burckstummer, T.; Stefanovic, A.; Krieger, S.; Bennett, K.L.; *et al.* IFIT1 is an antiviral protein that recognizes 5′-triphosphate RNA. *Nat. Immunol.* **2011**, *12*, 624–630. [CrossRef] [PubMed]
114. Vialat, P.; Billecocq, A.; Kohl, A.; Bouloy, M. The S segment of Rift Valley fever phlebovirus (Bunyaviridae) carries determinants for attenuation and virulence in mice. *J. Virol.* **2000**, *74*, 1538–1543. [CrossRef] [PubMed]
115. Struthers, J.K.; Swanepoel, R. Identification of a major non-structural protein in the nuclei of Rift Valley fever virus-infected cells. *J. Gen. Virol.* **1982**, *60*, 381–384. [CrossRef] [PubMed]
116. Yadani, F.Z.; Kohl, A.; Prehaud, C.; Billecocq, A.; Bouloy, M. The carboxy-terminal acidic domain of Rift Valley fever virus NSs protein is essential for the formation of filamentous structures but not for the nuclear localization of the protein. *J. Virol.* **1999**, *73*, 5018–5025. [PubMed]
117. Muller, R.; Saluzzo, J.F.; Lopez, N.; Dreier, T.; Turell, M.; Smith, J.; Bouloy, M. Characterization of Clone-13, a naturally attenuated avirulent isolate of Rift Valley fever virus, which is altered in the small segment. *Am. J. Trop. Med. Hyg.* **1995**, *53*, 405–411. [PubMed]
118. Billecocq, A.; Spiegel, M.; Vialat, P.; Kohl, A.; Weber, F.; Bouloy, M.; Haller, O. NSs protein of Rift Valley fever virus blocks interferon production by inhibiting host gene transcription. *J. Virol.* **2004**, *78*, 9798–9806. [CrossRef] [PubMed]
119. Le May, N.; Dubaele, S.; Proietti De Santis, L.; Billecocq, A.; Bouloy, M.; Egly, J.M. TFIIH transcription factor, a target for the Rift Valley hemorrhagic fever virus. *Cell* **2004**, *116*, 541–550. [CrossRef]
120. Kalveram, B.; Lihoradova, O.; Ikegami, T. NSs protein of Rift Valley fever virus promotes posttranslational downregulation of the TFIIH subunit p62. *J. Virol.* **2011**, *85*, 6234–6243. [CrossRef] [PubMed]
121. Pichlmair, A.; Kandasamy, K.; Alvisi, G.; Mulhern, O.; Sacco, R.; Habjan, M.; Binder, M.; Stefanovic, A.; Eberle, C.A.; Goncalves, A.; *et al.* Viral immune modulators perturb the human molecular network by common and unique strategies. *Nature* **2012**, *487*, 486–490. [CrossRef] [PubMed]
122. Hermand, D. F-box proteins: More than baits for the SCF? *Cell Div.* **2006**, *1*, 30. [CrossRef] [PubMed]
123. Kainulainen, M.; Habjan, M.; Hubel, P.; Busch, L.; Lau, S.; Colinge, J.; Superti-Furga, G.; Pichlmair, A.; Weber, F. Virulence factor NSs of Rift Valley fever virus recruits the F-box protein FBXO3 to degrade subunit p62 of general transcription factor TFIIH. *J. Virol.* **2014**, *88*, 3464–3473. [CrossRef] [PubMed]
124. Cyr, N.; de la Fuente, C.; Lecoq, L.; Guendel, I.; Chabot, P.R.; Kehn-Hall, K.; Omichinski, J.G. A Omega XaV motif in the Rift Valley fever virus NSs protein is essential for degrading p62, forming nuclear filaments and virulence. *Proc. Natl. Acad. Sci. USA* **2015**, *112*, 6021–6026. [CrossRef] [PubMed]
125. Copeland, A.M.; van Deusen, N.M.; Schmaljohn, C.S. Rift Valley fever virus NSs gene expression correlates with a defect in nuclear mRNA export. *Virology* **2015**, *486*, 88–93. [CrossRef] [PubMed]
126. Le May, N.; Mansuroglu, Z.; Leger, P.; Josse, T.; Blot, G.; Billecocq, A.; Flick, R.; Jacob, Y.; Bonnefoy, E.; Bouloy, M. A SAP30 complex inhibits IFN-beta expression in Rift valley fever virus infected cells. *PLoS Pathog.* **2008**, *4*, e13. [CrossRef] [PubMed]
127. Kainulainen, M.; Lau, S.; Samuel, C.E.; Hornung, V.; Weber, F. NSs virulence factor of Rift Valley fever virus engages the F-box proteins FBXW11 and beta-TRCP1 to degrade the antiviral protein kinase PKR. *J. Virol.* **2016**, *90*, 6140–6147. [CrossRef] [PubMed]
128. Mudhasani, R.; Tran, J.P.; Retterer, C.; Kota, K.P.; Whitehouse, C.A.; Bavari, S. Protein kinase R degradation is essential for Rift Valley fever virus infection and is regulated by SKP1-CUL1-F-box (SCF)FBXW11-NSs E3 Ligase. *PLoS Pathog.* **2016**, *12*, e1005437. [CrossRef] [PubMed]
129. Anderson, G.W., Jr.; Slayter, M.V.; Hall, W.; Peters, C.J. Pathogenesis of a phleboviral infection (Punta Toro virus) in golden Syrian hamsters. *Arch. Virol.* **1990**, *114*, 203–212. [CrossRef] [PubMed]
130. Perrone, L.A.; Narayanan, K.; Worthy, M.; Peters, C.J. The S segment of Punta Toro virus (Bunyaviridae, *Phlebovirus*) is a major determinant of lethality in the Syrian hamster and codes for a type I interferon antagonist. *J. Virol.* **2007**, *81*, 884–892. [CrossRef] [PubMed]
131. Lihoradova, O.A.; Indran, S.V.; Kalveram, B.; Lokugamage, N.; Head, J.A.; Gong, B.; Tigabu, B.; Juelich, T.L.; Freiberg, A.N.; Ikegami, T. Characterization of Rift Valley fever virus MP-12 strain encoding NSs of Punta Toro virus or sandfly fever Sicilian virus. *PLoS Negl. Trop. Dis.* **2013**, *7*, e2181. [CrossRef] [PubMed]
132. Kalveram, B.; Ikegami, T. Toscana virus NSs protein promotes degradation of double-stranded RNA-dependent protein kinase. *J. Virol.* **2013**, *87*, 3710–3718. [CrossRef] [PubMed]
133. Di Bonito, P.; Nicoletti, L.; Mochi, S.; Accardi, L.; Marchi, A.; Giorgi, C. Immunological characterization of Toscana virus proteins. *Arch. Virol.* **1999**, *144*, 1947–1960. [CrossRef] [PubMed]

134. Savellini, G.G.; Weber, F.; Terrosi, C.; Habjan, M.; Martorelli, B.; Cusi, M.G. Toscana virus induces interferon although its NSs protein reveals antagonistic activity. *J. Gen. Virol.* **2011**, *92*, 71–79. [CrossRef] [PubMed]
135. Gori-Savellini, G.; Valentini, M.; Cusi, M.G. Toscana Virus NSs Protein Inhibits the Induction of Type I Interferon by Interacting with RIG-I. *J. Virol.* **2013**, *87*, 6660–6667. [CrossRef] [PubMed]
136. Savellini, G.G.; Gandolfo, C.; Cusi, M.G. Truncation of the C-terminal region of Toscana Virus NSs protein is critical for interferon-beta antagonism and protein stability. *Virology* **2015**, *486*, 255–262. [CrossRef] [PubMed]
137. Brisbarre, N.M.; Plumet, S.; de Micco, P.; Leparc-Goffart, I.; Emonet, S.F. Toscana virus inhibits the interferon beta response in cell cultures. *Virology* **2013**, *442*, 189–194. [CrossRef] [PubMed]
138. Qu, B.; Qi, X.; Wu, X.; Liang, M.; Li, C.; Cardona, C.J.; Xu, W.; Tang, F.; Li, Z.; Wu, B.; *et al.* Suppression of the interferon and NF-kappaB responses by severe fever with thrombocytopenia syndrome virus. *J. Virol.* **2012**, *86*, 8388–8401. [CrossRef] [PubMed]
139. Ning, Y.J.; Wang, M.; Deng, M.; Shen, S.; Liu, W.; Cao, W.C.; Deng, F.; Wang, Y.Y.; Hu, Z.; Wang, H. Viral suppression of innate immunity via spatial isolation of TBK1/IKKepsilon from mitochondrial antiviral platform. *J. Mol. Cell Biol.* **2014**, *6*, 324–337. [CrossRef] [PubMed]
140. Santiago, F.W.; Covaleda, L.M.; Sanchez-Aparicio, M.T.; Silvas, J.A.; Diaz-Vizarreta, A.C.; Patel, J.R.; Popov, V.; Yu, X.J.; Garcia-Sastre, A.; Aguilar, P.V. Hijacking of RIG-I signaling proteins into virus-induced cytoplasmic structures correlates with the inhibition of type I interferon responses. *J. Virol.* **2014**, *88*, 4572–4585. [CrossRef] [PubMed]
141. Wu, X.D.; Qi, X.; Qu, B.Q.; Zhang, Z.R.; Liang, M.F.; Li, C.; Cardona, C.J.; Li, D.X.; Xing, Z. Evasion of antiviral immunity through sequestering of TBK1/IKK epsilon/IRF3 into viral inclusion bodies. *J. Virol.* **2014**, *88*, 3067–3076. [CrossRef] [PubMed]
142. Wu, X.D.; Qi, X.; Liang, M.F.; Li, C.; Cardona, C.J.; Li, D.X.; Xing, Z. Roles of viroplasm-like structures formed by nonstructural protein NSs in infection with severe fever with thrombocytopenia syndrome virus. *Faseb J. Off. Publ. Fed. Am. Soc. Exp. Biol.* **2014**, *28*, 2504–2516. [CrossRef] [PubMed]
143. Gack, M.U.; Shin, Y.C.; Joo, C.H.; Urano, T.; Liang, C.; Sun, L.J.; Takeuchi, O.; Akira, S.; Chen, Z.J.; Inoue, S.S.; *et al.* TRIM25 RING-finger E3 ubiquitin ligase is essential for RIG-I-mediated antiviral activity. *Nature* **2007**, *446*, 916–920. [CrossRef] [PubMed]
144. Ning, Y.J.; Feng, K.; Min, Y.Q.; Cao, W.C.; Wang, M.; Deng, F.; Hu, Z.; Wang, H. Disruption of type I interferon signaling by the nonstructural protein of severe fever with thrombocytopenia syndrome virus via the hijacking of STAT2 and STAT1 into inclusion bodies. *J. Virol.* **2015**, *89*, 4227–4236. [CrossRef] [PubMed]
145. Simons, J.F.; Hellman, U.; Pettersson, R.F. Uukuniemi virus S RNA Segment—Ambisense coding strategy, packaging of complementary strands into virions, and homology to members of the genus *Phlebovirus*. *J. Virol.* **1990**, *64*, 247–255. [PubMed]
146. Rezelj, V.V.; Overby, A.K.; Elliott, R.M. Generation of mutant Uukuniemi viruses lacking the nonstructural protein NSs by reverse genetics indicates that NSs is a weak interferon antagonist. *J. Virol.* **2015**, *89*, 4849–4856. [CrossRef] [PubMed]
147. Brennan, B.; Li, P.; Zhang, S.; Li, A.; Liang, M.; Li, D.; Elliott, R.M. Reverse genetics system for severe fever with thrombocytopenia syndrome virus. *J. Virol.* **2015**, *89*, 3026–3037. [CrossRef] [PubMed]
148. Ikegami, T.; Won, S.; Peters, C.J.; Makino, S. Rift Valley fever virus NSs mRNA is transcribed from an incoming anti-viral-sense S RNA segment. *J. Virol.* **2005**, *79*, 12106–12111. [CrossRef] [PubMed]
149. Billecocq, A.; Gauliard, N.; le May, N.; Elliott, R.M.; Flick, R.; Bouloy, M. RNA polymerase I-mediated expression of viral RNA for the rescue of infectious virulent and avirulent Rift Valley fever viruses. *Virology* **2008**, *378*, 377–384. [CrossRef] [PubMed]
150. Habjan, M.; Penski, N.; Spiegel, M.; Weber, F. T7 RNA polymerase-dependent and -independent systems for cDNA-based rescue of Rift Valley fever virus. *J. Gen. Virol.* **2008**, *89*, 2157–2166. [CrossRef] [PubMed]
151. Ikegami, T.; Won, S.; Peters, C.J.; Makino, S. Rescue of infectious Rift Valley fever virus entirely from cDNA, analysis of virus lacking the NSs gene, and expression of a foreign gene. *J. Virol.* **2006**, *80*, 2933–2940. [CrossRef] [PubMed]

© 2016 by the authors. Licensee MDPI, Basel, Switzerland. This article is an open access article distributed under the terms and conditions of the Creative Commons Attribution (CC BY) license (http://creativecommons.org/licenses/by/4.0/).

Review

Making Bunyaviruses Talk: Interrogation Tactics to Identify Host Factors Required for Infection

Amber M. Riblett [1] and Robert W. Doms [2,*]

[1] Department of Microbiology, Perelman School of Medicine, University of Pennsylvania, Philadelphia, PA 19104, USA; ariblett@mail.med.upenn.edu

[2] Department of Pathology and Laboratory Medicine, Children's Hospital of Philadelphia, Philadelphia, PA 19104, USA

* Correspondence: domsr@email.chop.edu; Tel.: +1-215-590-4446

Academic Editors: Jane Tao and Pierre-Yves Lozach
Received: 19 March 2016; Accepted: 6 May 2016; Published: 13 May 2016

Abstract: The identification of host cellular genes that act as either proviral or antiviral factors has been aided by the development of an increasingly large number of high-throughput screening approaches. Here, we review recent advances in which these new technologies have been used to interrogate host genes for the ability to impact bunyavirus infection, both in terms of technical advances as well as a summary of biological insights gained from these studies.

Keywords: bunyavirus; high-throughput screening; host-pathogen interaction; haploid genetic screening; RNAi screening; yeast two-hybrid; affinity purification mass spectrometry

1. Introduction

Viruses rely on a large number of host cellular factors and pathways as they enter, traffic through, replicate, assemble, and exit the cell. Identification of host cell factors needed for virus replication, such as cell surface receptors, has provided tremendous insights into viral lifecycles and pathogenesis. Traditionally, relatively reductionist approaches have been taken to identify specific interactions between viral and host cell molecules. More recently, rapid advances in high-throughput screening technologies based upon small molecules, loss-of-function libraries, and interactome characterization have informed our understanding of nearly every stage of virus replication cycles and identified valuable targets for antiviral therapeutics. Since the technical aspects for most of these techniques have been extensively reviewed, we aim here to present a summary of their use within the bunyavirus field, a brief comparison of their relative advantages and disadvantages, technical considerations that apply to screening with bunyaviruses, and recent advances in screening approaches that may be of general interest. The emergence and spread of newly-identified bunyaviruses in recent years, as well as important progress toward a more detailed understanding of bunyavirus structure and genetics, has renewed interest in this large and diverse family of viruses. Simultaneous advances in genome-wide screening techniques, and their well-demonstrated power to identify novel host cellular factors that either support or restrict viral infection, present an exciting tool for probing many aspects of bunyavirus biology.

2. Genetic Approaches to Identify Bunyavirus Host Factors

RNA interference (RNAi) technology was the first of a new generation of high-throughput screening approaches applied to the study of virus-host interactions. Examples of its use include the pioneering screens by Cherry *et al.* to uncover a role for host organelle-reshaping and ribosomal proteins in *Drosophila* C virus (DCV) replication [1,2], a series of 2008 studies from multiple labs that identified many host factors necessary for human immunodeficiency virus (HIV)-1 replication [3–5], and the characterization in 2009 by Brass *et al.* of interferon-induced transmembrane (IFITM) proteins as restriction factors for influenza, West Nile, and dengue viruses [6]. For this screening technique, the incorporation of small interfering RNAs (siRNAs) into the RNA-induced silencing complex (RISC) effects the cleavage of target cellular mRNA and consequent knockdown of gene product expression. These siRNAs can be either directly introduced into the cell, or derived from supplied precursors: long double-stranded RNAs (dsRNAs) or short hairpin RNAs (shRNAs) that are then processed by cellular machinery. The availability of increasingly robust genome-wide libraries for RNAi screening has greatly increased its popularity as a high-throughput, unbiased screening platform.

Within the bunyavirus field, a 2013 RNAi screen by Hopkins *et al.* in *Drosophila* cells used dsRNAs targeting more than 13,000 genes, identifying 124 that restricted infection by the phlebovirus Rift Valley fever virus (RVFV), with genes involved in DNA replication, the cell cycle, and mRNA metabolic processing being significantly enriched [7]. Among these were the catalytic component of the mRNA decapping machinery mRNA-decapping enzyme 2 (Dcp2), as well as two decapping activators, DEAD (Asp-Glu-Ala-Asp) Box Helicase 6 (DDX6) and U6 snRNA-associated Sm-like protein LSm7 (LSM7). Bunyaviruses "cap-snatch" the 5′ ends of host mRNAs, and the authors showed that RVFV specifically cap-snatches the 5′ ends of Dcp2-targeted mRNAs, as did La Crosse virus (LACV), a member of the *Orthobunyavirus* genus. The year after, Meier *et al.* performed a screen using Uukuniemi virus (UUKV) in HeLa cells expressing the surface lectin CD209, which is an attachment factor for UUKV in dendritic cells [8]. Two independent genome-wide siRNA libraries were used from two manufacturers: one with four unpooled siRNAs per gene and one with four pooled siRNAs per gene. In both screens the vesicle-soluble NSF attachment protein receptor (vSNARE) vesicle-associated membrane protein 3 (VAMP3) was identified as a host factor required for the entry of UUKV. The importance of VAMP3 was also indicated by virtue of its being a target for the endogenous microRNA miR-142-3p, a microRNA identified as impacting infection after analysis of the seed sequences of the siRNAs used for screening. The authors examined incoming UUKV virions trafficking through the endocytic pathway and noted increasing colocalization of virions with VAMP3 as they moved within vesicles through the cytoplasm. At 20 min after internalization, maximum colocalization between UUKV virions and VAMP3 was observed within vesicles positive for lysosomal-associated membrane protein 1 (LAMP1), a marker for late endosomes and lysosomes. In VAMP3-depleted cells, incoming virions failed to reach these LAMP1-positive vesicles, indicating that their trafficking was arrested at an earlier endosomal compartment. In contrast, depletion of VAMP3 did not affect the entry of Semliki Forest virus (SFV), which penetrates from early endosomes [9], or of influenza A virus (IAV), which fuses from late endosomes [9–11]. This suggests that the entry defect of UUKV in the absence of VAMP3 is not due to a lack of endosomal acidification, and that late steps of viral entry of UUKV are distinct from those of IAV. Interestingly, VAMP3 has been shown to be required for the fusion of multivesicular bodies with autophagosomes [12], although it is unclear whether this activity may have any bearing on its role in bunyavirus entry. These data informed our understanding of the host cellular machinery required for maturation of endosomal compartments and for the fusion of late-penetrating viruses within the acidic environment of late endosomes.

The arrival of haploid screening in human cells, described by Carette *et al.* in 2009, offered a loss-of-function forward genetic approach as a powerful alternative to traditional siRNA-based depletion screens [13,14]. In these screens, null alleles are generated in mammalian haploid cells using insertional mutagenesis, and the resulting cellular library is challenged by a selective agent such as a virus or toxin. Surviving cells, which presumably lack a gene required by the selective agent as a consequence of retroviral insertion, are pooled and deep sequencing is used to map the insertion sites of the mutagenizing lentivirus. Statistically enriched insertion sites within the surviving (selected) population compared to the original mutant library yield a list of genes whose disruption confers a resistance phenotype. This approach identified the homotypic fusion and vacuole protein sorting (HOPS) complex and the endo/lysosomal cholesterol transporter protein Nieman-Pick 1 (NPC1) as essential host factors for Ebola virus (EBOV) entry, and uncovered the receptor-switching process of

Lassa virus (LASV) as it engages first its α-dystroglycan receptor at the cell surface and then later its intracellular receptor, LAMP1 [15–17]. These studies have provided potential antiviral targets, as well as insight into the molecular determinants of host tropism, for these important human pathogens.

In 2014, Petersen *et al.* used a recombinant vesicular stomatitis virus (rVSV), in which the Andes virus (ANDV) glycoproteins are expressed on the VSV core, to identify cellular host factors required for ANDV entry [18]. This rVSV-ANDV was used to challenge a human haploid mutant library and multiple members of the sterol regulatory pathway were identified as impacting ANDV entry. This dependence upon cholesterol was validated using live wild-type ANDV, a member of the New World hantaviruses that are causative agents of hantavirus pulmonary syndrome (HPS). Cholesterol requirement during viral entry was verified through the use of Chinese hamster ovary (CHO) knockout cell lines, pharmacological inhibitors, siRNA depletion, and transcription activator-like effector nuclease (TALEN) disruption of members of the cholesterol pathway, as well as by direct depletion of cholesterol in the cellular membranes. Virus binding at the cell surface was unaffected, but an internalization defect was observed within cells that lack a functional sterol regulatory pathway. The following year, Kleinfelter *et al.* independently confirmed these findings and extended the cholesterol-dependence phenotype to members of both the Old World and New World hantavirus clades [19]. Cholesterol depletion was shown to significantly delay virus internalization, and to inhibit the ability of virions to fuse with cellular membranes. This finding is intriguing, as the pH requirement for ANDV implicates it as a late-penetrating virus, but the liposome fusion results from Kleinfelter *et al.* suggest that ANDV may require a greater cholesterol concentration than what is present in the membranes of late endosomes. Detailed mechanistic studies will be needed to reconcile this, and to determine whether hantaviruses somehow modulate endosomal cholesterol composition, fuse specifically at cholesterol-rich microdomains, or whether cholesterol plays some other role during virus-membrane fusion.

We recently employed the haploid genetic screening technology using RVFV and identified a role for heparan sulfate proteoglycans as attachment factors for RVFV on some but not all cell types [20]. Within the surviving RVFV-resistant population, there was a significant enrichment of inactivating insertions into genes encoding proteins involved in nearly every step of the heparan sulfate biosynthesis pathway, as well as multiple members of the conserved oligomeric Golgi (COG) complex. The COG complex is required for normal Golgi function, and it has been shown that perturbing the COG complex leads to a defect in *O*-linked glycosylation [21,22]. Infection of heparan sulfate-deficient cells was also inhibited for a panel of pathogenic primary RVFV isolates, indicating that the use of glycosaminoglycans by RVFV was not a trait acquired during repeated passaging and laboratory strain attenuation. The screen also identified the previously-uncharacterized gene *PTAR1*, and disruption of this gene led to decreased levels of heparan sulfate on the cell surface and conferred resistance to RVFV infection. This was consistent with the results published by Blomen *et al.* showing that cells lacking protein prenyltransferase alpha subunit repeat containing 1 (PTAR1) have a defect in glycosylation [23].

3. Recent Advances in Genetic Screening Techniques

Bunyavirus research going forward will be greatly aided by many exciting developments in loss-of-function screening technology. In addition to the near-haploid human cell line HAP1, haploid cell lines have been generated from fish, mouse, monkey, and rat embryonic stem cells [24–28]. A fully haploid human cell line has also been derived by genome editing using the clustered regularly interspaced short palindromic repeats (CRISPR) RNA-guided endonuclease Cas9 to excise the fragment of chromosome 15 that was integrated onto chromosome 19 and was preventing the HAP1 cell line from being fully haploid [29]. This updated, engineered haploid cell line, termed eHAP, will likely replace the HAP1 line in the generation of new mutagenesis libraries.

CRISPR/Cas9 technology has now also been applied to high-throughput functional genomic screening. This DNA-editing technique was adapted from the type II CRISPR bacterial adaptive immune system in which the endonuclease Cas9 is recruited to the DNA of invading pathogens by two RNA components: a CRISPR RNA (crRNA) that contains a DNA fragment complementary to the foreign target, and a trans-activating CRISPR RNA (tracrRNA) which acts as a scaffold. The crRNA and tracrRNA can be fused to form a single guide RNA (sgRNA), greatly simplifying the process of synthesizing and delivering custom CRISPR/Cas9 machinery in order to disrupt a gene of interest. The Cas9-induced cleavage triggers the cell's double-strand break repair response, leading either to indel mutations, or (if supplied) the introduction of a sequence of interest. For a detailed technical review of CRISPR/Cas systems and their utility for genome engineering, see reference [30].

Generation of sgRNA libraries providing genome-wide targeting by CRISPR/Cas9 has opened the door to a new method of high-throughput screening to identify host factors required for infection. In one recent study, a CRISPR sgRNA library was used to identify genes required for the induction of cell death by West Nile virus (WNV) [31]. Lentiviral vector delivery of both the sgRNAs and the Cas9 endonuclease have been developed, and are being optimized for efficient delivery [32].

4. Small Molecule Screening

The lack of available antivirals for bunyaviruses has renewed interest in the screening of small molecule inhibitors, including the repurposing of clinically-approved pharmacologics. In 2016, Islam *et al.* used a high-throughput drug screen to identify compounds, which potently inhibited RVFV infection, based upon a replication-competent recombinant virus lacking the gene encoding the nonstructural protein S (NSs) and bearing a fluorescent reporter [33]. This study yielded six compounds (out of approximately 28,000 screened) that exhibited inhibitory activity at low concentrations with minimal cytotoxicity. Follow-up studies will be required to determine the mechanism of action of these compounds and their potential suitability as therapeutic agents against RVFV and perhaps other bunyaviruses.

Advances in inhibitor drug screening have included methods to study the interactions between compounds that may be able to synergistically restrict viral infection. In 2012, Tan *et al.* described multiplex screening for interacting compounds (MuSIC), an analysis of all of the possible pairs of 1000 commercially available compounds that were approved by the U.S. Food and Drug Administration (FDA) or clinically tested [34]. The authors identified anti-inflammatory drugs as a group that synergistically enhanced anti-HIV activity and informed drug-interaction network formation. Such screening methods may uncover previously uncharacterized therapeutic options within the pool of clinically-tested or -approved drugs, which is particularly attractive for bunyavirus diseases, most of which lack vaccines and therapeutic options.

5. Biochemical Approaches: Viral Proteins as Bait for Host Factors

Valuable insight into the host-pathogen relationship can also be gleaned from interrogating physical interactions between viral and cellular proteins. The most widely-used applications for probing protein-protein interactions are yeast two-hybrid (Y2H) and affinity purification followed by mass spectrometry (AP/MS) techniques. Y2H screens utilize a reporter gene whose expression depends upon the activity of a transcription factor whose modular binding and activation domains have been fused, respectively, to bait and prey proteins. The protein of interest whose interacting partners are to be probed is the bait, and the prey proteins are typically libraries of proteins (or protein fragments) covering the genome of the organism of interest. These hybrid proteins are then introduced into cells, and if the bait and prey proteins interact, the binding and activation domains come into close enough proximity to reconstitute transcription factor activity and effect the expression of the reporter gene. For AP/MS, a bait protein of interest is pulled down via affinity for an antibody against either the protein itself or a tag to which it has been fused. Tandem affinity purification (TAP) systems are attractive approaches for their ability to reduce non-specific interactions. The classic TAP tag comprises a Protein A tag and calmodulin binding peptide (CBP) tag separated by a recognition sequence that is specific to the Tobacco etch virus (TEV) protease. Protein complexes are purified by first capturing with the terminal IgG-binding Protein A tag, then using the TEV protease to cleave and release bound complexes and expose the CBP, followed by a second affinity purification step of immobilization on calmodulin. This dual-affinity approach reduces co-purifying non-specific interactions.

A variety of protein-protein interaction approaches have been employed with the bunyavirus NSs, which is known to be critical for viral defense against the host's type I interferon response. Léonard and colleagues performed Y2H screening of a HeLa cDNA library using the Bunyamwera virus (BUNV) NSs protein as bait [35]. They identified Mediator of RNA polymerase II transcription subunit 8 (MED8), a component of the Mediator complex, as a target of NSs during infection, and mapped this interaction to the C-terminal domain of NSs. Mediator is a key regulator of RNA polymerase II transcriptional activity, and C-terminal NSs truncation mutants were unable to effect host cell protein shutoff. Additionally, whereas wild-type BUNV is able to inhibit transcription of interferon-β (IFN-β) mRNA, infection with a recombinant BUNV lacking the MED8 interaction domain of NSs resulted in strong induction of the IFN-β promoter and thus rendered the virus sensitive to the host interferon response. The domain of NSs responsible for this MED8 interaction contains a motif that is highly conserved among orthobunyaviruses, suggesting that this interaction represents an important defense mechanism used by the virus to dismantle the host interferon response. In 2012, Rönnberg *et al.* used Y2H screening with a mouse embryo cDNA library with the hantaviruses Puumala virus (PUUV) and Tula virus (TULV) NSs proteins as bait [36]. From these two screens, 65 total host cellular proteins were identified as hantavirus interacting partners: 47 were associated with the PUUV NSs protein, while 21 were found to interact with the TULV NSs protein. The overlap between the two screens comprised three proteins: Acyl-coenzyme A-binding domain containing 3 protein (ACBD3), ARP5 actin-related protein 5 homolog (ACTR5), and keratin-14 (KRT14). ACBD3 was validated as an interacting partner of TULV NSs by fluorescence resonance energy transfer (FRET) and colocalization of the two proteins in the perinuclear area was observed by confocal microscopy. Bioinformatic analysis of the pooled interactome of 65 proteins revealed overlapping cellular pathways between the two hantavirus NSs proteins. This dataset provided insight into potential, previously-undescribed roles for NSs during infection, including regulation of apoptosis and interaction with proteins of the integrin complex.

An extensive survey of viral-host protein-protein interactions by Pichlmair *et al.* in 2012 used as bait a panel of 70 viral open reading frames (ORFs) selected for their roles in defending against the host innate immune response [37]. The bunyavirus ORFs included in the panel were the NSs of RVFV, LACV, and Sandfly fever Sicilian virus (SFSV). Viral ORFs were expressed within a HEK293 cell line and then affinity purified followed by liquid chromatography tandem-mass spectrometry (LC-MS/MS). The authors identified 579 interacting host proteins, which displayed an overrepresentation of proteins known to be involved in innate immunity, and specifically they noted an enrichment within the interacting partners of the negative-sense single-strand RNA for host proteins that may promote processing of viral RNA transcripts or prevent detection and degradation of these transcripts. In 2014, a follow-up study was published by Kainulainen *et al.* examining the interaction between RVFV NSs and the host F-box protein FBXO3 [38]. FBXO3, which is a component of an E3 ubiquitin ligase, was shown to be recruited by NSs to effect the degradation of p62, a subunit of the general transcription factor II Human (TFIIH). Depletion of FBXO3 failed to fully rescue interferon induction in RVFV-infected cells, did not affect the ability of NSs to degrade the interferon-induced antiviral effector dsRNA-dependent protein kinase R (PKR), and did not significantly impact viral replication. The authors therefore concluded that this FBXO3-mediated degradation of p62 is partially, although not completely, responsible for the ability of NSs to suppress the host interferon response. These findings highlight the capacity of protein-protein interaction studies for uncovering host factors that might not have been detected by gene-disruption or gene-depletion screening strategies, which usually depend upon robust viral replication or host cell survival phenotypes.

6. The Next Generation of Biochemical Screening Techniques

To identify potential cellular interacting partners during bunyavirus infection, it is now possible to circumvent the requirement that proteins associate strongly with bait proteins during affinity purification. Martell *et al.* introduced in 2012 a new genetically-encoded reporter molecule that can be used for proximity labeling followed by MS to detect nearby proteins, as well as electron microscopy [39,40]. The authors engineered a monomeric variant of ascorbate peroxidase, which they have termed APEX, that is active in all cellular compartments (including the cytosol), a major advantage over the horseradish peroxidase (HRP) tag typically used. This APEX tag can oxidize biotin-phenol (in the presence of a hydrogen peroxide catalyst) into phenoxyl radicals, and these short-lived radical species react with electron-rich amino acids present in proteins that are fewer than 20 nm away. This results in the biotin-labeling of endogenous proteins adjacent to the APEX-tagged protein of interest, and these can be identified by streptavidin purification followed by digestion and MS analysis. An improved version of this peroxidase, termed APEX2, was recently obtained by yeast display evolution and exhibits increased activity, stability, and sensitivity [41].

Another proximity labeling approach developed in 2012 by Roux and colleagues is named proximity-dependent biotin identification (BioID) and it employs a promiscuous mutant of the *E. coli* biotin ligase BirA fused to a bait protein of interest [42]. As with the APEX labeling technique, neighboring proteins that have been biotinylated within the cell can be affinity purified and identified. BioID has been used to characterize the constituents and architecture of the nuclear pore complex and to identify the interactome of the Ewing sarcoma fusion oncoprotein EWS-Fli-1 [43,44]. This approach has also been used to study host-pathogen interactions during bacterial and viral infection. Mojica *et al.* fused the BirA to SINC, a type III secreted effector from *Chlamydia psittaci*, and showed that it targets the nuclear envelope of both infected and neighboring cells [45]. In 2015, Le Sage and colleagues used HIV-1 Gag protein fused to BirA to identify 47 associated proteins that were biotinylated by the fusion protein when it was transfected into Jurkat cells [46]. Two of the putative host factors identified, DDX17 and RPS6, were validated as interacting partners of Gag by co-immunoprecipitation experiments. A substantially smaller biotin ligase, BioID2, was recently described to have higher activity and to improve the function and localization of the resultant fusion protein [47]. These new proximity labeling technologies represent exciting additions to the bunyavirus screening toolbox.

Please refer to Figure 1 for a summary of the host cellular factors that have been identified by the bunyavirus screens discussed in this review.

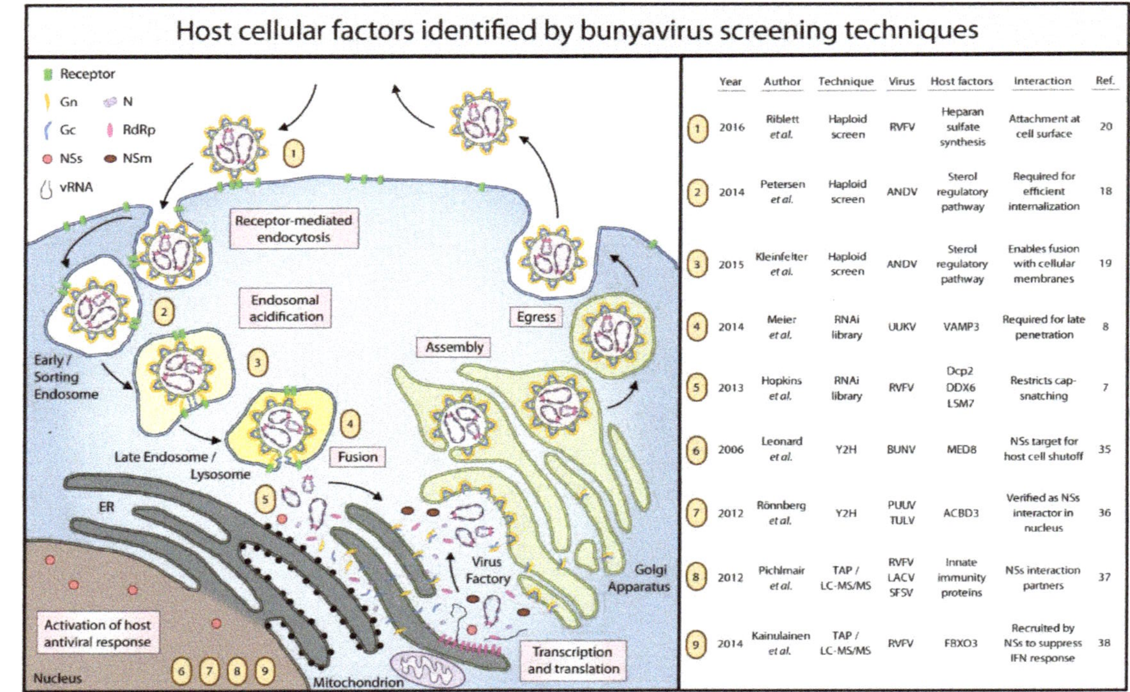

Figure 1. Summary of bunyavirus host factors identified by high-throughput screening techniques. RdRp: RNA-dependent RNA polymerase; NSs: non-structural protein S; NSm: non-structural protein M; vRNA: viral RNA; RVFV: Rift Valley fever virus; ANDV: Andes virus; UUKV: Uukuniemi virus; BUNV: Bunyamwera virus; PUUV: Puumala virus; TULV: Tula virus; LACV: La Crosse encephalitis virus; SFSV: Sandfly fever Sicilian virus; VAMP3: vesicle-associated membrane protein 3; Dcp2: mRNA-decapping enzyme 2; DDX6: DEAD (Asp-Glu-Ala-Asp) Box Helicase 6; LSM7: U6 snRNA-associated Sm-like protein LSm7; MED8: Mediator of RNA polymerase II transcription subunit 8; ACBD3: acyl-coenzyme A-binding domain containing 3 protein; FBXO3: F-Box Protein.

7. Diverse Screening Approaches are Highly Complementary

As RNAi screening became more popular, it also became evident that the technology suffered from issues of reproducibility and a high rate of false discovery. Results of the three genome-wide siRNA screens performed with HIV in 2008 [3–5], each of which had generated a list of approximately 300 genes supporting HIV infection in 293T or HeLa-derived cells, were subjected to in-depth meta-analysis by Bushman *et al.* in 2009, who reported that the percentage of overlap in gene hits between any two of the three screens was 6% at most [48]. Two genome-wide RNAi screens were performed in 2009 to uncover host factors required for hepatitis C virus (HCV) replication in human cells. Tai *et al.* [49], using an HCV subgenomic replicon, reported the identification of 96 genes that support HCV replication, and Li *et al.* [50], using infectious virus, then identified 262 genes impacting infection, only 15 of which overlapped with the previous screen's findings. In the last five years, two genome-wide RNAi screens using Sindbis virus (SINV) have been performed, one in *Drosophila* cells [51,52], and one in human cells [53]. The screen in *Drosophila* cells identified 57 genes supporting and 37 genes that restricted SINV infection, while the screen in human cells identified 56 genes supporting and 62 genes restricting infection—but there was very little overlap between the genes identified (compare [53] Tables S2 and S3 to human homologues of [52] Table S1).

Much of the reason for this lack of overlap between seemingly similar RNAi screens has been ascribed to the off-target effects of siRNAs and differences between technical aspects of the screening conditions. In a recent analysis of three genome-wide RNAi screens (one with UUKV and two with bacterial pathogens), Franceschini *et al.* concluded that the phenotypic effects of siRNA oligos were in fact predominantly due to off-target microRNA activity conferred by the seed region sequence, rather than the intended siRNA activity [54]. They found significantly higher phenotypic correlations when siRNA oligos from different vendors were grouped by seed sequence (nucleotides 2–8) than when they were grouped by intended target (full-length complementarity of all 21 nucleotides). The authors confirmed these findings by designing custom oligos containing seed sequences predicted to impact infection that were flanked by arbitrary sequences outside of the seed region, and demonstrated that overexpression of known human microRNAs phenocopied the effect of siRNA oligos with corresponding seed sequences. These findings beg a reexamination of the raw data that have been generated by previous RNAi screens, as well as an attentive consideration of microRNA effects during analysis of any future screens. In addition to the off-target activities of the oligos themselves (which can cause both false-positive and -negative results), differing gene expression levels between cell types, variable efficiencies of transfection protocols, and discordance between knockdown timing and the half-life of the target protein can all contribute to a high false-negative rate. Recent improvements in both design and analysis of RNAi screens have sought to address these problems, such as the Minimum Information About an RNAi Experiment (MIARE) reporting guidelines (http://miare.sourceforge.net) that have been established, and the utilization of the multiple orthologous RNAi reagents coupled with RNAi gene enrichment ranking (MORR-RIGER) method, which helps to reduce false negatives and filter off-target effects [55]. For a detailed discussion about the factors impacting RNAi screen success, recent technical updates, and current design and analysis strategies, see reference [56] and the references therein.

Like RNAi, insertional mutagenesis screening is a forward genetic approach, allowing for discovery of novel host factors in the absence of a presumed or suspected mechanism of action. Although the technique is relatively new and comparatively few studies employing this approach to study virus-host interactions have been published, it is clear that haploid screening offers some important advantages over RNAi screening. A significant advantage of haploid screening is the fact that the insertional mutagenesis strategy employed to generate the haploid libraries usually results in complete disruption of the gene product, rather than the transient partial depletion that results from RNAi targeting. This in turn greatly increases the signal-to-noise ratio of the data that are obtained. Generation of many independent mutants within the library that each bear separate integrations into the same gene locus also allows for rigorous statistical analysis to identify genes whose absence was selected for within the surviving mutant pool. The fact that this selection occurs in a cell line of human origin is also attractive because it increases the likelihood of finding biologically meaningful factors that participate in the host-pathogen interaction during the course of human disease.

It may be premature to evaluate the reproducibility of haploid genetic screens as published applications of this screening technique have utilized a diverse array of viruses and toxins, including EBOV, LASV, RVFV, enterovirus D68, and adeno-associated virus serotype 2 [15–17,20,57,58]. In addition, diphtheria and anthrax toxins, *Clostridium perfringens* TpeL toxin, *Pseudomonas aeruginosa* exotoxin A, and *Staphylococcus aureus* α-toxin [13,59–61] have been investigated with this approach. To our knowledge, ANDV is the only selective agent to have been used in two completely independent haploid genetic screens performed by different labs. The degree of overlap between these two screens, however, was striking. In the 2014 study by Petersen *et al.* four genes encoding members of the sterol regulatory pathway (*SREBF2, S1P, S2P,* and *SCAP*) were enriched for disrupting integrations well above any other genes [18] and the 2015 screen performed by Kleinfelter *et al.* reported that these exact same four genes were also their top hits, and that three other genes involved in cholesterol biosynthesis (*LSS, SQLE,* and *ACAT2*) were the next most frequently disrupted [19]. This identification of multiple members of a biological pathway has been seen in many of the haploid screens mentioned in this review, and it not only demonstrates the high level of mutagenesis coverage in the libraries that have been generated thus far, but it also increases the confidence that screening hits are biologically relevant.

The haploid screening technique is not without drawbacks. Due to the nature of disrupting mutagenesis in a haploid genetic background, this screening strategy is unlikely to identify host factors that are required for cell viability. Additionally, most haploid screens have relied upon cell death as a phenotypic read-out, a decision that greatly increases the throughput of the screen but that may prevent the identification of a gene whose disruption produces an intermediate phenotype in which virus infection is delayed or partially suppressed. We find it interesting that in a number of the published screens a single biological pathway is clearly identified by virtue of multiple retroviral gene insertions to the near exclusion of other hits. In the two ANDV haploid screens [18,19], cells that survived the viral challenge almost invariably had one of several genes involved in cholesterol biosynthesis disrupted, and in the RVFV haploid cell screen we performed, genes contributing to glycosaminoglycan synthesis and Golgi complex function were mutated in the surviving pool almost to the exclusion of any other mutations. In contrast, RNAi screens often implicate several biological pathways as being important for viral replication, as did the RVFV RNAi screen published by Hopkins *et al.* [7]. Variables that could impact the results of haploid cell screens could include the multiplicity and timing of infection as well as the length of time cells are cultured after virus challenge. Finally, most haploid screens have utilized mutant libraries generated in the human haploid cells HAP1, a line derived from the KBM-7 chronic myeloid leukemia cell line, which restricts its use to viruses that are capable of infecting these cells.

Interrogating host-pathogen protein-protein interactions through Y2H, AP/MS, or proximity labeling makes it possible to identify host factors based upon the *a priori* association of a viral and a cellular protein within the biological context of the host cellular environment. Many of the common phenotypic read-outs used during screening techniques, such as production of a reporter protein or host cell death, have the distinct disadvantage of restricting host factor discovery to those which impact a specific subset of stages during the viral replication cycle. High-throughput screens to identify cellular factors required for viral assembly and egress, for example, have proven difficult to design and, screens to identify host factors required for viral infections have largely focused on the rate-limiting stages of entry and replication. Another important advantage to protein-protein interaction screening is that it allows for the identification of host factors whose depletion or disruption may be cytotoxic, or even lethal. On the other hand, antibodies to affinity purify a viral protein are not always available, and the introduction of a tag or the precipitation conditions may perturb viral protein function or have other unforeseen consequences.

The use of multiple complementary screening techniques can serve to address and overcome the varying advantages and disadvantages presented by each of the techniques on their own. Performing multiple screens in parallel can help eliminate false-positive hits, even if the differences between the screens are relatively subtle technical changes such as use of different viral strains, cell types, or siRNA libraries. With each new published screen, the pool of datasets available to draw from also increases, which will allow for valuable comparisons of one's screening results with the reported hits from other related screens.

8. Future Perspectives: Expanding Cellular Targets and Bunyavirus Technical Resources

In addition to the screening techniques focused on genes and proteins, there has been renewed interest in developing high-throughput approaches to identify metabolites and lipids that are involved in viral infection. Analysis with LC-MS can be used to quantify changes in the metabolomic profile of infected cells relative to uninfected cells, providing insight into viral alteration of host metabolism as well as yielding potential therapeutic targets. This approach was used to quantify the levels of known metabolites at different time points during infection with human cytomegalovirus (HCMV), herpes simplex virus type-1 (HSV-1), and IAV, demonstrating each virus's ability to differentially remodel the host's metabolism during infection [62–64]. In the case of HCMV and IAV, pharmacological inhibition of fatty acid biosynthesis was shown to effectively restrict viral replication, demonstrating the power of such screens to inform the development (or repurposing) of therapeutics. In 2013, Morita *et al.* tested a library of bioactive lipids for an effect on IAV replication, and observed potent inhibition with the lipid mediator protectin D1 (PD1) [65]. Treatment with PD1 was able to protect against influenza in a mouse model, even if it was not supplied until severe disease had developed. Another important aspect of virus-host dynamics that could be examined for bunyaviruses is that of interactions between RNA and proteins during infection. Yeast three-hybrid (Y3H) screening provides a powerful tool for identifying proteins that bind to a specific RNA sequence. This technique, first described by SenGupta *et al.* [66], detects RNA-protein interactions by utilizing two-hybrid proteins whose proximity activates a reporter gene when both proteins bind to a hybrid RNA molecule. Y3H screening was used to identify human ribosomal proteins that bind to the 3′ untranslated region of HCV using a human cDNA library as prey and the viral RNA sequence as bait [67]. Covalent UV crosslinking during infection could also be used to capture and characterize the RNA-protein interactome in a manner similar to the technique described by Castello *et al.* in 2012 [68].

These and other recent advances in screening technology have the potential to significantly inform bunyavirus research, particularly in light of the continually expanding options available for generating bunyavirus reporter systems to enable high-throughput or automated screening. Among orthobunyaviruses, a replication-competent recombinant BUNV has been generated bearing a fluorescent or V5 tag on either Gc glycoprotein or L proteins, respectively [69,70]. In 2013, reverse genetics was described for Schmallenberg virus (SBV) and in 2015 a BHK cell line was developed that constitutively expressed the SBV N protein and a minigenome system was described for Oropouche virus (OROV) [71–73]. Efficient reverse genetics has also now been established for Akabane virus (AKAV), further expanding the options for bunyavirus screening approaches [74]. For the phlebovirus RVFV, there exists both a reverse genetics toolset as well as a BHK replicon cell line expressing the S and L segments of the genome [75,76]. We and others have also utilized pseudovirion systems, described in [77] and [78], in order to screen for host factors required during entry of bunyaviruses. These pseudotyped virions can be used at the biosafety level (BSL)-2 and allow for the convenient use of either cell death or a genetically-encoded reporter (e.g., luciferase or a fluorescent protein) to facilitate high-throughput, cell-based screening approaches. The future of bunyavirus screening techniques is bright, and the marriage of improved screening techniques with the increasing availability of virological tools promises to push forward our understanding of how these viruses interact with their host cells, and will help us develop targeted antiviral therapeutics.

Acknowledgments: We thank Jason Wojcechowskyj for a critical reading of the manuscript. We apologize to the authors of studies whose work we were unable to cite due to space constraints.

Author Contributions: A.M.R. and R.W.D. wrote the paper.

Conflicts of Interest: The authors declare no conflict of interest.

References

1. Cherry, S.; Doukas, T.; Armknecht, S.; Whelan, S.; Wang, H.; Sarnow, P.; Perrimon, N. Genome-wide RNAi screen reveals a specific sensitivity of IRES-containing RNA viruses to host translation inhibition. *Genes Dev.* **2005**, *19*, 445–452. [CrossRef] [PubMed]
2. Cherry, S.; Kunte, A.; Wang, H.; Coyne, C.; Rawson, R.B.; Perrimon, N. COPI activity coupled with fatty acid biosynthesis is required for viral replication. *PLoS Pathog.* **2006**, *2*, e120. [CrossRef] [PubMed]
3. Brass, A.L.; Dykxhoorn, D.M.; Benita, Y.; Yan, N.; Engelman, A.; Xavier, R.J.; Lieberman, J.; Elledge, S.J. Identification of host proteins required for HIV infection through a functional genomic screen. *Science* **2008**, *319*, 921–926. [CrossRef] [PubMed]
4. König, R.; Zhou, Y.; Elleder, D.; Diamond, T.L.; Bonamy, G.M.C.; Irelan, J.T.; Chiang, C.; Tu, B.P.; de Jesus, P.D.; Lilley, C.E.; *et al.* Global analysis of host-pathogen interactions that regulate early-stage HIV-1 replication. *Cell* **2008**, *135*, 49–60. [CrossRef] [PubMed]
5. Zhou, H.; Xu, M.; Huang, Q.; Gates, A.T.; Zhang, X.D.; Castle, J.C.; Stec, E.; Ferrer, M.; Strulovici, B.; Hazuda, D.J.; *et al.* Genome-scale RNAi screen for host factors required for HIV replication. *Cell Host Microbe* **2008**, *4*, 495–504. [CrossRef] [PubMed]
6. Brass, A.L.; Huang, I.C.; Benita, Y.; John, S.P.; Krishnan, M.N.; Feeley, E.M.; Ryan, B.J.; Weyer, J.L.; van der Weyden, L.; Fikrig, E.; *et al.* The IFITM proteins mediate cellular resistance to influenza A H1N1 virus, West Nile virus, and dengue virus. *Cell* **2009**, *139*, 1243–1254. [CrossRef] [PubMed]
7. Hopkins, K.C.; McLane, L.M.; Maqbool, T.; Panda, D.; Gordesky-Gold, B.; Cherry, S. A genome-wide RNAi screen reveals that mRNA decapping restricts bunyaviral replication by limiting the pools of dcp2-accessible targets for cap-snatching. *Genes Dev.* **2013**, *27*, 1511–1525. [CrossRef] [PubMed]
8. Meier, R.; Franceschini, A.; Horvath, P.; Tetard, M.; Mancini, R.; von Mering, C.; Helenius, A.; Lozach, P.-Y. Genome-wide small interfering RNA screens reveal VAMP3 as a novel host factor required for Uukuniemi virus late penetration. *J. Virol.* **2014**, *88*, 8565–8578. [CrossRef] [PubMed]
9. White, J.; Matlin, K.; Helenius, A. Cell fusion by Semliki Forest, influenza, and vesicular stomatitis viruses. *J. Cell Biol.* **1981**, *89*, 674–679. [CrossRef] [PubMed]
10. Maeda, T.; Ohnishi, S. Activation of influenza virus by acidic media causes hemolysis and fusion of erythrocytes. *FEBS Lett.* **1980**, *122*, 283–287. [CrossRef]
11. Huang, R.T.C.; Rott, R.; Klenk, H.D. Influenza viruses cause hemolysis and fusion of cells. *Virology* **1981**, *110*, 243–247. [CrossRef]
12. Fader, C.M.; Sánchez, D.G.; Mestre, M.B.; Colombo, M.I. TI-VAMP/VAMP7 and VAMP3/cellubrevin: Two v-SNARE proteins involved in specific steps of the autophagy/multivesicular body pathways. *Biochim. Biophys. Acta* **2009**, *1793*, 1901–1916. [CrossRef] [PubMed]
13. Carette, J.E.; Guimaraes, C.P.; Varadarajan, M.; Park, A.S.; Wuethrich, I.; Godarova, A.; Kotecki, M.; Cochran, B.H.; Spooner, E.; Ploegh, H.L.; *et al.* Haploid genetic screens in human cells identify host factors used by pathogens. *Science* **2009**, *326*, 1231–1235. [CrossRef] [PubMed]
14. Carette, J.E.; Guimaraes, C.P.; Wuethrich, I.; Blomen, V.A.; Sun, C.; Bell, G.; Yuan, B.; Muellner, M.K.; Nijman, M.; Ploegh, H.L.; *et al.* Global gene disruption in human cells to assign genes to phenotypes. *Nat. Biotechnol.* **2011**, *29*, 542–546. [CrossRef] [PubMed]
15. Carette, J.E.; Raaben, M.; Wong, A.C.; Herbert, A.S.; Obernosterer, G.; Mulherkar, N.; Kuehne, A.I.; Kranzusch, P.J.; Griffin, A.M.; Ruthel, G.; *et al.* Ebola virus entry requires the cholesterol transporter Niemann-Pick C1. *Nature* **2011**, *477*, 340–343. [CrossRef] [PubMed]
16. Jae, L.T.; Raaben, M.; Riemersma, M.; van Beusekom, E.; Blomen, V.; Velds, A.; Kerkhoven, R.M.; Carette, J.E.; Topaloglu, H.; Meinecke, P.; *et al.* Deciphering the glycosylome of dystroglycanopathies using haploid screens for Lassa virus entry. *Science* **2013**, *340*, 479–483. [CrossRef] [PubMed]

17. Jae, L.T.; Raaben, M.; Herbert, A.S.; Kuehne, A.I.; Wirchnianski, A.S.; Soh, T.K.; Stubbs, S.H.; Janssen, H.; Damme, M.; Saftig, P.; *et al.* Lassa virus entry requires a trigger-induced receptor switch. *Science* **2014**, *344*, 1506–1510. [CrossRef] [PubMed]
18. Petersen, J.; Drake, M.J.; Bruce, E.A.; Riblett, A.M.; Didigu, C.A.; Wilen, C.B.; Malani, N.; Male, F.; Lee, F.-H.; Bushman, F.D.; *et al.* The major cellular sterol regulatory pathway is required for Andes virus infection. *PLoS Pathog.* **2014**, *10*, e1003911. [CrossRef] [PubMed]
19. Kleinfelter, L.M.; Jangra, R.K.; Jae, L.T.; Herbert, A.S.; Mittler, E.; Stiles, K.M.; Wirchnianski, A.S.; Kielian, M.; Brummelkamp, T.R.; Dye, J.M. Haploid genetic screen reveals a profound and direct dependence on cholesterol for hantavirus membrane fusion. *mBio* **2015**, *6*, 1–14. [CrossRef] [PubMed]
20. Riblett, A.M.; Blomen, V.A.; Jae, L.T.; Altamura, L.A.; Doms, R.W.; Brummelkamp, T.R.; Wojcechowskyj, J.A. A haploid genetic screen identifies heparan sulfate proteoglycans supporting Rift Valley fever virus infection. *J. Virol.* **2016**, *90*, 1414–1423. [CrossRef] [PubMed]
21. Ungar, D.; Oka, T.; Brittle, E.E.; Vasile, E.; Lupashin, V.; Chatterton, J.E.; Heuser, J.E.; Krieger, M.; Waters, M.G. Characterization of a mammalian Golgi-localized protein complex, COG, that is required for normal Golgi morphology and function. *J. Cell Biol.* **2002**, *157*, 405–415. [CrossRef]
22. Kingsley, D.M.; Kozarsky, K.F.; Segal, M.; Krieger, M. Three types of low density lipoprotein receptor-deficient mutant have pleiotropic defects in the synthesis of N-linked, O-linked, and lipid-linked carbohydrate chains. *J. Cell Biol.* **1986**, *102*, 1576–1585. [CrossRef] [PubMed]
23. Blomen, V.A.; Májek, P.; Jae, L.T.; Bigenzahn, J.W.; Nieuwenhuis, J.; Staring, J.; Sacco, R.; van Diemen, F.R.; Olk, N.; Stukalov, A.; *et al.* Gene essentiality and synthetic lethality in haploid human cells. *Science* **2015**, *350*, 1092–1096. [CrossRef] [PubMed]
24. Yi, M.; Hong, N.; Hong, Y. Generation of medaka fish haploid embryonic stem cells. *Science* **2009**, *326*, 430–433. [CrossRef] [PubMed]
25. Elling, U.; Taubenschmid, J.; Wirnsberger, G.; O'Malley, R.; Demers, S.P.; Vanhaelen, Q.; Shukalyuk, A.I.; Schmauss, G.; Schramek, D.; Schnuetgen, F.; *et al.* Forward and reverse genetics through derivation of haploid mouse embryonic stem cells. *Cell Stem Cell* **2011**, *9*, 563–574. [CrossRef] [PubMed]
26. Leeb, M.; Wutz, A. Derivation of haploid embryonic stem cells from mouse embryos. *Nature* **2011**, *479*, 131–134. [CrossRef] [PubMed]
27. Yang, H.; Liu, Z.; Ma, Y.; Zhong, C.; Yin, Q.; Zhou, C.; Shi, L.; Cai, Y.; Zhao, H.; Wang, H.; *et al.* Generation of haploid embryonic stem cells from Macaca fascicularis monkey parthenotes. *Cell Res.* **2013**, *23*, 1187–1200. [CrossRef] [PubMed]
28. Li, W.; Li, X.; Li, T.; Jiang, M.G.; Wan, H.; Luo, G.Z.; Feng, C.; Cui, X.; Teng, F.; Yuan, Y.; *et al.* Genetic modification and screening in rat using haploid embryonic stem cells. *Cell Stem Cell* **2014**, *14*, 404–414. [CrossRef] [PubMed]
29. Essletzbichler, P.; Konopka, T.; Santoro, F.; Chen, D.; Gapp, B.V.; Kralovics, R.; Brummelkamp, T.R.; Nijman, S.M.B.; Bürckstümmer, T. Megabase-scale deletion using CRISPR/Cas9 to generate a fully haploid human cell line. *Genome Res.* **2014**, *24*, 2059–2065. [CrossRef] [PubMed]
30. Wright, A.V.; Nunez, J.K.; Doudna, J.A. Biology and applications of CRISPR systems: Harnessing nature's toolbox for genome engineering. *Cell* **2016**, *164*, 29–44. [CrossRef] [PubMed]
31. Ma, H.; Dang, Y.; Wu, Y.; Jia, G.; Anaya, E.; Zhang, J.; Abraham, S.; Choi, J.G.; Shi, G.; Qi, L.; *et al.* A CRISPR-based screen identifies genes essential for West-Nile-virus-induced cell death. *Cell Rep.* **2015**, *12*, 673–683. [CrossRef] [PubMed]
32. Sanjana, N.E.; Shalem, O.; Zhang, F. Improved vectors and genome-wide libraries for CRISPR screening. *Nat. Methods* **2014**, *11*, 006726. [CrossRef] [PubMed]
33. Islam, M.K.; Baudin, M.; Eriksson, J.; Öberg, C.; Habjan, M.; Weber, F.; Överby, A.K.; Ahlm, C.; Evander, M. High-throughput screening using a whole-cell virus replication reporter gene assay to identify inhibitory compounds against Rift Valley fever virus infection. *J. Biomol. Screen.* **2016**, *21*, 354–362. [CrossRef] [PubMed]
34. Tan, X.; Hu, L.; Luquette, L.J.; Gao, G.; Liu, Y.; Qu, H.; Xi, R.; Lu, Z.J.; Park, P.J.; Elledge, S.J. Systematic identification of synergistic drug pairs targeting HIV. *Nat. Biotechnol.* **2012**, *30*, 1125–1130. [CrossRef] [PubMed]
35. Léonard, V.H.J.; Kohl, A.; Hart, T.J.; Elliott, R.M. Interaction of Bunyamwera orthobunyavirus NSs protein with Mediator protein MED8: A mechanism for inhibiting the interferon response. *J. Virol.* **2006**, *80*, 9667–9675. [CrossRef] [PubMed]

36. Rönnberg, T.; Jääskeläinen, K.; Blot, G.; Parviainen, V.; Vaheri, A.; Renkonen, R.; Bouloy, M.; Plyusnin, A. Searching for cellular partners of hantaviral nonstructural protein NSs: Y2H screening of mouse cDNA library and analysis of cellular interactome. *PLoS ONE* **2012**, *7*. [CrossRef] [PubMed]
37. Pichlmair, A.; Kandasamy, K.; Alvisi, G.; Mulhern, O.; Sacco, R.; Habjan, M.; Binder, M.; Stefanovic, A.; Eberle, C.-A.; Goncalves, A.; *et al.* Viral immune modulators perturb the human molecular network by common and unique strategies. *Nature* **2012**, *487*, 486–490. [CrossRef] [PubMed]
38. Kainulainen, M.; Habjan, M.; Hubel, P.; Busch, L.; Lau, S.; Colinge, J.; Superti-Furga, G.; Pichlmair, A.; Weber, F. Virulence factor NSs of Rift Valley fever virus recruits the F-box protein FBXO3 to degrade subunit p62 of general transcription factor TFIIH. *J. Virol.* **2014**, *88*, 3464–3473. [CrossRef] [PubMed]
39. Martell, J.D.; Deerinck, T.J.; Sancak, Y.; Poulos, T.L.; Mootha, V.K.; Sosinsky, G.E.; Ellisman, M.H.; Ting, A.Y. Engineered ascorbate peroxidase as a genetically encoded reporter for electron microscopy. *Nat. Biotechnol.* **2012**, *30*, 1143–1148. [PubMed]
40. Rhee, H.; Zou, P.; Udeshi, N.D.; Martell, J.D.; Mootha, V.K.; Carr, S.A.; Ting, A.Y. Proteomic mapping of mitochondria in living cells via spatially restricted enzymatic tagging. *Science* **2013**, *339*, 1328. [CrossRef] [PubMed]
41. Lam, S.S.; Martell, J.D.; Kamer, K.J.; Deerinck, T.J.; Ellisman, M.H.; Mootha, V.K.; Ting, A.Y. Directed evolution of APEX2 for electron microscopy and proximity labeling. *Nat. Methods* **2014**, *12*, 51–54. [CrossRef] [PubMed]
42. Roux, K.J.; Kim, D.I.; Raida, M.; Burke, B. A promiscuous biotin ligase fusion protein identifies proximal and interacting proteins in mammalian cells. *J. Cell Biol.* **2012**, *196*, 801–810. [CrossRef] [PubMed]
43. Kim, D.I.; KC, B.; Zhu, W.; Motamedchaboki, K.; Doye, V.; Roux, K.J. Probing nuclear pore complex architecture with proximity-dependent biotinylation. *Proc. Natl. Acad. Sci. USA* **2014**, *111*, E2453–E2461. [CrossRef] [PubMed]
44. Elzi, D.J.; Song, M.; Hakala, K.; Weintraub, S.T.; Shiio, Y. Proteomic analysis of the EWS-Fli-1 interactome reveals the role of the lysosome in EWS-Fli-1 turnover. *J. Proteome Res.* **2014**, *13*, 3783–3791. [CrossRef] [PubMed]
45. Mojica, S.A.; Hovis, K.M.; Frieman, M.B.; Tran, B.; Hsia, R.C.; Ravel, J.; Jenkins-Houk, C.; Wilson, K.L.; Bavoil, P.M. SINC, a type III secreted protein of Chlamydia psittaci, targets the inner nuclear membrane of infected cells and uninfected neighbors. *Mol. Biol. Cell* **2015**, *26*, 1918–1934. [CrossRef] [PubMed]
46. Le Sage, V.; Cinti, A.; Valiente-Echeverría, F.; Mouland, A.J. Proteomic analysis of HIV-1 Gag interacting partners using proximity-dependent biotinylation. *Virol. J.* **2015**, *12*, 138. [CrossRef] [PubMed]
47. Kim, D.I.; Jensen, S.C.; Noble, K.A.; Kc, B.; Roux, K.H.; Motamedchaboki, K.; Roux, K.J. An improved smaller biotin ligase for BioID proximity labeling. *Mol. Biol. Cell* **2016**, *27*, 1188–1196. [CrossRef] [PubMed]
48. Bushman, F.D.; Malani, N.; Fernandes, J.; D'Orso, I.; Cagney, G.; Diamond, T.L.; Zhou, H.; Hazuda, D.J.; Espeseth, A.S.; Konig, R.; *et al.* Host cell factors in HIV replication: Meta-analysis of genome-wide studies. *PLoS Pathog.* **2009**, *5*, e1000437. [CrossRef] [PubMed]
49. Tai, A.W.; Benita, Y.; Peng, L.F.; Kim, S.; Sakamoto, N.; Xavier, R.J.; Chung, R.T. A functional genomic screen identifies cellular cofactors of hepatitis C virus replication. *Cell Host Microbe* **2009**, *5*, 298–307. [CrossRef] [PubMed]
50. Li, Q.; Brass, A.L.; Ng, A.; Hu, Z.; Xavier, R.J.; Liang, T.J.; Elledge, S.J. A genome-wide genetic screen for host factors required for hepatitis C virus propagation. *Proc. Natl. Acad. Sci. USA* **2009**, *106*, 16410–16415. [CrossRef] [PubMed]
51. Rose, P.P.; Hanna, S.L.; Spiridigliozzi, A.; Wannissorn, N.; Beiting, D.P.; Ross, S.R.; Hardy, R.W.; Bambina, S.A.; Heise, M.T.; Cherry, S. Natural resistance-associated macrophage protein is a cellular receptor for Sindbis virus in both insect and mammalian hosts. *Cell Host Microbe* **2011**, *10*, 97–104. [CrossRef] [PubMed]
52. Panda, D.; Rose, P.P.; Hanna, S.L.; Gold, B.; Hopkins, K.C.; Lyde, R.B.; Marks, M.S.; Cherry, S. Genome-wide RNAi screen identifies SEC61A and VCP as conserved regulators of Sindbis virus entry. *Cell Rep.* **2013**, *5*, 1737–1748. [PubMed]
53. Ooi, Y.S.; Stiles, K.M.; Liu, C.Y.; Taylor, G.M.; Kielian, M. Genome-wide RNAi screen identifies novel host proteins required for alphavirus entry. *PLoS Pathog.* **2013**, *9*, e1003835. [CrossRef] [PubMed]

54. Franceschini, A.; Meier, R.; Casanova, A.; Kreibich, S.; Daga, N.; Andritschke, D.; Dilling, S.; Rämö, P.; Emmenlauer, M.; Kaufmann, A.; *et al.* Specific inhibition of diverse pathogens in human cells by synthetic microRNA-like oligonucleotides inferred from RNAi screens. *Proc. Natl. Acad. Sci. USA* **2014**, *111*, 4548–4553. [CrossRef] [PubMed]
55. Zhu, J.; Davoli, T.; Perriera, J.M.; Chin, C.R.; Gaiha, G.D.; John, S.P.; Sigiollot, F.D.; Gao, G.; Xu, Q.; Qu, H.; *et al.* Comprehensive identification of host modulators of HIV-1 replication using multiple orthologous RNAi reagents. *Cell Rep.* **2014**, *9*, 752–766. [CrossRef] [PubMed]
56. Mohr, S.E.; Smith, J.A.; Shamu, C.E.; Neumüller, R.A.; Perrimon, N. RNAi screening comes of age: Improved techniques and complementary approaches. *Nat. Rev. Mol. Cell Biol.* **2014**, *15*, 591–600. [CrossRef] [PubMed]
57. Baggen, J.; Jan, H.; Staring, J.; Jae, L.T.; Liu, Y.; Guo, H.; Slager, J.J.; de Bruin, J.W.; van Vliet, A.L.W.; Blomen, V.A.; *et al.* Enterovirus D68 receptor requirements unveiled by haploid genetics. *Proc. Natl. Acad. Sci. USA* **2015**, *113*, 1–6. [CrossRef] [PubMed]
58. Pillay, S.; Meyer, N.L.; Puschnik, A.S.; Davulcu, O.; Diep, J.; Ishikawa, Y.; Jae, L.T.; Wosen, J.E.; Nagamine, C.M.; Chapman, M.S.; *et al.* An essential receptor for adeno-associated virus infection. *Nature* **2016**, *530*, 108–112. [CrossRef] [PubMed]
59. Schorch, B.; Song, S.; van Diemen, F.R.; Bock, H.H.; May, P.; Herz, J.; Brummelkamp, T.R.; Papatheodorou, P.; Aktories, K. LRP1 is a receptor for Clostridium perfringens TpeL toxin indicating a two-receptor model of clostridial glycosylating toxins. *Proc. Natl. Acad. Sci. USA* **2014**, *111*, 6431–6436. [CrossRef] [PubMed]
60. Tafesse, F.G.; Guimaraes, C.P.; Maruyama, T.; Carette, J.E.; Lory, S.; Brummelkamp, T.R.; Ploegh, H.L. GPR107, a G-protein-coupled receptor essential for intoxication by Pseudomonas aeruginosa exotoxin A, localizes to the Golgi and is cleaved by furin. *J. Biol. Chem.* **2014**, *289*, 24005–24018. [CrossRef] [PubMed]
61. Popov, L.M.; Marceau, C.D.; Starkl, P.M.; Lumb, J.H.; Shah, J.; Guerrera, D.; Cooper, R.L.; Merakou, C.; Bouley, D.M.; Meng, W.; *et al.* The adherens junctions control susceptibility to Staphylococcus aureus α-toxin. *Proc. Natl. Acad. Sci. USA* **2015**, *112*, 201510265. [CrossRef] [PubMed]
62. Munger, J.; Bajad, S.U.; Coller, H.A.; Shenk, T.; Rabinowitz, J.D. Dynamics of the cellular metabolome during human cytomegalovirus infection. *PLoS Pathog.* **2006**, *2*, 1165–1175. [CrossRef] [PubMed]
63. Munger, J.; Bennett, B.D.; Parikh, A.; Feng, X.-J.; McArdle, J.; Rabitz, H.A.; Shenk, T.; Rabinowitz, J.D. Systems-level metabolic flux profiling identifies fatty acid synthesis as a target for antiviral therapy. *Nat. Biotechnol.* **2008**, *26*, 1179–1186. [CrossRef] [PubMed]
64. Vastag, L.; Koyuncu, E.; Grady, S.L.; Shenk, T.E.; Rabinowitz, J.D. Divergent effects of human cytomegalovirus and herpes simplex virus-1 on cellular metabolism. *PLoS Pathog.* **2011**, *7*. [CrossRef] [PubMed]
65. Morita, M.; Kuba, K.; Ichikawa, A.; Nakayama, M.; Katahira, J.; Iwamoto, R.; Watanebe, T.; Sakabe, S.; Daidoji, T.; Nakamura, S.; *et al.* The lipid mediator protectin D1 inhibits influenza virus replication and improves severe influenza. *Cell* **2013**, *153*, 112–125. [CrossRef] [PubMed]
66. SenGupta, D.J.; Zhang, B.; Kraemer, B.; Pochart, P.; Fields, S.; Wickens, M. A three-hybrid system to detect RNA-protein interactions *in vivo*. *Proc. Natl. Acad. Sci. USA* **1996**, *93*, 8496–8501. [CrossRef] [PubMed]
67. Wood, J.; Frederickson, R.M.; Fields, S.; Patel, A.H. Hepatitis C virus 3'X region interacts with human ribosomal proteins. *J. Virol.* **2001**, *75*, 1348–1358. [CrossRef] [PubMed]
68. Castello, A.; Fischer, B.; Eichelbaum, K.; Horos, R.; Beckmann, B.M.; Strein, C.; Davey, N.E.; Humphreys, D.T.; Preiss, T.; Steinmetz, L.M.; *et al.* Insights into RNA biology from an atlas of mammalian mRNA-binding proteins. *Cell* **2012**, *149*, 1393–1406. [CrossRef] [PubMed]
69. Shi, X.; van Mierlo, J.T.; French, A.; Elliott, R.M. Visualizing the replication cycle of Bunyamwera orthobunyavirus expressing fluorescent protein-tagged Gc glycoprotein. *J. Virol.* **2010**, *84*, 8460–8469. [CrossRef] [PubMed]
70. Shi, X.; Elliott, R.M. Generation and analysis of recombinant Bunyamwera orthobunyaviruses expressing V5 epitope-tagged L proteins. *J. Gen. Virol.* **2009**, *90*, 297–306. [CrossRef] [PubMed]
71. Elliott, R.M.; Blakqori, G.; van Knippenberg, I.C.; Koudriakova, E.; Li, P.; McLees, A.; Shi, X.; Szemiel, A.M. Establishment of a reverse genetics system for Schmallenberg virus, a newly emerged orthobunyavirus in Europe. *J. Gen. Virol.* **2013**, *94*, 851–859. [CrossRef]
72. Zhang, Y.; Wu, S.; Song, S.; Lv, J.; Feng, C.; Lin, X. Preparation and characterization of a stable BHK-21 cell line constitutively expressing the Schmallenberg virus nucleocapsid protein. *Mol. Cell. Probes* **2015**, *29*, 244–253. [CrossRef] [PubMed]

73. Acrani, G.O.; Tilston-Lunel, N.L.; Spiegel, M.; Weidmann, M.; Dilcher, M.; da Silva, D.E.A.; Nunes, M.R.T.; Elliott, R.M. Establishment of a minigenome system for oropouche virus reveals the S genome segment to be significantly longer than reported previously. *J. Gen. Virol.* **2015**, *96*, 513–523. [CrossRef] [PubMed]
74. Takenaka-Uema, A.; Sugiura, K.; Bangphoomi, N.; Shioda, C.; Uchida, K.; Kato, K.; Haga, T.; Murakami, S.; Akashi, H.; Horimoto, T. Development of an improved reverse genetics system for Akabane bunyavirus. *J. Virol. Methods* **2016**, *232*, 16–20. [CrossRef] [PubMed]
75. Ikegami, T.; Won, S.; Peters, C.J.; Makino, S. Rescue of infectious Rift Valley fever virus entirely from cDNA, analysis of virus lacking the NSs gene, and expression of a foreign gene. *J. Virol.* **2006**, *80*, 2933–2940. [CrossRef] [PubMed]
76. Kortekaas, J.; Oreshkova, N.; Cobos-Jimenez, V.; Vloet, R.P.M.; Potgieter, C.A.; Moormann, R.J.M. Creation of a nonspreading Rift Valley fever virus. *J. Virol.* **2011**, *85*, 12622–12630. [CrossRef] [PubMed]
77. Ray, N.; Whidby, J.; Stewart, S.; Hooper, J.W.; Bertolotti-Ciarlet, A. study of Andes virus entry and neutralization using a pseudovirion system. *J. Virol. Methods* **2010**, *163*, 416–423. [CrossRef] [PubMed]
78. Higa, M.M.; Petersen, J.; Hooper, J.; Doms, R.W. Efficient production of Hantaan and Puumala pseudovirions for viral tropism and neutralization studies. *Virology* **2012**, *423*, 134–142. [CrossRef] [PubMed]

© 2016 by the authors. Licensee MDPI, Basel, Switzerland. This article is an open access article distributed under the terms and conditions of the Creative Commons Attribution (CC BY) license (http://creativecommons.org/licenses/by/4.0/).

MDPI AG
St. Alban-Anlage 66
4052 Basel, Switzerland
Tel. +41 61 683 77 34
Fax +41 61 302 89 18
http://www.mdpi.com

Viruses Editorial Office
E-mail: viruses@mdpi.com
http://www.mdpi.com/journal/viruses

www.ingramcontent.com/pod-product-compliance
Lightning Source LLC
Chambersburg PA
CBHW051838210326
41597CB00033B/5702

* 9 7 8 3 0 3 8 4 2 3 9 2 8 *